This book describes for the first time in a modern text the fundamental principles on which solid state electrochemistry is based. In this sense, it is distinct from other books in the field which concentrate on a description of materials. The text provides an essential foundation of understanding for postgraduates or others entering the field for the first time and may also be of value in advanced undergraduate courses.

Chemistry of Solid State Materials

Solid state electrochemistry

Chemistry of Solid State Materials

Series Editors:
Bruce Dunn, Department of Materials Science and Engineering, UCLA
John W. Goodby, School of Chemistry, University of Hull
A. R. West, Department of Chemistry, University of Aberdeen

Titles in this series:
David Segal, *Chemical synthesis of advanced ceramic materials*
Philippe Colomban (ed.), *Proton conductors*
Alan Wilson and John Nicholson, *Acid base reaction cements*
Raul Valenzuela, *Magnetic ceramics*
Peter Bruce (ed.), *Solid state electrochemistry*

Solid state electrochemistry

edited by Peter G. Bruce
Department of Chemistry, University of St Andrews, Scotland

PUBLISHED BY THE PRESS SYNDICATE OF THE UNIVERSITY OF CAMBRIDGE
The Pitt Building, Trumpington Street, Cambridge CB2 1RP, United Kingdom

CAMBRIDGE UNIVERSITY PRESS
The Edinburgh Building, Cambridge CB2 2RU, United Kingdom
40 West 20th Street, New York, NY 10011-4211, USA
10 Stamford Road, Oakleigh, Melbourne 3166, Australia

First published 1995
First paperback edition 1997

Typeset in Linotron Times 10/13 pt

A catalogue record for this book is available from the British Library

Library of Congress Cataloguing in Publication data
Solid state electrochemistry/edited by Peter G. Bruce.
p. cm. – (Chemistry of solid state materials ;)
Includes bibliographical references.
ISBN 0-521-40007-4
1. Electrochemistry. 2. Solid state chemistry. I. Bruce, Peter
G. II. Series.
QD555.5.S65 1995
541′.002—dc20 93-42634 CIP

ISBN 0 521 40007 4 hardback
ISBN 0 521 59949 0 paperback

Transferred to digital printing 2003

KW

Dedicated to Margaret, David and Caroline, for their friendship, support and patience

Contents

Preface

Significant advances in our understanding of solid state electrochemistry have taken place over the last 30 years. The subject has grown rapidly and is, as illustrated in Chapter 1, much closer to the more established electrochemistry in liquid electrolytes than at any time since Faraday's pioneering work in both fields some 150 years ago. Although several books have appeared dealing with ionically conducting materials from a structural point of view, very few texts are available which describe the physical electrochemistry of solids. The present textbook aims to present the fundamentals of solid state electrochemistry with a strong emphasis on the physical aspects. It is directed primarily towards postgraduate students and other scientists and engineers entering the field for the first time, as well as those active in the areas of batteries, fuel cells, sensors and electrochromic devices, topics for which solid state electrochemistry makes a major contribution. Although the reader requires little prior knowledge of solid state electrochemistry, the subject is treated at a relatively advanced level and therefore significant sections of the book should be of interest to all electrochemists, as well as those already active in the electrochemistry of solids. The structural and material aspects of the subject are not ignored, indeed some knowledge of the most important solid electrolytes and intercalation electrodes is an essential foundation on which to build an understanding of the physical properties.

Despite strenuous efforts by this editor to avoid overlap of topics at all stages in the production of the book, it is inevitable in a multi-author text that some topics will be addressed more than once. In general where overlap has been permitted to remain the treatment of the topics is quite distinct, either in depth or approach. For example, both Chapters 2 and 3 deal with crystalline solid electrolytes: West provides an excellent introduction to the field and describes the key materials in Chapter 2, while the contrasting and distinctive approach to the description of ion transport by Goodenough provides the reader with a consistent and advanced framework for thinking about electronic and ionic transport, which many will find stimulating.

In a subject with the breadth of solid state electrochemistry, and with the inevitable constraints on the length of a textbook, some topics have been omitted. In particular, less emphasis on proton conduction has been given than on the transport of other ions, in part because of the existence of a recent book on this topic in the same series, *Proton Conductors*, Ed. Philippe Colomban, Cambridge University Press (1992). Also the important topic of intercalation into graphite has been largely omitted. Several excellent texts on this subject are already available.

Finally, I am indebted to all the authors who have contributed to the book. In the present climate of research it is increasingly difficult to find time to write for a work such as this. They have borne my pressure with patience and good humour.

St Andrews — Peter G. Bruce

1 Introduction

PETER G. BRUCE
Department of Chemistry, University of St Andrews

I formerly described a substance, sulphuret of silver, whose conducting power was increased by heat; and I have since then met with another as strongly affected in the same way: this is fluoride of lead. When a piece of that substance, which had been fused and cooled, was introduced into the circuit of a voltaic battery, it stopped the current. Being heated, it acquired conducting powers before it was visibly red-hot in daylight; and even sparks could be taken against it whilst still solid.
M. Faraday; *Philosophical Transactions of the Royal Society of London* (1838)

1.1 A brief history of solid state electrochemistry

Solid state electrochemistry may be divided into two broad topics.

(*a*) Solid electrolytes, which conduct electricity by the motion of ions, and exhibit negligible electronic transport. Included in this group are crystalline and amorphous inorganic solids as well as ionically conducting polymers.
(*b*) Intercalation electrodes, which conduct both ions and electrons. Again there are numerous examples based mainly on inorganic solids and polymers.

The field of solid state electrochemistry is not new. It has its origins, as does so much of electrochemistry, with Michael Faraday who discovered that PbF_2 and Ag_2S were good conductors. He therefore established both the first solid electrolyte and the first intercalation electrode (Faraday, 1838). Faraday had the wisdom to appreciate the benefits of unifying science rather than compartmentalising it. Regrettably, in the years after Faraday's pioneering work on electrochemistry this wisdom was largely lost. The two subjects of solid and liquid electrochemistry grew apart and developed separately until recent times. The dominance of the latter was probably due to the ease with which a wide range of liquid electrolytes could be prepared and purified. Solid state electrochemistry developed steadily up to the late 1960s. Conducting polymers were, in that period, unknown; the subject dealt exclusively with inorganic solids and glasses. Warburg (1884) demonstrated that Na^+ ions could be transported through glass and together with Tegetmeier (Warburg and Tegetmeier, 1888) carried out the first transference number measurement in solids. The dawn of the twentieth century saw the first technological application of ion transport in solids, when no less an electrochemist than Nernst (1900) proposed a new form of electric light, the 'Nernst glower'. He described how ZrO_2 when doped with a small amount of Y_2O_3 would emit a bright white light on the passage of a current at high temperature, due to its ability to conduct oxide ions. This remains one of the few non-electrochemical applications of solid electrolytes. Until the 1960s solid state ionics was confined largely to the study of oxide ion conductors such as doped ZrO_2 and Ag^+ ion conductors such as AgI, which above 147 °C adopts a structure that sustains very high ionic conductivity. In fact Tubandt and Lorenz (1914) found that the conductivity of solid AgI is higher just below its melting point than that of the molten salt! Insertion electrodes were also under investigation in the early years, specifically Ag_2S. The theory of transport in such mixed ionic and electronic conductors owes much to the elegant work of Carl Wagner (1956) whose academic descendant, W. Weppner is a contributor to this book.

The latter half of the 1960s and the early years of the 1970s saw an explosion of interest in solid state electrochemistry. In 1966 Kummer and Weber at the Ford Motor Company announced the development of a new type of battery which would be light in weight and deliver significant power. The sodium/sulphur cell consists of a solid sodium ion conductor, known as sodium beta alumina, separating electrodes of molten sodium and sulphur. This battery relies on the unique aspect of a *solid* electrolyte separating *liquid* electrodes. The Na^+ ion conductor sodium beta alumina, often referred to as β-Al_2O_3, was first discovered embedded in the linings of glass making furnaces. It had formed at high temperatures due to reaction between soda from the glass melt and the bricks of the furnace walls, which contained the common α-Al_2O_3 phase. The oil crisis of the early 1970s focused attention on the development of batteries and fuel cells (Steele, 1992) for electric traction, and other contenders joined sodium/sulphur; most relied on the use of either solid electrolytes or intercalation electrodes. The intensive search for, and study of, new intercalation electrodes led to an appreciation that the long-established battery technologies such as zinc/manganese dioxide and lead/lead dioxide also rely on the process of intercalation. In the 1990s powerful environmental concerns have, to some extent, replaced the oil crisis as a technological driving force in the field of solid state ionics, both from the point of view of batteries and gas sensors. Of no less importance is the development of compact low power batteries for portable electronic equipment, including heart pacemakers, mobile telephones, laptop computers, etc. and electrochromic devices such as SMART windows, which are electrochemical cells that can change their opacity by the passage of a small amount of charge. Many of these applications are discussed in Chapter 11.

The desire to realise technological goals has spurred the discovery of many new solid electrolytes and intercalation compounds based on crystalline and amorphous inorganic solids. In addition an entirely new class of ionic conductors has been discovered by P. V. Wright (1973) and M. B. Armand, J. M. Chabagno and M. Duclot (1978). These polymer electrolytes can be fabricated as soft films of only a few microns, and their flexibility permits interfaces with solid electrodes to be formed which remain intact when the cells are charged and discharged. This makes possible the development of *all-solid-state* electrochemical devices.

Along with the explosion of new electrochemical solids and the intense effort in the field of applications, there have been significant advances in our understanding of the fundamental physical principles of solid state electrochemistry since the 1960s. As is the case for other branches of electrochemistry, fundamental understanding and technological progress have frequently occurred together, often in the same laboratories. Electrochemistry in general is the richer for the synergy between theory and applications. In the 1990s solid and liquid electrochemistry are, after a separation of 150 years, growing together. Electronically conducting polymer electrodes provide the most obvious link. Much early work on these materials was carried out using the models of solution electrochemistry which were frequently inappropriate, when, in fact, polymer electrodes are much more akin to insertion electrodes based on inorganic solids (Chapter 8), where the concepts of charge transfer at the insertion electrode/solution interface and the coupled diffusion of dopant ions and electrons within the electrode, are well established. Conversely, polymer electrolytes are in many senses closer to aprotic liquid electrolytes than to inorganic solids, as acknowledged by the inclusion of polymer electrolytes in the latest general electrochemistry textbooks (e.g. Koryta, Dvořák and Kavan, 1993). There is also much that can be learned about the most fundamental aspects of electrochemistry in general by studying solid state systems. New insights into the nature of the electrical double layer and the mechanism by which electron transfer reactions are activated can be obtained since solid electrolytes do not possess freely rotating or librating dipoles. If this book can play a part in bringing solid and liquid phase electrochemistry together, as well as widening the access of solid state electrochemistry to scientists at large, it will have more than justified the effort in its production.

The authors of the succeeding chapters in this book have, in large measure, provided a sufficiently clear presentation of their topics that this introductory chapter can be much shorter than would otherwise have been necessary. However, for those new to the field, a concise overview of solid state electrochemistry may be of value and is presented in the following sections.

1.2 *Crystalline electrolytes* (Chapters 2 and 3)

These materials provide an essentially rigid framework with channels along which one of the ionic species of the solid can migrate. Ion transport involves hopping from site to site along these channels. Whereas all ionic solids conduct, only those with very specific structural features are capable of exhibiting conductivities comparable to liquid electrolytes. For example, one such solid electrolyte, $RbAg_4I_5$, possesses a conductivity of 0.27 S cm^{-1} at 25 °C (Owens and Argue, 1970) comparable to many liquid electrolytes.

Crystalline solid electrolytes which conduct Ag^+, Cu^+, Tl^+, Li^+, Na^+, K^+, H^+, O^{2-} and F^- as well as many divalent and trivalent cations are all readily available. The most important applications are of oxide ion conductors for solid oxide fuel cells and oxygen gas sensors. The latter are already widely used to monitor vehicle exhaust gases in catalytic converters.

1.3 *Glass electrolytes* (Chapter 4)

Rather specific structural features appear to be necessary for high ionic conductivity in crystalline solids, and as a result it may seem surprising that glasses can support high ionic conductivity. For example, a Li^+ conductivity of 0.16 mS cm^{-1} at 25 °C for a glass with the composition

0.7 Li_2S–0.3 P_2S_5 has been reported, which compares favourably with the best crystalline Li^+ ion conductors at the same temperature. Again a hopping mechanism for ion transport is believed to operate, although there is still debate concerning the detailed mechanism (Chapter 4). Glasses capable of transporting Li^+, Na^+, K^+, Cs^+, Rb^+, Ag^+ and F^- (Angell, 1989) have been prepared.

1.4 **Polymer electrolytes** (Chapters 5 and 6)

These materials are introduced in Chapter 5 and only brief mention of them is necessary here. It is important to appreciate that polymer electrolytes, which consist of salts, e.g. NaI, dissolved in solid cation coordinating polymers, e.g. $(CH_2CH_2O)_n$, conduct by quite a different mechanism from crystalline or glass electrolytes. Ion transport in polymers relies on the dynamics of the framework (i.e. the polymer chains) in contrast to hopping within a rigid framework. Intense efforts are being made to make use of these materials as electrolytes in all solid state lithium batteries for both microelectronic medical and vehicle traction applications.

1.5 **Intercalation electrodes** (Chapters 7, 8 and 9)

Intercalation is the process of inserting an atom (or ion with its charge compensating electrons) into a solid and removing the atom from the solid. The archetypal example is the insertion of Li between the layers of graphite. This particular insertion process has gained great technological interest as a replacement for lithium metal electrodes in rechargeable lithium batteries. Most of the work has concentrated on lithium intercalation because batteries with a high energy density may be fabricated. Li^+ or I^- ions may be intercalated into π-conjugated polymers such as polyacetylene $(CH)_x$, the ionic charge being balanced respectively by the addition or removal of electrons on the polymer chain. These polymer intercalation electrodes find application in batteries and electrochromic devices.

1.6 **Interfaces** (Chapter 10)

To date the greatest emphasis in solid state electrochemistry has been placed on ionics, i.e. the study of ion transport in the bulk phases. This situation is changing. All electrochemical devices rely on the performance of the interfaces between electrolytes and electrodes, at least as much as on the performance of the bulk phases. As a result interfacial studies are growing rapidly in importance. This has only served to emphasise how little is presently understood concerning the fundamental processes at such interfaces. The solid state interface in particular exhibits unique behaviour. Our present knowledge in this most important area is described by Armstrong and Todd.

References

Angell, C. A. (1989) in *High Conductivity Solid Ionic Conductors*, Ed. T. Takahashi, World Scientific, Singapore p. 89.

Armand, M. B., Chabagno, J. M. and Duclot, M. (1978) *2nd Int. Conference on Solid Electrolytes, Extended Abstracts 20–22*, St. Andrews.

Faraday, M. (1838) *Philosophical Transactions of the Royal Society of London*, p. 90. Richard and John Taylor, London.

Koryta, J., Dvořák, J. and Kavan, L. (1993) *Principles of Electrochemistry*, Second Edition, Wiley, Chichester.

Kummer, J. T. and Weber, N. (1966) US Patent 3 458 356.

Nernst, W. (1900) *Z. Elektrochem.*, **6**, 41.

Owens, B. B. and Argue, G. R. (1970) *J. Electrochem Soc.*, **117**, 898.

Steele, B. C. H. (1992) *Materials Sci. Eng.*, **1313**, 79.

Tubandt, C. and Lorenz, E. (1914) *Z. Physik Chem.*, **87**, 513.

Wagner, C. (1956) *Z. Electrochem.*, **60**, 4.

Warburg, E. (1884) *Ann. Physik u Chem. N. F.*, **21**, 662.

Warburg, E. and Tegetmeier, F. (1888) *Ann. Physik u Chem. N. F.*, **32**, 455.

Wright, P. V. (1973) *Polymer*, **14**, 589.

Zhang, Z. and Kennedy, J. H. (1990) *Solid State Ionics*, **38**, 217.

2 Crystalline solid electrolytes I: General considerations and the major materials

A. R. WEST
Department of Chemistry, University of Aberdeen

2.1 Introduction

High ionic conductivity in crystalline solids is a widely recognised, although still relatively rare, phenomenon. Most ionic solids are electrical insulators unless they exhibit electronic conductivity. They begin to show significant levels of ionic conductivity only at high temperatures, as the melting point is approached. Materials in the family of crystalline *solid electrolytes* (also called *superionic conductors*, *fast ion conductors* or *optimised ionic conductors*), however, exhibit high conductivity in one of their ionic sublattices – the *mobile ion sublattice* – at temperatures well below melting and often as low as room temperature.

The first half of this chapter concentrates on the mechanisms of ion conduction. A basic model of ion transport is presented which contains the essential features necessary to describe conduction in the different classes of solid electrolyte. The model is based on the isolated hopping of the mobile ions; in addition, brief mention is made of the influence of ion interactions between both the mobile ions and the immobile ions of the solid lattice (ion hopping) and between different mobile ions. The latter leads to either ion ordering or the formation of a more dynamic structure, the *ion atmosphere*. It is likely that in solid electrolytes, such ion interactions and cooperative ion movements are important and must be taken into account if a quantitative description of ionic conductivity is to be attempted. In this chapter, the emphasis is on presenting the basic elements of ion transport and comparing ionic conductivity in different classes of solid electrolyte which possess different gross structural features. Refinements of the basic model presented here are then described in Chapter 3.

The second half of this chapter deals with the most important solid electrolytes and also includes discussion of their structures and properties.

2.2 Conduction mechanisms

Ionic conductivity occurs by means of ions hopping from site to site through a crystal structure, therefore it is necessary to have partial occupancy of energetically equivalent or near-equivalent sites.† Two broad classes of *conduction mechanism* – *vacancy* and *interstitial migration* – may be distinguished. In vacancy migration, a number of sites that would be occupied in the ideal, defect-free structure are in fact, empty perhaps due to either a thermally generated *Schottky defect* (a cation and anion vacancy pair) formation or the presence of charged impurities. An ion adjacent to a vacancy may be able to hop into it leaving its own site vacant. This process is regarded as vacancy migration, although, of course, it is the ions and not the vacancies that hop. An example of vacancy migration in NaCl is shown schematically in Fig. 2.1(*a*).

Interstitial sites are defined as those that would usually be empty in an ideal structure. Occasionally in real structures, ions may be displaced from their lattice sites into interstitial sites (*Frenkel defect* formation). Once this happens, the ions in interstitial sites can often hop into adjacent interstitial sites. These hops may be one stage in a long range conduction process. A schematic example is shown in Fig. 2.1(*b*): a small number of Na^+ ions are displaced into the tetrahedral interstitial sites and can subsequently hop into adjacent tetrahedral sites. It should be noted, however, that while a small number of Frenkel defects may form in NaCl, conduction is primarily by means of vacancies whereas in some other structures, e.g. AgCl, Frenkel defects do predominate.

The above two mechanisms may be regarded as isolated ion hops. Sometimes, especially in solid electrolytes, *cooperative ion migration* occurs. An example is shown in Fig. 2.1(*c*) for the so-called *interstitialcy* or *knock-on* mechanism. A Na^+ ion, A, in an interstitial site in the 'conduction plane' of β-alumina (see later) cannot move unless it persuades one of the three surrounding Na^+ ions, B, C or D, to move first. Ion A is shown moving in direction 1 and, at the same time, ion B hops out of its lattice site in either of the directions, 2 or 2′. It is believed that interstitial Ag^+ ions in AgCl also migrate by an interstitialcy mechanism, rather than by a direct interstitial hop.

In crystalline electrolytes, *conduction pathways* for the mobile ions

† A particular site must clearly be either full (i.e. contain an ion) or empty. By partial occupancy, we mean that only some of the sites in a particular crystallographic set are occupied.

permeate the 'immobile ion sublattice' in one, two or three dimensions, depending on the structure of the material. Thus, in β-alumina, Fig. 2.1(*c*), Na^+ ions can migrate only in two dimensions. The sites containing the mobile ions are not fully occupied and are connected, via open windows or *bottlenecks*, to adjacent sites that are also partially occupied or empty. In crystalline electrolytes, the sites for the mobile ions are clearly defined by the structure of the immobile sublattice (unlike melts, for instance, where there is no fixed set of sites). Ionic conduction occurs, therefore, by means of a series of definite hops between adjacent sites in the

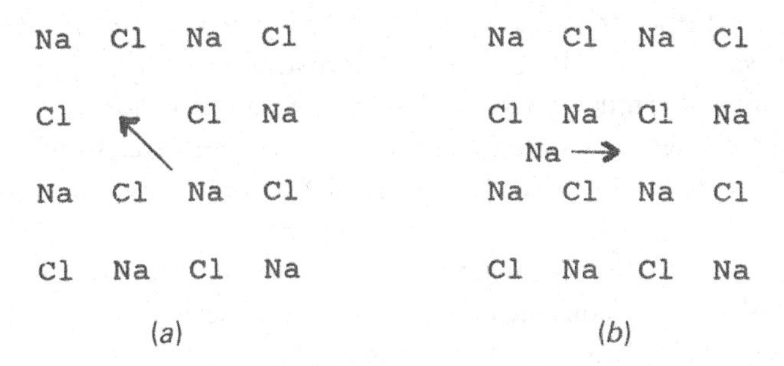

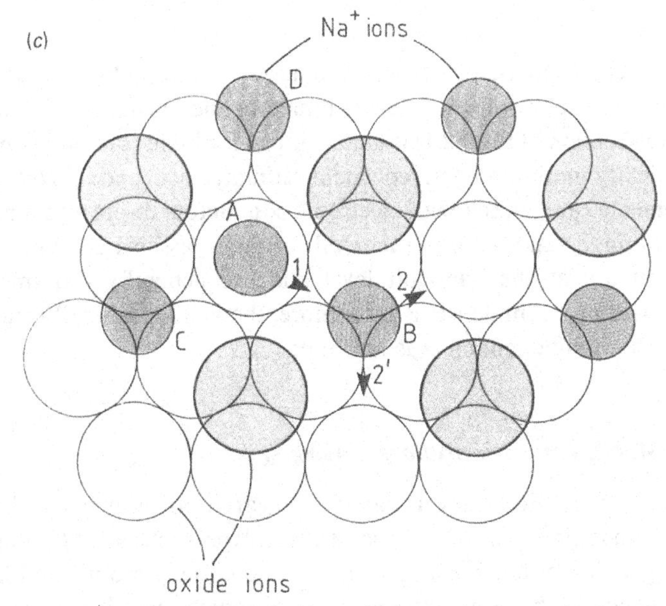

Fig. 2.1 (*a*) Vacancy, (*b*) interstitial and (*c*) interstitialcy conduction mechanisms. In (*c*), Na^+ ion A can move only by first ejecting Na^+ ion B from its site.

conduction pathways. For most of the time, the 'mobile' ions are located in a particular site, where they undergo thermal vibrations within the site. Just occasionally, they escape from their site and hop quickly into an adjacent site where they may then reside for a considerable time before either moving on or hopping back into their original site.

This notion of occasional ion hops, apparently at random, forms the basis of *random walk theory* which is widely used to provide a semi-quantitative analysis or description of ionic conductivity (Goodenough, 1983; see Chapter 3 for a more detailed treatment of conduction). There is very little evidence in most solid electrolytes that the ions are instead able to move around without thermal activation in a true liquid-like motion. Nor is there much evidence of a free-ion state in which a particular ion can be activated to a state in which it is completely free to move, i.e. there appears to be no ionic equivalent of free or nearly free electron motion.

A simple yet valuable starting point for treating ionic conductivity, σ_i, is as the product of the concentration, c_i, of mobile species (interstitial ions or vacancies), their charge, q and their mobility, u_i:

$$\sigma_i = c_i q u_i. \tag{2.1}$$

This same equation is, of course, also used to rationalise the general electronic behaviour of metals, semiconductors and insulators. The quantitative application of Eqn (2.1) is handicapped for ionic conductors by the great difficulty in obtaining independent estimates of c_i and u_i. *Hall effect* measurements can be used with electronic conductors to provide a means of separating c_i and u_i but the Hall voltages associated with ionic conduction are at the nanovolt level and are generally too small to measure with any confidence. Furthermore, the validity of Hall measurements on hopping conductors is in doubt.

2.3 Mobile ion concentrations: doping effects

The parameter c_i, Eqn (2.1), is capable of variation by many orders of magnitude in ionic solids. In good solid electrolytes such as Na β''-alumina and $RbAg_4I_5$, all of the Na^+/Ag^+ ions are potentially mobile and hence c is optimised. At the other extreme, in pure, stoichiometric salts such as NaCl, ionic conduction depends on the presence of crystal defects, whether vacancies or interstitials and the concentration of these is vanishingly small at, for instance, room temperature.

An important practical way of increasing the value of c_i is by means of doping with *aliovalent* (or *heterovalent*) ions. This involves partial replacement of ions of one type by ions of different formal charge. In order to retain charge balance, either interstitial ions or vacancies must be generated at the same time. If the interstitials or vacancies are able to migrate, dramatic increases in conductivity can result.

For aliovalent doping of cations, there are four fundamental ionic mechanisms for achieving charge balance. (There are also electronic compensation mechanisms leading to electron/hole creation and possibly to electronic conduction; these are not considered here.) These four ionic mechanisms are shown in Fig. 2.2, together with an example of each. Doping with a higher valent cation necessitates the creation of either cation vacancies (1) or anion interstitials (2), whereas doping with lower valent cations leads to the creation of either interstitial cations (3) or

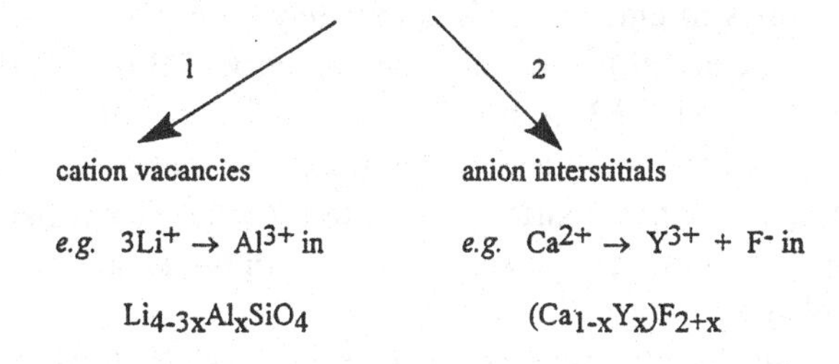

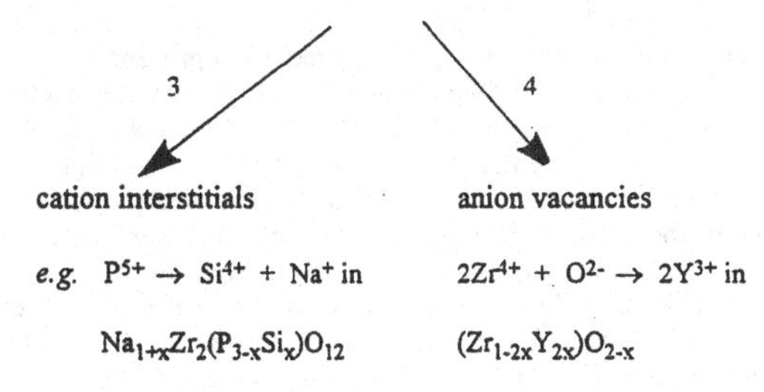

Fig. 2.2 Solid solution formation by doping with aliovalent ions.

anion vacancies (4). In the examples and formulae shown, the number of vacancies or interstitials increases with x. Usually in a particular material, there is a practical limit as to how many vacancies/interstitials can be introduced while still retaining a homogeneous solid solution phase. In many cases, this limit is small, $\ll 1\%$ but in others it may be large, 10–20%, giving rise to massive defect concentrations.

Occasionally, it is possible to vary the composition to such an extent that it is possible either to fill completely a set of interstitial sites or to empty completely a particular set of lattice sites. When this happens, random walk theory predicts that at the half-stage, when the concentrations of filled and empty sites are equal, the ionic conductivity should pass through a maximum because the product of the concentration of mobile species, c_i, and sites to which they may migrate $(1 - c_i)$ is at a maximum.

A clear example of this effect is found in the $Li_{4-3x}Al_xSiO_4$ solid solutions (Garcia, Torres-Trevino and West, 1990), scheme 1, Fig. 2.2. Single phase solid solutions form over the entire range $x = 0$–0.5. At $x = 0$, in stoichiometric Li_4SiO_4, all the Li^+ sites are full and the conductivity is low. As x increases, one particular set of Li^+ sites in the crystal structure starts to empty and is completely empty at $x = 0.5$, i.e. at $Li_{2.5}Al_{0.5}SiO_4$. The effect on the ionic conductivity of this variation is dramatic, as shown in Fig. 2.3. A broad conductivity maximum occurs around $x = 0.25$, at which composition the mobile Li^+ sites are half-full ($n_c = 0.5$, where n_c is the occupancy of the site by ions). To either side, the conductivity becomes very small as $x \rightarrow 0$ ($n_c \rightarrow 1$) and $x \rightarrow 0.5$ ($n_c \rightarrow 0$) respectively.†

In most solid electrolyte systems, it is not possible to vary the composition sufficiently so as to have the complete spectrum of mobile ion concentrations, from $n_c = 0$ to $n_c = 1$. Instead, the properties are usually limited to one or other of the 'wings' in the type of behaviour

† In this book and in general, the concentration, c_i, is used to represent the number of mobile species, ions or vacancies, per unit volume whereas n_c is always the fraction of the crystallographically equivalent sites that are occupied by ions. Often if $n_c > 0.5$ we talk of ion migration and c refers to ion concentration, whereas if $n_c < 0.5$ we talk of vacancy migration and c refers to the concentration of vacancies. However, this is simply a convenient way of thinking about ion transport in solids by focusing on the minority species. It is equally possible to describe conduction always in terms of ions or of vacancies provided account is taken of the fact that both the concentration of mobile species and sites to which they may migrate are important. The importance of c and $(1 - c)$ is emphasised and placed within a unified framework in Chapter 3. The concentration of ions c_i is related to the occupancy n_c by $n_c = c_i/C$, where C is the concentration of the sites.

shown in Fig. 2.3, depending on whether vacancies or interstitials are being introduced.

2.4 Materials with disordered sublattices: α-AgI

Thus far we have considered, starting from an ideal, defect-free structure, how to introduce defects using schemes such as that outlined in Fig. 2.2. In favourable structures, the defects may be mobile, leading to high ionic conductivity. There is a small group of materials in which defect creation by doping is unnecessary since, in the parent stoichiometric crystal, there is already extensive disorder in the mobile ion sublattice above 0 K. A classic example of this is provided by the high temperature, α polymorph of AgI stable about 146 °C; similar effects occur at room temperature in various complex salts such as $RbAg_4I_5$.

α-AgI has a body centred cubic arrangement of I^- ions, Fig. 2.4(*a*),

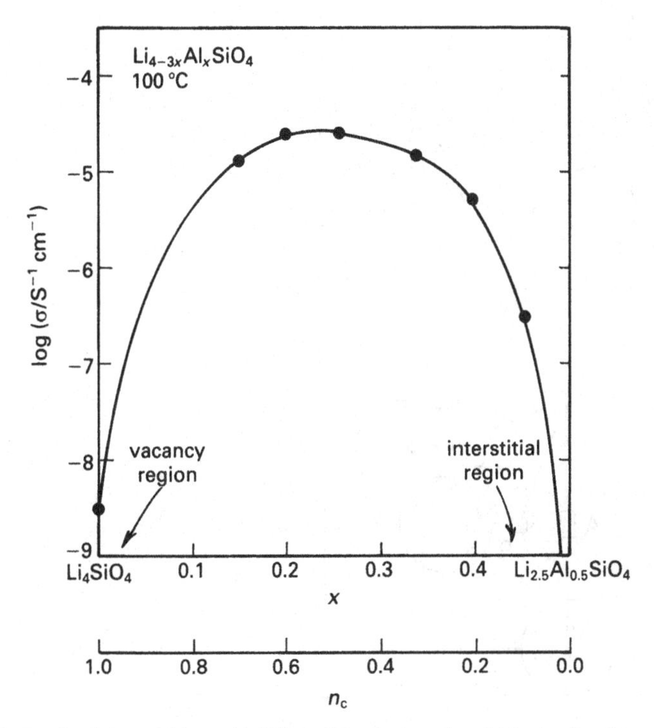

Fig. 2.3 Conductivity of $Li_{4-3x}Al_xSiO_4$ solid solutions in which the fraction of occupied mobile ion sites, n_c, varies from 1 (at $x = 0$) to 0 (at $x = 0.5$).

which form a rigid anion sublattice. The Ag^+ ions (on average 2 per unit cell) are distributed over a very large number of sites with coordination numbers 2, 3 and 4 (Strock, 1934; Hoshino, 1957; Geller, 1967; Wright and Fender, 1977), although the latest evidence indicates that the ions spend most of their time in the four-coordinate sites (Yoshiasa, Maeda, Ishii and Koto, 1990). Some of these sites, located on one face of the unit cell, are shown in Fig. 2.4(*b*). Two-coordinate sites, A, occur midway along cube cell edges; displaced from these and in the direction of the cube face centres are distorted tetrahedral sites, B. Since adjacent pairs of tetrahedral sites – B, B′ – share a common 'face' of three I^- ions, a three-coordinate site, C, is found in the middle of such faces.

The Ag^+ ions are arranged at random or 'statistically' over these various interstitial sites with a preference for the four- and to a lesser

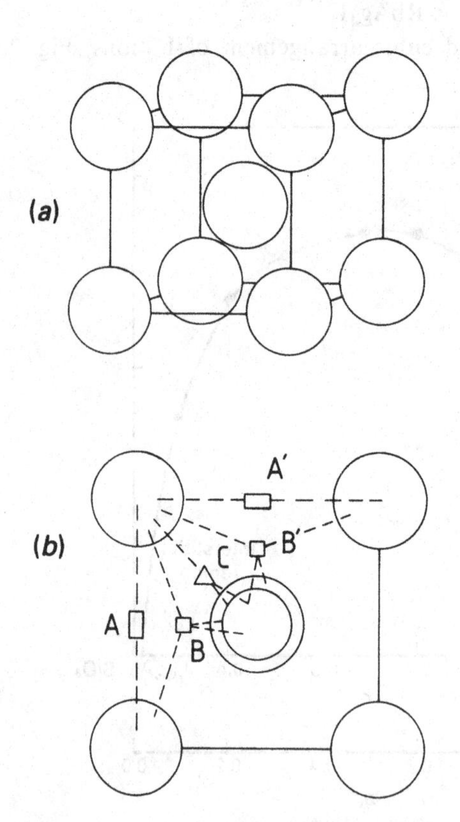

Fig. 2.4 (*a*) Body centred lattice of I^- ions in α-AgI. (*b*) Sites available to Ag^+ ions in the conduction pathway.

extent three-coordinate sites. Clearly, it is easy for them to move between sites, as reflected in the very low activation energy for conduction (see later). We may say that, from energetic considerations, the individual two-, three- and four-coordinate sites do not form deep *potential traps* for the Ag^+ ions; instead, the Ag^+ ions occupy rather shallow potential wells and adjacent wells are connected by low potential barriers. Thus, ions spend most of their time in the four-coordinate sites, passing through two- and three-coordinate sites during long range migration (Yoshiasa *et al.*, 1990).

The disordered nature of the α-AgI structure may be regarded as intermediate between that of a typical ionic solid in which every lattice site is occupied and a typical ionic liquid in which both anions and cations are disordered. This is borne out by entropy calculations based on heat capacity data (O'Keeffe and Hyde, 1976). These show a large increase in entropy at the β (or γ) to α transition and a similar increase in entropy on melting of the α polymorph at 557 °C:

$$\beta\text{-AgI} = \xrightarrow[147\,^\circ\text{C}]{\Delta S\,=\,14.5\ \text{J K}^{-1}\,\text{mol}^{-1}} \alpha\text{-AgI} \xrightarrow[557\,^\circ\text{C}]{\Delta S\,=\,11.3\ \text{J K}^{-1}\,\text{mol}^{-1}} \text{liquid}$$

Effectively, therefore, the $\beta \rightarrow \alpha$ transition may be regarded as 'melting' or disordering of the Ag^+ ion sublattice and the $\alpha \rightarrow$ liquid transition as melting of the I^- ion sublattice.

α-AgI is, in many ways, an 'ideal' electrolyte. It is solid due to the rigid sublattice of I^- ions. The carrier concentration is high since all of the Ag^+ ions are potentially mobile. It has a low activation barrier to migration and consequently, the conductivity is high, $\sim 1\ \text{S cm}^{-1}$ at 147 °C. It therefore combines the advantages of a strong liquid-electrolyte level of conductivity with having only one species mobile (unlike liquid electrolytes where everything moves) and having the mechanical properties of a solid. An additional advantage arises because α-AgI is stoichiometric, does not require dopants to achieve its high conductivity and, indeed, is insensitive to the presence of dopants. One major disadvantage is that α-AgI is stable only at higher temperatures, >147 °C. Similar features are found at room temperature, however, in phases such as $RbAg_4I_5$ (Bradley and Greene, 1966; Owens and Argue, 1967).

2.5 Ion trapping effects

The above situation, in which Ag^+ ions are disordered over, and readily move between, a large number of interstitial sites contrasts with the properties of many of the dopant-induced mobile ion systems, such as indicated in Fig. 2.2. This is because the aliovalent dopant may 'trap' or form complexes with the associated vacancies or interstitials. To see this, consider the case of the oxide ion conducting, lime-stabilised zirconia system given by the general formula $(Zr_{1-x}Ca_x)O_{2-x}$. The replacement mechanism is

$$Zr^{4+} + O^{2-} \rightleftharpoons Ca^{2+}$$

When we consider the defect charges on the species involved, using the *Kröger–Vink notation* in which the superscripts $^{\bullet}$, $'$ and $^{\times}$ refer to positive, negative and neutral species, the above equation may be rewritten:

$$Zr^{\times} + O^{\times} \rightleftharpoons Ca_{Zr}'' + V_O^{\bullet\bullet}.$$

The substitution of Ca onto a Zr site leaves a residual charge of 2− on that site, whereas the oxygen vacancy, V_O, that is created carries an effective double positive charge. Since the aliovalent impurity and the anion vacancy carry opposite effective charges, they are likely to attract each other strongly, forming dipoles, quadrupoles or larger clusters. In order for the vacancy to move it must first break free from the cluster and this adds an additional barrier to the activation energy for conduction.

The occurrence of such ion trapping is clearly undesirable since it inevitably leads to a decrease in conductivity. In practice, in materials that contain potential traps such as charged aliovalent impurities/dopants, the conductivity values of a particular sample may actually decrease with time as the mobile ions gradually become trapped. Such *ageing effects* greatly limit the usefulness of a solid electrolyte in any device that needs to have a long working-life.

2.6 Potential energy profiles

It may be useful to think of the conduction pathway for a mobile ion as a series of potential wells and barriers. An example of a schematic energy profile is shown for α-AgI in Fig. 2.5(*a*), for sites connected in the sequence A–B–C–B′–A′, Fig. 2.4(*b*). Sites B are of somewhat lower energy than the A and C sites and form the preferred residences of Ag^+ ions. However, the barriers for hopping between the different sites are all low.

The relatively shallow potential energy profile for Ag^+ ions in α-AgI, Fig. 2.5(*a*), may be contrasted with that shown schematically for Na^+ vacancies, in NaCl, Fig. 2.5(*b*) and Ag^+ interstitials in AgCl, Fig. 2.5(*c*).

2.6 Potential energy profiles

In NaCl, a large energy barrier, ΔH_m must be overcome for vacancy migration to occur; there is an interstitial site (Frenkel defect) in the conduction pathway which may be regarded as a transition state but it has such a shallow potential well that Na^+ ions do not reside in it for significant times. In AgCl, creation of a Frenkel defect (an interstitial Ag^+ ion) also requires a large amount of energy $\Delta H_g/2$ (see Chapter 3) but once formed, such defects have a certain amount of stability. Migration between adjacent interstitial sites is relatively easy, since the barrier, ΔH_m to interstitial migration, by means of the interstitialcy mechanism, is considerably less than $\Delta H_g/2$.

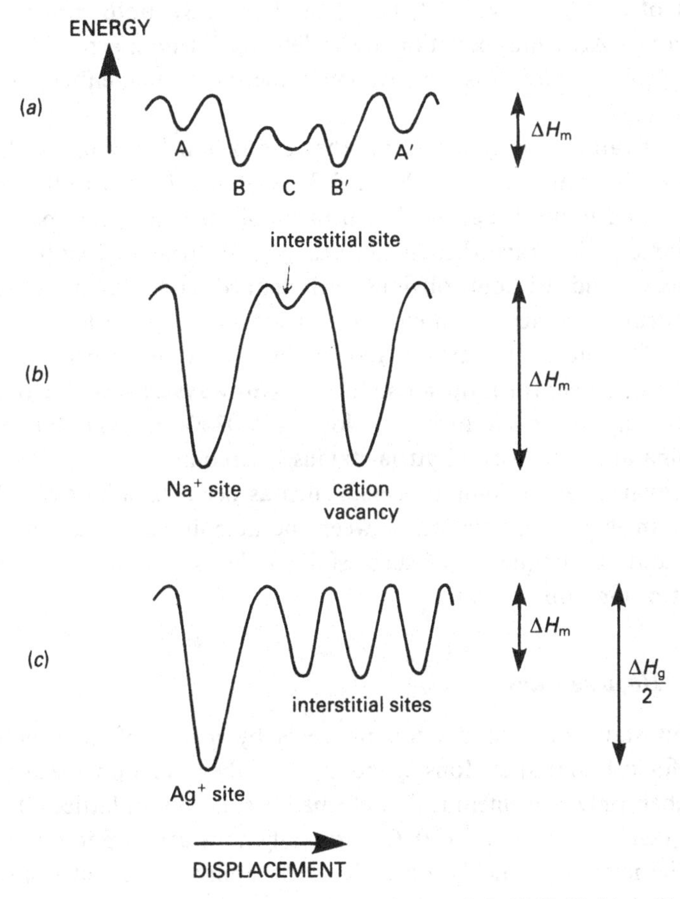

Fig. 2.5 Potential energy profile for ion migration in (*a*) α-AgI, (*b*) NaCl, (*c*) AgCl.

2.7 The activation energy for conduction

The activation energy for conduction, ΔH_m, is the major factor controlling the ionic mobility, u. The Arrhenius expression for conductivity is either

$$\sigma = A\exp(-\Delta H_m/RT) \quad (2.2)$$

or

$$\sigma T = A_T\exp(-\Delta H_m/RT) \quad (2.3)$$

The prefactor A or A_T contains many terms, including the number of mobile ions. Of the two equations, Eqn (2.3) is derived from random walk theory and has some theoretical justification; Eqn (2.2) is not based on any theory but is simpler to use since data are plotted as log σ vs T^{-1} instead of as log σT vs T^{-1}, based on Eqn (2.3). Both forms of the conductivity Arrhenius equation are widely used; within errors the value of ΔH_m that is obtained is approximately the same using either equation in many cases.

The activation energy represents the ease of ion hopping, as already indicated above and shown in Fig. 2.5. It is related directly to the crystal structure and in particular, to the openness of the conduction pathways. Most ionic solids have densely packed crystal structures with narrow bottlenecks and without obvious well-defined conduction pathways. Consequently, the activation energies for ion hopping are large, usually 1 eV ($\sim$96 kJ mole^{-1}) or greater and conductivity values are low. In solid electrolytes, by contrast, open conduction pathways exist and activation energies may be much lower, as low as 0.03 eV in AgI, 0.15 eV in β-alumina and $\sim$0.90 eV in yttria-stabilised zirconia.

In activated ion hopping processes such as occur in solid electrolytes, there is an inverse correlation between the magnitude of the activation energy and the frequency of successful ion hops; this leads us to the concept of ion hopping rates.

2.8 Hopping rates

All solid state ionic conduction proceeds by means of hops between well-defined lattice sites. Ions spend most of their time on specific sites where their only movement is that of small oscillations at lattice vibrational frequencies (10^{12}–10^{13} Hz). Occasionally, ions can hop into adjacent sites. The ions hop quickly, on a timescale approaching but somewhat longer than that of a single lattice vibration. This is because the hop distance, typically 1–2 Å, is an order of magnitude greater than the atomic displacement during a lattice vibration.

There are two times to consider, therefore, in order to characterise ionic conduction. One is the actual time, t_j, taken to jump between sites; this is of the order of 10^{-11}–10^{-12} s and is largely independent of the material. The other time is the site residence time, t_r, which is the time (on average) between successful hops. The site residence times can vary enormously, from nanoseconds in the good solid electrolytes to geological times in the ionic insulators. Ion hopping rates, ω_p, are defined as the inverse of the site residence times, i.e.

$$\omega_p = t_r^{-1}. \tag{2.4}$$

Hopping rates are traditionally obtained from mechanical relaxation techniques such as internal friction or ultrasonic attenuation measurements (Almond and West, 1988). In these, the sample is squeezed or stressed at a certain frequency and ions may hop so as to relieve the stress. This effect is well documented in studies of atomic diffusion in metals, where it is known as the *Snoek effect*. When the conditions are such that the frequency of the applied stress coincides with the ion hopping rate, a maximum in the absorption or attenuation occurs. Since hopping rates vary with temperature (in a manner dependent on the activation energy for hopping or conduction), the method usually used to determine ω_p involves a scan of temperature at fixed frequency. This gives an estimate of the temperature at which the ion hopping rate equals the applied frequency, Fig. 2.6. If the measurements are repeated over a range of set

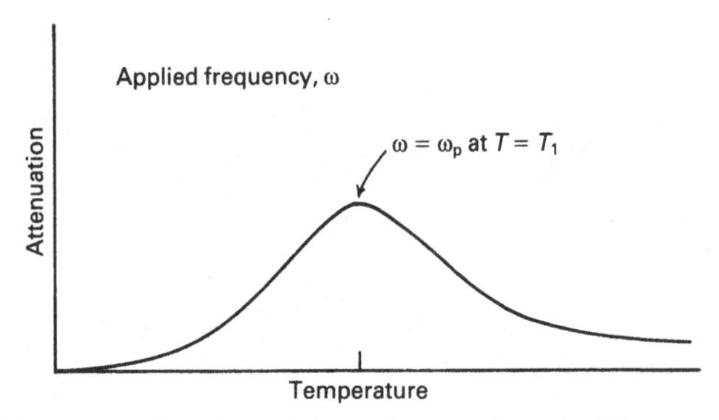

Fig. 2.6 Temperature dependence of ultrasonic attenuation at fixed frequency ω for an ionic conductor.

frequencies, the temperature dependence, and hence the activation energy, of ion hopping rates may be obtained.

The ion hopping rate is an apparently simple parameter with a clear physical significance. It is the number of hops per second that an ion makes, on average. As an example of the use of hopping rates, measurements on Na β-alumina indicate that many, if not all the Na^+ ions can move and at rates that vary enormously with temperature, from, for example, 10^3 jumps per second at liquid nitrogen temperatures to 10^{10} jumps per second at room temperature. Mobilities of ions may be calculated from Eqn (2.1) provided the number of carriers is known, but it is not possible to measure ion mobilities directly.

2.9 *The ac conductivity spectrum: local motions and long range conduction*

The picture outlined above of ionic conduction in crystals is one of rapid hops between adjacent sites, separated by long residence times in which the ions are confined to oscillations within particular sites. The residence times depend on (amongst other things) the activation energy for hopping. Activation energy is a complex parameter that includes not only a physical barrier, in which an ion has to squeeze through a narrow bottleneck, but also a longer range electrostatic barrier between the mobile ions. This arises whenever an ion hops out of a regular lattice site, Fig. 2.7, arrow 1. Hopping generates local departures from electroneutrality, which may also be viewed as the creation of dipoles, Fig. 2.7(*b*). Such departures from local electroneutrality act as a drag on further ion hops. Local electro-

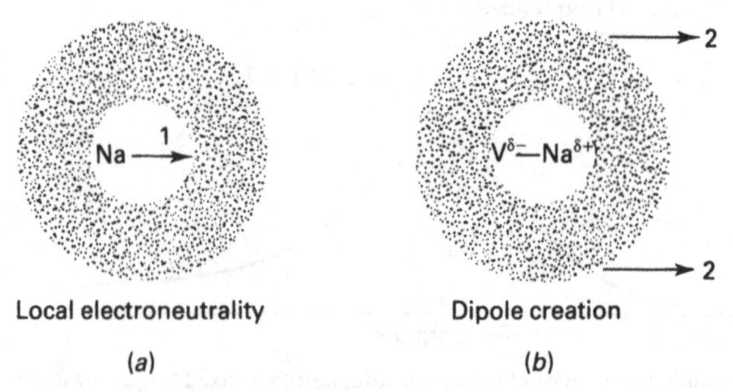

Fig. 2.7 Generation of dipoles due to ion displacements.

neutrality can be restored, however, by means of a redistribution in the positions of the surrounding ions, arrow 2, Fig. 2.7; after this process is complete, the ion under consideration is able to move again.

This effect is reminiscent of a similar phenomenon in liquid electrolytes known as the *Debye–Falkenhagen effect* (Debye and Falkenhagen, 1928). In it, an ion is not free to move unhindered but has an associated *ion atmosphere* (Debye and Hückel, 1923; Onsager, 1927). For long range conduction to occur, the ion must drag along its ion atmosphere. One key difference from ionic conduction in crystalline solids is that both anions and cations are able to move in liquids. Both are likely to be involved in reorganisation of the ion atmosphere therefore. In ionic solids, however, only one type of ion is involved, both in conduction and in the reorganisation of its surrounding ion atmosphere.

In materials that have high ionic conductivity, effects such as the above are undoubtedly very important. They show up particularly in materials that have a high concentration of mobile ions and in experimental values of the ac conductivity measured as a function of frequency. In materials with a high carrier concentration, mobile ions are inevitably quite close together, separated by at most a few angstroms. Consequently, ions cannot hop in isolation but are influenced by the distribution of mobile ions in their vicinity. This contrasts with the behaviour of dilute defect systems with low carrier concentrations. In these, the mobile ions are well separated from each other and their conduction can largely be treated in terms of isolated hops.

Some experimental data that illustrate these points are shown for Na β-alumina single crystals (Grant, Hodge, Ingram and West, 1977; Strom and Ngai, 1981) in Fig. 2.8. The ac conductivity is given as a function of frequency. At low frequencies, a plateau region (A) is obtained and corresponds to the conductivity value that would be measured by dc methods. In this plateau region, Na^+ ions have time to migrate over large distances, by making many successful hops during one half-cycle of the applied ac field. At the other end of the conductivity spectrum, at, e.g. 10^{10}–10^{11} Hz, a short plateau region is again observed, B. On this timescale, ions have time to make only a single hop during one half-cycle of the ac field. At intermediate frequencies, a broad conductivity dispersion is present in which the conductivity decreases with decreasing frequency, C. This is the region in which an ion spends a considerable amount of time 'waiting' for the surrounding ions to reorganise themselves.

Various theoretical attempts have been made to provide a quantitative interpretation of the dispersion region (Funke, 1986; Funke and Hoppe, 1990). While the situation is still not fully resolved, it is now clear that such a dispersion, which has been observed in a wide range of crystalline as well as glassy ionic conductors, is associated with ion–ion relaxation effects. The conductivity dispersion, $\sigma(\omega)$, is usually linear in a plot of log σ vs log ω, which means that it can be represented by a power law expression:

$$\sigma(\omega) = A\omega^s \qquad 0 < s < 1. \tag{2.5}$$

This behaviour is one example of a wide range of phenomena which are manifestations of *Jonscher's Universal Law of Dielectric Response* (Jonscher, 1977, 1983).

We earlier defined the hopping rate parameter, ω_p. It corresponds approximately to the frequency at which the conductivity dispersion commences, arrowed in Fig. 2.8.

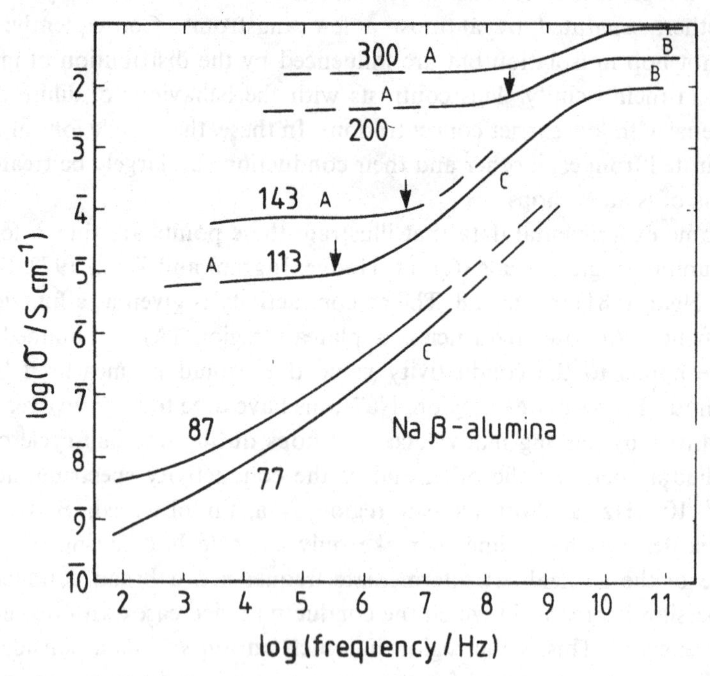

Fig. 2.8 Ionic conductivity of Na β-alumina single crystals. Numbers on the plot represent temperatures in Kelvin.

2.10 Survey of solid electrolytes: general comments

Solid electrolyte behaviour has been reported in a wide range of materials and is now known for a considerable number of mobile ions. Some key examples for each ion are listed in Table 2.1.

Silver ion conductors are numerically the most common and generally have the highest conductivities. Many are chalcogenides or complex halides, often derived from the α-AgI structure, Fig. 2.4. There is also a considerable number of monovalent copper ion conductors, although these are less well studied than the Ag^+ materials. The bonding between Ag(Cu) and the immobile chalcogenide/iodide sublattice appears to have considerable covalent character. Thus, the coordination numbers of Ag(Cu) are often low, e.g. 2, 3 or 4, which is a characteristic of covalent bonding. Such covalent bonding may have a key role in reducing the activation energy for conduction, by acting to stabilise the two-coordinate intermediate sites in the conduction pathways, Fig. 2.4. Coordination numbers of 2 are common for the d^{10} ions Ag^+, Cu^+ and occur in, for example, Ag_2O, Cu_2O. Stabilisation of two-coordinate intermediate sites would effectively reduce the bottleneck for ionic migration and hence reduce the activation energy, ΔH_m, as shown schematically in Fig. 2.5(*a*).

Sodium ion conduction appears to be common because of the well-known properties of the beta-aluminas and, to a lesser extent, the NASICONs (see Section 2.12.1), Table 2.1. There are, however, relatively few other examples of high Na^+ ion conductivity, especially at room temperature. In contrast to Ag, the usual coordination number of Na is high, often 7–9, and the sites may be distorted. The bonding of Na in such structures is much more ionic than that of Ag, therefore.

The high level of Na^+ ion conductivity in the beta-aluminas is probably a fortuitous consequence of having a crystal structure with open conduction pathways and a large number of partially occupied sites. The beta-alumina structure (β or β'' polymorphs, see later) is a marvellous host structure for a wide range of cations. Thus, as well as the Na forms, which are thermodynamically stable and can be prepared by solid state reaction at high temperature, a wide variety of cations can be exchanged for Na^+ by ion exchange in a molten salt. These metastable, ion exchanged materials are also, usually, good solid electrolytes and some examples are given in Table 2.1. A most unusual feature of the β''-aluminas is that both

Table 2.1. *Some solid electrolytes*

Solid electrolyte	Mobile ion	Conductivity/S cm^{-1}, temperature/°C	Activation energy/eV	Reference
Na β-alumina	Na^+	1.4×10^{-2}, 25	0.15	Whittingham and Huggins, 1972
NASICON $Na_3Zr_2PSi_2O_{12}$	Na^+	1×10^{-1}, 300	temperature-dependent	Goodenough, Hong and Kafalas, 1976; Kreuer, Kohler and Maier, 1989
Ag β-alumina	Ag^+	6.7×10^{-3}, 25	0.16	Whittingham and Huggins, 1972
K β-alumina	K^+	6.5×10^{-5}, 25	0.27	Whittingham and Huggins, 1972
$RbAg_4I_5$	Ag^+	0.25, 25	0.07	O'Keeffe and Hyde, 1976; Bradley and Greene, 1966; Owens and Argue, 1967
Li_3N (H-doped)	Li^+	6×10^{-3}, 25	0.20	Lapp, Skaarup and Hooper, 1983
$Li_{3.6}Ge_{0.6}V_{0.4}O_4$	Li^+	4×10^{-5}, 18	0.44	Kuwano and West, 1980
$Rb_4Cu_{16}I_7Cl_{13}$	Cu^+	0.34, 25	0.07	Takahashi, Yamamoto, Yamada and Hayashi, 1979
CuTeBr	Cu^+	1×10^{-5}, 25	0.11	von Alpen, Fenner, Marcoll and Rabenau, 1977
Pb β''-alumina	Pb^{2+}	4.6×10^{-3}, 40	variable	Seevers, DeNuzzio, Farrington and Dunn, 1983
$SrCe_{0.95}Yb_{0.05}O_{3-\delta}H_x$	H^+	8×10^{-3}, 900		Iwahara, Uchida and Tanaka, 1986
$H_3PW_{12}O_{40} \cdot 29H_2O$	H^+	0.17, 25	0.14	Nakamura, Kodama, Ogino and Miyake, 1979
LaF_3	F^-	3×10^{-6}, 27	$\sim$0.45	Roos, Aalders, Schoonman, Arts and de Wijn, 1983
PbF_2	F^-	1.0, 460		Benz, 1975
$(Bi_{1.67}Y_{0.33})O_3$	O^{2-}	1×10^{-2}, 550	0.80	Takahashi and Iwahara, 1978
$Ce_{0.8}Gd_{0.2}O_{1.9}$	O^{2-}	5×10^{-2}, 727	variable	Kudo and Obayashi, 1976

divalent and trivalent ions can be introduced into the structure by ion exchange, where they also conduct readily.

Lithium ion conductors are much sought after, because of the high voltages and power densities that may be obtained with lithium batteries, but as yet, none have been found that are highly conducting, easy to make and stable in a range of environments. Thus, lithium beta-alumina has very high conductivity (although still less than that of Na^+, Ag^+ beta-alumina) but is very difficult to prepare pure and free from water. The Li^+ ion is smaller and more polarising than the Na^+ ion; high Li^+ ion conductivity appears to require more polarisable environments than Na^+, therefore, and there is some evidence for this in that certain sulphides and iodides have higher conductivity than corresponding oxides or fluorides.

A special type of crystal structure that appears to favour moderately high Li^+ ion conduction is the tetragonal packed anion array which is found in various Li_4SiO_4-derivative materials. There is also a group of composite materials, e.g. LiI/Al_2O_3 composites, that have moderately high Li^+ ion conductivity associated with the interfacial region between LiI and Al_2O_3 grains.

Anion conduction, particularly oxide and fluoride ion conduction, is found in materials with the fluorite structure. Examples are CaF_2 and ZrO_2 which, when doped with aliovalent impurities, Fig. 2.2, schemes 2 and 4, are F^- and O^{2-} ion conductors, respectively, at high temperature. The δ polymorph of Bi_2O_3 has a fluorite-related structure with a large number of oxide vacancies. It has the highest oxide ion conductivity found to date at high temperatures, $>660\,°C$.

The reasons why the fluorite structure should be such a good host for anion conduction are still not clear. The cations in CaF_2 or the cubic polymorph of ZrO_2 form a face centred cubic array and the anions are distributed over the tetrahedral sites, which are presumably fully occupied in, for example, CaF_2. There is evidence that, especially in the doped materials, defect clusters may form in preference to, for instance, simple vacancies and these may have a key role in the conduction process. Conductivity in this structure type is not restricted to anions. In the inverse, antifluorite structure possessed by the alkali oxides, Na_2O, Li_2O and in the cation excess antifluorite structure of Na_3PO_4 and its derivatives, conduction of alkali ions occurs. The structure of Na_3PO_4 may be represented as $Na[Na_2X]$: $X = PO_4$ where that part in brackets, Na_2X, possesses the antifluorite structure.

2.11 The beta-aluminas

2.11.1 Stoichiometry

The beta-aluminas are ceramic oxides containing mainly Na_2O and Al_2O_2, often with small amounts of MgO and/or Li_2O (Kummer, 1972; Kennedy, 1977; Collongues, Thery and Boilot, 1978; Moseley, 1985). There are two 'parent' phases, designated β and β''. The β phase has a formula that is often quoted as being ideally $Na_2O.11Al_2O_3$, or $NaAl_{11}O_{17}$. In practice, it always contains more soda than this and the composition is around $Na_2O.8Al_2O_3$, or $Na_{1.33}Al_{11}O_{17.17}$.

The β'' phase is richer still in soda and its formula is usually given as being in the range $Na_2O.(5\text{–}7)Al_2O_3$. It appears not to be thermodynamically stable, however, unless certain stabiliser ions, particularly Li^+ and/or Mg^{2+}, are present. Its formula then appears to lie, ideally, on the so-called 'spinel join' between $NaAl_5O_8$, $LiAl_5O_8$ and $MgAl_2O_4$. In the case of lithia-stabilised β''-alumina the formula is $(Na_{1-x}Li_x)Al_5O_8$, $0.18 < x < 0.28$, although in addition, a range of solid solutions containing excess Al^{3+} ions also forms the β'' structure (Duncan and West, 1988).

In magnesia-stabilised β''-alumina, the formula is $Na_{1-x}Mg_{2x}Al_{5-x}O_8$ and x is essentially fixed at 0.175. An alternative way of representing it is in terms of a derivation from the stoichiometry $NaAl_{11}O_{17}$, with the formula $Na_{1+z}Mg_zAl_{11-z}O_{17}$. Various values of z are quoted in the literature, between 0.5 and 0.8. The value 0.747 which is in this range coincides with that obtained using the alternative, spinel-like formula given above, with $x = 0.175$.

2.11.2 Structure

The beta-alumina structures show a strong resemblance to the spinel structure. They are layered structures in which densely packed blocks with spinel-like structure alternate with open 'conduction planes' containing the mobile Na^+ ions. The β and β'' structures differ in the detailed stacking arrangement of the spinel blocks and conduction planes, Fig. 2.9.

In β'', (*a*), the oxide packing arrangement comprises close packed layers in a cubic close packed, ... ABC ..., stacking sequence, which continues throughout the structure. The spinel blocks are four oxide layers thick and contain Al^{3+} ions (and stabilising Li^+, Mg^{2+} ions) distributed over tetrahedral and octahedral sites, as in spinel, $MgAl_2O_4$. These sites are too small to accommodate the Na^+ ions which, instead, reside in the conduction planes. Up to $\frac{3}{4}$ of the oxide ions are missing from the conduction planes but in their place the Na^+ ions are located. The remaining $\frac{1}{4}$ of the oxide ions, the so-called bridging oxygens, serve to prop apart adjacent spinel blocks and prevent the structure from collapsing. Part of a conduction plane is shown in Fig. 2.1(*c*). Underneath is a close packed oxide layer (open circles), forming one end of a spinel block; the shaded oxide ions are the bridging oxide ions of the conduction plane; sites A, B, C, D indicate possible sites for Na^+ ions. On top (not shown) is the first oxide layer of the next spinel block. Since the stacking sequence of the oxide layers, including the conduction planes, is ABC in β'', the two oxide layers on either side of the conduction plane are staggered relative to each other.

The stacking sequence in β-alumina is shown in Fig. 2.9(*b*). It again has

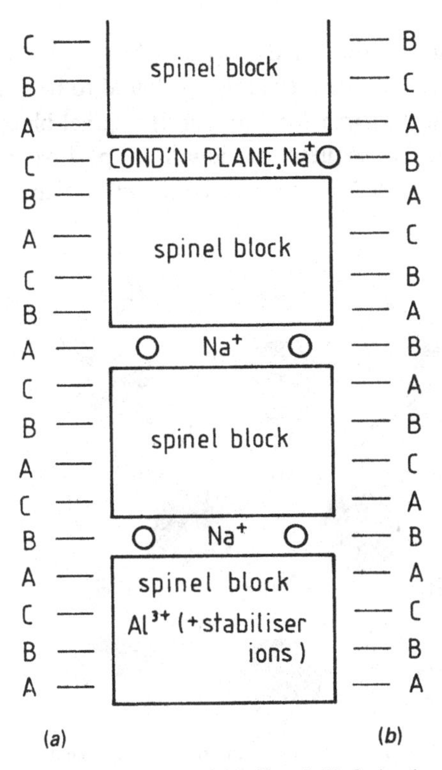

Fig. 2.9 Schematic structures of (*a*) β'' and (*b*) β alumina.

spinel blocks, four oxide layers thick, in an ABCA sequence. The conduction planes coincide with a mirror plane of symmetry, however, and hence the oxide stacking sequence is reversed on passing to the next spinel block. This means that the oxide layers on either side of the conduction plane would be superposed in the projection shown in Fig. 2.1(*c*).

In both β- and β″-alumina, the conduction planes contain a non-integral number of Na^+ ions and there are considerably more sites available than Na^+ ions to fill them. The structures are often described in terms of the supposedly ideal 1:11 stoichiometry with the formula $NaAl_{11}O_{17}$. In such a case, of the three out of four oxide ions that are missing from the conduction planes, only one half of their vacant sites would contain a Na^+ ion, as indicated schematically in Fig. 2.10. In practice, excess Na^+ ions are almost always present (e.g. ion A in Fig. 2.1(*c*)) but rarely in sufficient quantities to fill all the available vacancies; this then gives rise to the high carrier concentrations in these phases.

The excess of Na^+ ions (above the $NaAl_{11}O_{17}$ stoichiometry) requires some means of charge balance and the mechanism is found to be different in the β and β″ structures. In β″, some Al^{3+} ions in the spinel blocks are replaced by Li^+/Mg^{2+}; the consequent reduction in positive charge compensates that of the excess Na^+ ions in the conduction planes. In β,

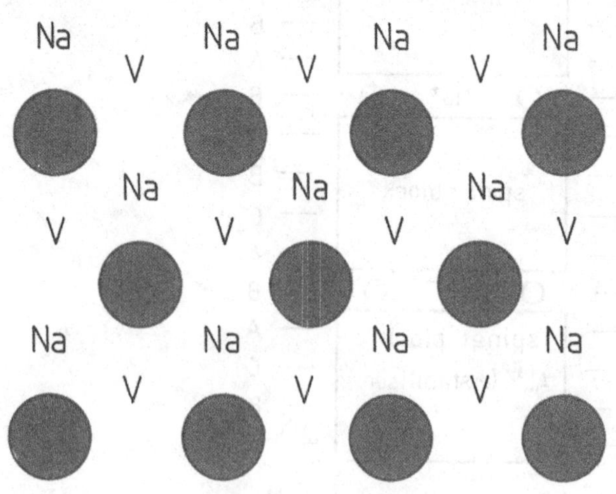

Fig. 2.10 Ordered arrangement of Na^+ ions in 'stoichiometric', soda-deficient β-alumina, $NaAl_{11}O_{17}$.

however, extra oxygen ions (interstitial oxygens) are present in the conduction planes, along with the extra Na^+ ions and in the approximate ratio 1:2 required for charge balance.

The different oxide stacking sequences in β and β'', Fig. 2.9, and in particular, the presence of mirror symmetry in the conduction plane of β but not β'', lead to differences in detail in the nature of the sites for Na^+ ions in the conduction plane. Such differences together with the different charge compensation mechanisms cause the electrical properties of β and β'' to differ somewhat, and in particular lead to rather different conduction mechanisms.

In β-alumina, Na^+ ions appear to move by means of a knock-on or interstitialcy mechanism in which it is convenient to regard the excess Na^+ ions as occupying interstitial sites. When these sites are nearly empty, as in crystals of composition close to the $NaAl_{11}O_{17}$ stoichiometry, then the conductivity is much reduced. In β''-alumina, by contrast, it is more appropriate to regard conduction as a vacancy process in which the limiting composition without vacancies would correspond to $NaAl_5O_8$.

2.11.3 Properties

The beta-aluminas are two-dimensional Na^+ ion conductors; Na^+ ions can move readily within the conduction planes but cannot penetrate the dense spinel blocks. Conductivity data for Na β-alumina in the form of an Arrhenius plot are shown in Fig. 2.11. The behaviour of β-alumina is beautifully simple; a straight line plot is obtained over a very wide range of temperatures and conductivities. The activation energy is low, 0.15 eV. Data for β'' are more complex, giving a curved Arrhenius plot (Wang, 1982). At high temperatures, $>300\,°C$, the activation energy is very low, <0.1 eV, indicating almost liquid-like disorder of the Na^+ ions, but at lower temperatures it increases to $\sim$0.25–0.30 eV. This latter effect is attributed to the onset of ordering in the positions of the Na^+ ions: the ions tend to prefer certain sites, in ordered fashion, which gives rise to domains of ordered structure. In order for Na^+ ions to move, a disordering enthalpy is required and this contributes to the increased activation energy. This effect is similar to the ion trapping effects discussed above; in the present case, however, rather than associating themselves with aliovalent dopants, the mobile ions order themselves into a regular array which is of lower free energy than the disordered state.

The Na^+ ions in both β- and β''-alumina can be replaced by a wide range of other ions, simply by immersing the crystals in a molten salt containing the desired cation (Kummer, 1972). Partition equilibria of the exchanging ions are established between the crystals and melt, and it may be necessary to repeat the process with fresh melt in order to obtain complete exchange. In this way, most monovalent ions have been substituted for Na^+ in the β structure; conductivity data for some of these are also given in Fig. 2.11(*b*). The conductivity of Ag^+ β-alumina is very similar to that of Na^+ β; for the other cations, the activation energy rises and conductivity decreases as the size of the substituting cation increases.

The ion exchange properties of β''-alumina are remarkable and perhaps unique in that many divalent and trivalent cations can be exchanged into

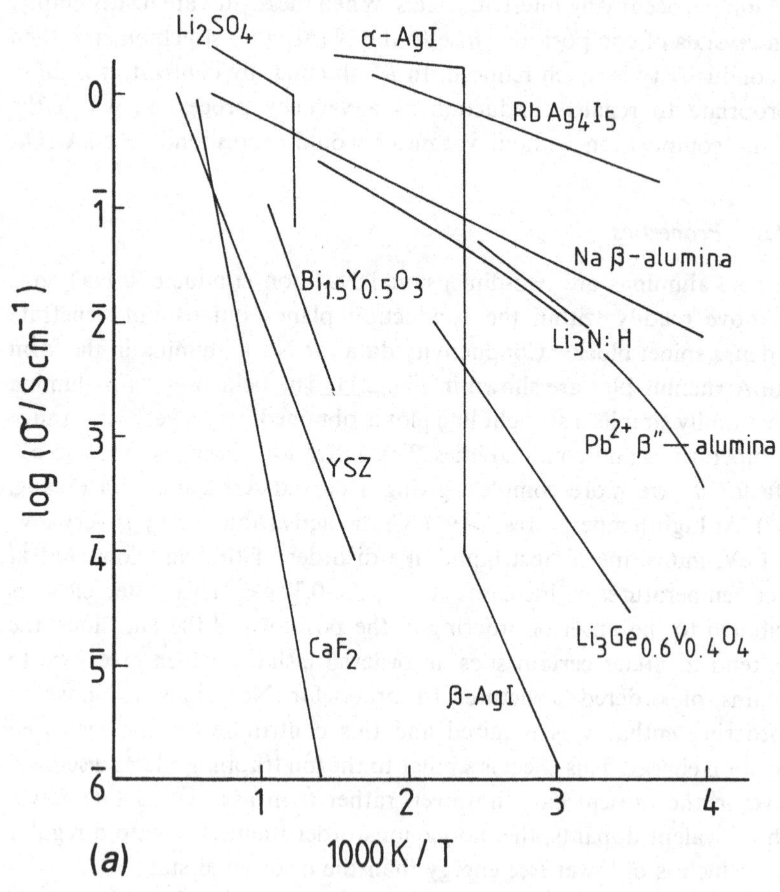

Fig. 2.11 Conductivity Arrhenius plots, shown on two separate figures for clarity.

the crystals (Farrington, Dunn and Thomas, 1989). Not only that, but they have high mobility, as shown in Fig. 2.11(*a*). Thus, the conductivity of Pb^{2+} β''-alumina is almost as high as that of the parent Na β''-alumina.

2.12 Other alkali ion conductors

2.12.1 NASICON

NASICON, the acronym for Na superionic conductor, is a non-stoichiometric framework zirconophosphosilicate (Kreuer *et al.*, 1989). It is

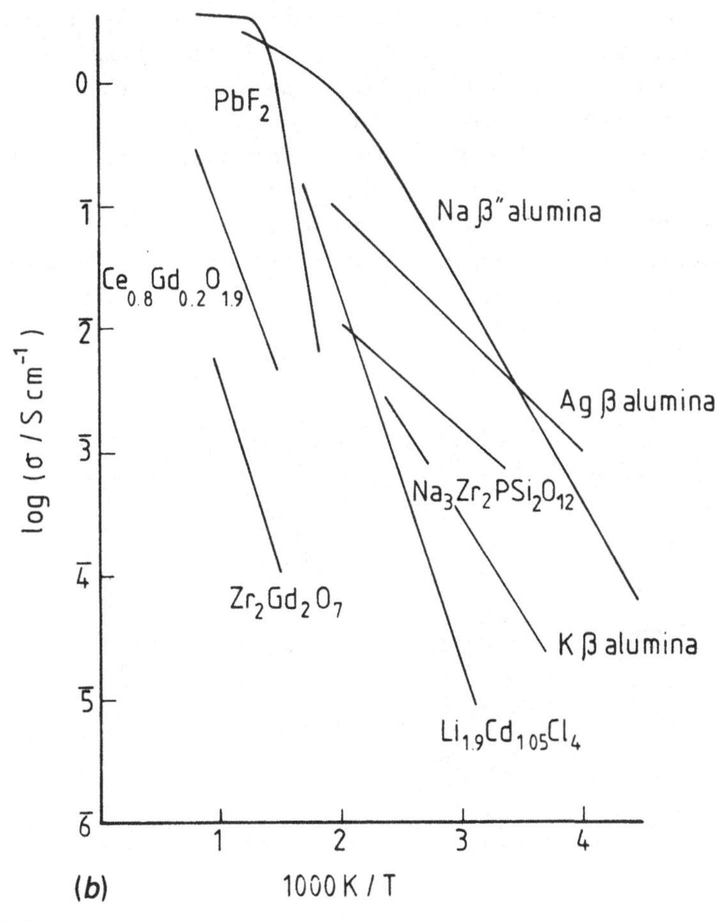

Fig. 2.11

primarily a solid solution between $NaZr_2(PO_4)_3$ and $Na_4Zr_2(SiO_4)_3$, in which the replacement mechanism, $P \rightleftharpoons Si + Na$, gives rise to the general formula

$$Na_{1+x}Zr_2(P_{1-x}Si_xO_4)_3 \qquad 0 < x < 3.$$

The conductivity, due to Na^+ ions, passes through a maximum at intermediate x. It is optimised at $x \sim 2$, where the values approach those of Na β''-alumina, especially at high temperature, >300 °C, Fig. 2.11. At the solid solution limits, $x = 0$ and 3, the conductivity is very low, for the same reasons given in the discussion of Fig. 2.3. The crystal structure of NASICON is a framework, built of (Si, P)O_4 tetrahedra and ZrO_6 octahedra which link up in such a way as to provide a relatively open, three-dimensional network of sites and conduction pathways for the Na^+ ions, Fig. 2.12(*a*). Two Na^+ sites are available, Na1 and Na2. The former is a six-coordinate site while the latter is an irregular eight-coordinate site. These sites are partially occupied at intermediate x.

As well as the solid solution formula given above, there have been suggestions that compositions off the join may also give single-phase NASICON. Part of the problem in determining solid solution stoichiometries and limits in materials such as NASICON arises because of the

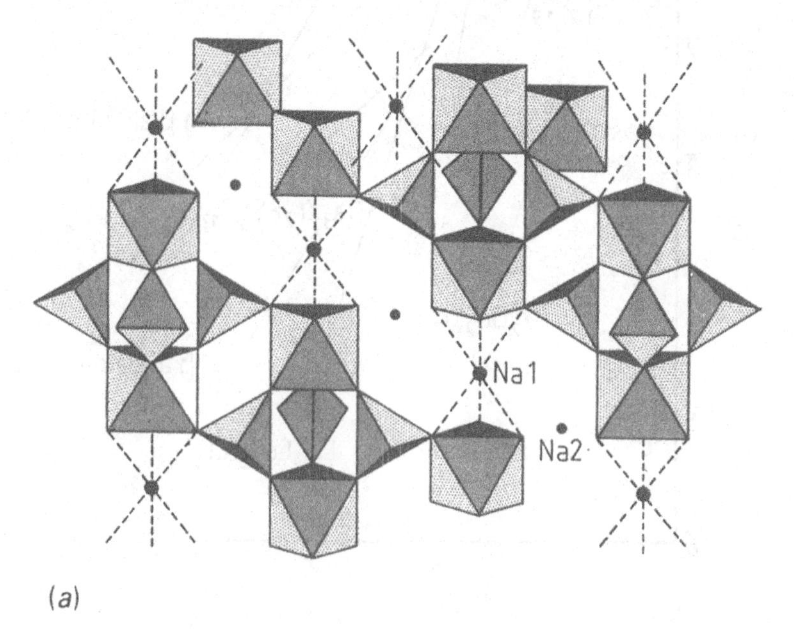

Fig. 2.12 (*a*) Crystal structure of NASICON.

difficulty in attaining equilibrium in reaction mixtures that contain both inert, refractory oxides such as ZrO_2 and volatile oxides such as Na_2O and especially P_2O_5. Thus, high temperatures are required to speed up reaction rates but, unless the samples are heated in sealed containers, some of the more volatile constituents may be lost before they have time to react fully.

2.12.2 LISICON

The lithium analogues of NASICON are, in fact, quite different materials (Irvine and West, 1989). They are solid solutions based on stoichiometric phases such as γ-Li_2ZnGeO_4 or γ-$Li_3(P, As, V)O_4$, but containing

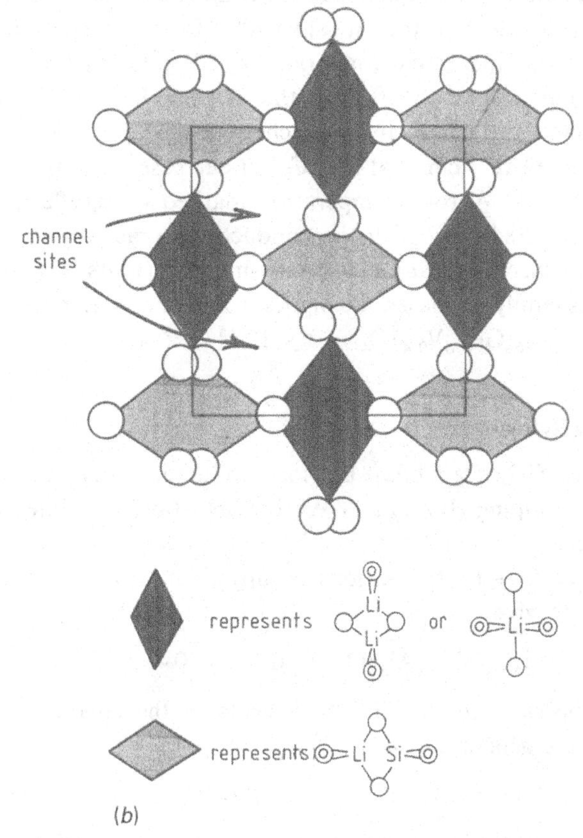

Fig. 2.12 (*Continued*) (*b*) Tetragonal packed oxide array in α-$Li_{2.5}Al_{0.5}SiO_4$ showing location of distorted channel sites.

interstitial Li^+ ions. These are introduced by the aliovalent substitutions, $Zn \rightleftharpoons 2Li$ in Li_2ZnGeO_4 or $P \rightleftharpoons Si + Li$ in Li_3PO_4, to give solid solution formulae:

$$Li_{2+2x}Zn_{1-x}GeO_4 \qquad 0.3 < x < 0.8 \text{ (limits vary with temperature)}$$

and

$$Li_{3+x}(P_{1-x}Si_x)O_4 \qquad 0 < x < 0.4.$$

The crystal structures of the end-members with $x = 0$ are so-called 'γ-tetrahedral structures', with distorted hexagonal close packed oxide arrays and cations distributed over various tetrahedral sites. In the solid solutions, Li^+ ions are found, by powder neutron diffraction, to occupy partially various tetrahedral and octahedral interstitial sites, which link up to form an essentially three-dimensional conduction pathway.

The conductivities increase dramatically with x, from essentially insulating behaviour in the stoichiometric, $x = 0$, end-members and reach a maximum around $x = 0.5$. In the original LISICON system, $Li_{2+2x}Zn_{1-x}GeO_4$, the conductivity reaches very high values at high temperatures, e.g. $0.1\ \Omega^{-1}\ cm^{-1}$ at 300 °C, but decreases rapidly at lower temperatures, Fig. 2.11. At room temperature, marked ageing effects occur (Bruce and West, 1984), in which the conductivity gradually decreases with time. Ageing effects appear unimportant in the $Li_3(P, As, V)O_4$-based materials. In this family of phases, the highest room temperature conductivity is found in $Li_{3.6}(Ge_{0.6}V_{0.4})O_4$, $\sim 3 \times 10^{-5}\ \Omega^{-1}\ cm^{-1}$.

2.12.3 *Li_4SiO_4 derivatives*

Stoichiometric Li_4SiO_4 is a modest Li^+ ion conductor but is a very good host material for doping (Irvine and West, 1989). Both Li^+ interstitials and Li^+ vacancies can be created resulting in high conductivities. In the substitution, $Si \rightleftharpoons Al + Li$, the Al ions occupy Si sites with Li entering interstitial sites, to give

$$Li_{4+x}(Si_{1-x}Al_x)O_4 \qquad 0 < x < 0.4.$$

Alternatively, the substitution, $3Li \rightleftharpoons Al$, leads to the creation of Li^+ vacancies, with the general formula

$$Li_{4-3x}Al_xSiO_4 \qquad 0 < x < 0.5.$$

Conductivity data for the latter system, showing a dramatic variation with x, are given in Fig. 2.3.

The crystal structure of Li_4SiO_4 and its solid solutions is particularly favourable for Li^+ ion conduction because it has an oxide array which approximates to tetragonal packing, Fig. 2.12. Permeating the structure are rather open channels containing various distorted sites of coordination numbers 4, 5 and 6; these sites are partially occupied in the high conductivity solid solutions (Smith and West, 1991). In addition to Li^+ ion migration along the channels, Li^+ ions can also apparently move through the channel walls and so the materials are three-dimensional conductors.

An interesting feature of the conduction mechanism in these materials and the LISICONS is that it is, at least partially, an interstitialcy mechanism. Both structure types contain examples of face-sharing tetrahedral sites, Fig. 2.13. Such sites are much too close together for both to be occupied simultaneously. Crystal structure refinements show that often, on average, one site of each pair contains a Li^+ ion but the occupancy appears to be random. This means that, during conduction, one site of each pair may contain a Li^+ ion but this is ejected when an incoming

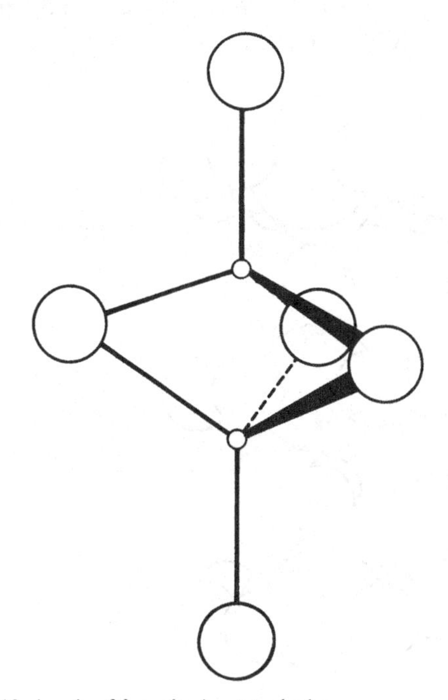

Fig. 2.13 A pair of face-sharing tetrahedra.

Li^+ ion enters the adjacent, face-sharing site. It is clear that, in these materials, Li^+ ions do not move by means of isolated, random hops; in the LISICONS there is clear evidence for clustering of lithium ions and indeed, migration may involve a process of continual reorganisation of the clusters (Bruce and Abrahams, 1991).

2.12.4 Li_3N

The crystalline material with the highest Li^+ ion conductivity found to date is H-doped lithium nitride (Lapp *et al.*, 1983), Fig. 2.11. It is essentially a vacancy conductor because the substituting hydrogen atoms in the formula $Li_{3-x}H_xN$ are tightly bound as NH^{2-} groups. These are located in such a way as to leave vacancies in the Li^+ ion conduction pathway.

Li_3N has a layer structure, Fig. 2.14, in which sheets of stoichiometry 'Li_2N' are separated by additional Li^+ ions, which act to bridge the N atoms in adjacent Li_2N sheets. In H-doped Li_3N, 1–2% of the Li^+ ions in the Li_2N sheets are vacant, giving rise to a high in-plane conductivity of 1×10^{-3} S cm^{-1} at 25 °C, but with a much lower conductivity perpendicular to the sheets, 1×10^{-5} S cm^{-1}.

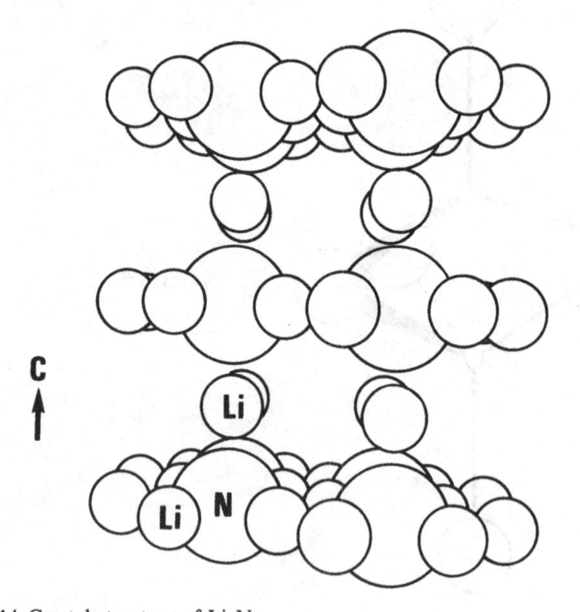

Fig. 2.14 Crystal structure of Li_3N.

Li_3N has a low decomposition potential, 0.445 V, which limits its usefulness as a practical solid electrolyte in batteries. Various Li_3N-derivative phases have been synthesised, such as cubic Li_5NI_2 with an antifluorite-derivative structure and a related phase in the system, Li_3N–LiI–LiOH. Some of these have high conductivity and the latter materials in particular are comparable to those of H-doped Li_3N. They also have a much higher decomposition voltage, ~1.5 V, but do suffer from increased sensitivity to chemical attack.

2.12.5 *Miscellaneous materials*

Complex lithium halide spinels (Kanno, Takeda and Yamamoto, 1982; Lutz, Schmidt and Haeuseler, 1981), based on Li_2CdCl_4 and Li_2MgCl_4 have remarkably high Li^+ ion conductivity for close packed structures, Fig. 2.11. These are complicated materials; however, they have essentially inverse spinel structures but may exist also in various distorted forms. Some of them undergo a phase transition to defect rock salt structures at high temperatures; some are non-stoichiometric.

Simple lithium halides are very poor lithium ion conductors, but composites made from mixtures of LiI and high surface area, insulating oxides such as Al_2O_3 or SiO_2 have conductivities as high as 3×10^{-5} S cm^{-1} at 25 °C (Liang, 1973; Shahi, Wagner and Owens, 1983). These are unusual materials since there is no apparent chemical reaction between LiI and Al_2O_3, nor evidence of the Al_2O_3 acting as a dopant to increase the conductivity of LiI and yet the conductivity is much higher than that of LiI alone. It is believed that the high conductivity is associated with the LiI/Al_2O_3 interfacial regions; the mechanism of charge carrier generation is described in Chapter 3.

The high temperature, α polymorph of Li_2SO_4 has a very high Li^+ ion conductivity, >1 S cm^{-1}, between 575 °C and the melting point, 870 °C. Many attempts have been made to dope Li_2SO_4 and stabilise the highly conducting α polymorph to lower temperatures but these have met with limited success (Lunden, 1987); transformation to the low conductivity β polymorph, or other low conductivity phases, always occurs when the temperature decreases below 500 °C.

The conduction mechanism in α-Li_2SO_4 is of considerable interest and is a matter of some controversy (Lunden and Thomas, 1989). The uncertainty focuses on whether the sulphate groups are fixed or whether they can rotate and, if they can rotate, whether the lithium ion conductivity may be enhanced by a paddle-wheel effect. The scientific evidence available points to the presence of rotational disorder of the sulphate groups; whether this is a dynamic disorder and what effect this would have on the Li^+ ion conductivity is not certain.

2.13 *Oxide ion conductors*

The most well-studied and useful materials to date are those with fluorite-related structures, especially ones based on ZrO_2, ThO_2, CeO_2 and Bi_2O_3 (Steele, 1989). To achieve high oxide ion conductivity in ZrO_2, CeO_2 and ThO_2, aliovalent dopants are required that lead to creation of oxide vacancies, Fig. 2.2, scheme 4. The dopants are usually alkaline earth or trivalent rare earth oxides.

CeO_2 and ThO_2 have the cubic fluorite structure, Fig. 2.15, and can be doped with large amounts of, for example, Ca, La or Gd to give extensive ranges of cubic solid solutions. ZrO_2 is cubic only above $\sim$2400 °C, however, and requires $\sim$8% of dopant to stabilise the cubic form to room temperature (as in YSZ, yttria-stabilised zirconia).

Bi_2O_3 is different since, with its cation:anion stoichiometry of 2:3, it already has a large number of anion vacancies in the cubic δ form stable above 730 °C. This accounts for its exceptional conductivity at these temperatures. Dopants are added to Bi_2O_3 with the purpose of reducing the temperature of the α (monoclinic) $\rightarrow$ δ transition and stabilising the cubic form to lower temperatures. Thus, in the yttria-doped $(Bi_{2-x}Y_x)O_3$, the cubic form is stabilised to low temperatures for $x > 0.25$. Although the high temperature conductivity is lower than that in doped Bi_2O_3, Fig. 2.11,

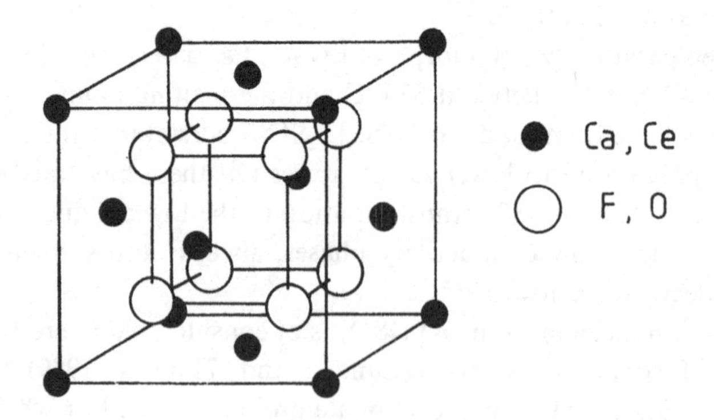

Fig. 2.15 The fluorite structure of CaF_2 or CeO_2.

the conductivity is still very high, e.g. 10^{-2} S cm^{-1} at 500 °C which is about 20 times that of YSZ at 500 °C.

A key factor in the possible applications of oxide ion conductors is that, for use as an electrolyte, their electronic *transport number* should be as low as possible. While the stabilised zirconias have an oxide ion transport number of unity in a wide range of atmospheres and oxygen partial pressures, the Bi_2O_3-based materials are easily reduced at low oxygen partial pressures. This leads to the generation of electrons, from the reaction $2O^{2-} \rightleftharpoons O_2 + 4e$, and hence to a significant electronic transport number. Thus, although Bi_2O_3-based materials are the best oxide ion conductors, they cannot be used as the solid electrolyte in, for example, fuel cell or sensor applications. Similar, but less marked, effects occur with ceria-based materials, due to the tendency of Ce^{4+} ions to become reduced to Ce^{3+}.

The activation energy for oxide ion conduction in the various zirconia-, thoria- and ceria-based materials is usually at least 0.8 eV. A significant fraction of this is due to the association of oxide vacancies and aliovalent dopants (ion trapping effects). Calculations have shown that the association enthalpy can be reduced and hence the conductivity optimised, when the ionic radius of the aliovalent substituting ion matches that of the host ion. A good example of this effect is seen in Gd-doped ceria in which Gd^{3+} is the optimum size to substitute for Ce^{4+}; these materials are amongst the best oxide ion conductors, Fig. 2.11.

Some pyrochlore, $A_2B_2O_7$, phases are moderately good oxide ion conductors. The pyrochlore structure may be regarded as a fluorite derivative in which $\frac{1}{8}$ of the oxygens are missing but since the oxygen sublattice, ideally, is fully ordered, it is necessary to introduce defects to achieve high conductivity.

Intrinsic Frenkel disorder, in which some of the oxygens are displaced into normally unoccupied sites, is responsible for the oxide ion conduction in, for example, $Zr_2Gd_2O_7$, Fig. 2.11. The interstitial oxygen concentration is rather low, however, and is responsible for the low value of the preexponential factor and for the rather low (by δ-Bi_2O_3 standards!) conductivity.

The perovskite structure is capable of high anion conductivity when oxide vacancies are introduced, as in, for example, $La_{1-x}Sr_xCoO_{3-x/2}$ or in the perovskite-related superconductor phases, La_2CuO_4 and $YBa_2Cu_3O_7$. The oxide ion transport number is not unity since such materials are often electronic conductors as well, due to the presence of mixed-valence cations. They are called *mixed conductors* and, while they cannot be used as solid electrolytes in most devices due to electronic short-circuits, they do have other applications in catalysis and as reversible electrodes.

2.14 Fluoride ion conductors

Fluoride ion conduction is quite common, especially in materials with the fluorite structure, Fig. 2.15. The first solid electrolyte to be discovered was a fluoride ion conductor, PbF_2, whose high temperature properties were investigated by Faraday. With increasing temperature, PbF_2 changes from a poor ionic conductor, $\sim 1 \times 10^{-7}$ S cm^{-1} at 20 °C to an exceptional conductor with essentially liquid-like F^- motion ~ 1 S cm^{-1} at 400 °C, Fig. 2.11 (Benz, 1975). Similar trends but with lower conductivities and/or at higher temperatures are shown by CaF_2, SrF_2, BaF_2 and by some chlorides, e.g. $SrCl_2$.

Aliovalent doping can also be used to enhance low temperature conductivities. Both interstitial F^- ions and F^- vacancies can be generated in, for example, CaF_2 by doping with Na and La, respectively (Kudo and Fueki, 1990).

2.15 Proton conductors

Because of the small size and high polarising power of the bare proton, it is unlikely to conduct in the same way as other ions (Poulsen, 1989; Colomban, 1992). Various mechanisms have been proposed, including conduction via species such as NH_4^+ and H_3O^+, the passage of protons along H-bonded networks and a combination of H^+ transferring between adjacent water molecules, linked to rotation of H_3O^+ groups as a means of proton transfer to the next water molecule. Most of the proton conductors discovered initially were hydrates or other thermally unstable phases, but that has now changed with the discovery of proton conduction at high temperatures, 500–1000 °C in perovskite-like $SrCeO_3$.

Various hydrated acid salts are proton conductors in the range 25–100 °C. These include HUP (hydrated uranyl phosphate) $HUO_2PO_4 . 4H_2O$, HUAs, and Keggin-type heteropolyanion structures such as $H_3(PMo_{12}O_{40}) . nH_2O$ and $H_3(PW_{12}O_{40}) . 29H_2O$, all of which have room temperature conductivities in the range 10^{-1}–10^{-4} S cm^{-1}, especially when measured in a humid atmosphere.

The beta-alumina structures (β,β''-alumina and the analogous gallates) can be prepared as H_3O^+ or $NH_4{}^+$ derivatives by ion exchange and some of these are good proton conductors at temperatures up to 200–400 °C, until they decompose by loss of H_2O/NH_3.

Proton conduction at high temperatures occurs in certain perovskites such as doped strontium cerate, $Sr\,Ce_{0.95}Yb_{0.05}O_{3-x}$. In air, this material is primarily an electronic conductor due to the mixed valence of Ce. In the presence of moisture, water is absorbed by the reaction with positive holes to generate protons:

$$H_2O + 2h^+ \rightleftharpoons 2H^+ + \tfrac{1}{2}O_2.$$

The small number of protons introduced gives rise to a high conductivity, e.g. 10^{-2} S cm^{-1} at 900 °C.

References

Almond, D. P. and West, A. R. (1988) *Solid State Ionics*, **26**, 265.
Benz, R. (1975) *Z. Phys. Chem. N.F.*, **95**, 28.
Bradley, J. N. and Greene, P. D. (1966) *Trans. Far. Soc.*, **62**, 2069.
Bruce, P. G. and Abrahams, I. (1991) *J. Solid State Chem.*, **95**, 74.
Bruce, P. G. and West, A. R. (1984) *J. Solid State Chem.*, **53**, 430.
Collongues, R., Thery, J. and Boilot, J. P. (1978) in *Solid Electrolytes*, eds. P. Hagenmuller and W. van Gool, Academic Press, New York, p. 253.
Colomban, P. (ed) (1992) *Proton conductors: Solids, Membranes and Gels – Materials and Devices*, Cambridge University Press, Cambridge.
Debye, P. and Falkenhagen, H. (1928) *Phys. Z.*, **29**, 121, 401.
Debye, P. and Hückel, E. (1923) *Phys. Z.*, **24**, 185, 305.
Duncan, G. K. and West, A. R. (1988) *Solid State Ionics*, **28–30**, 338.
Farrington, G. C., Dunn, B. and Thomas, J. O. (1989) in *High Conductivity Solid Ionic Conductors*, ed. T. Takahashi, World Scientific, Singapore, p. 32.7.
Funke, K. (1986) *Solid State Ionics*, **18/19**, 183.
Funke, K. and Hoppe, R. (1990) *Solid State Ionics*, **40/41**, 200.
Garcia, A., Torres-Trevino, G. and West, A. R. (1990) *Solid State Ionics*, **40/41**, 13.
Geller, S. A. (1967) *Science*, **157**, 310.
Goodenough, J. B. (1983) in *Progress in Solid Electrolytes* CANMET, eds. T. A. Wheat, A. Ahmad and A. K. Kuriakase.
Goodenough, J. B., Hong, H. Y. P. and Kafalas, J. A. (1976) *Mat. Res. Bull.*, **11**, 173, 203.
Grant, R. J., Hodge, I. M., Ingram, M. D. and West, A. R. (1977) *Nature*, **266**, 42.
Hoshino, S. (1957) *J. Phys. Soc. Jap.*, **12**, 315.
Irvine, J. T. S. and West, A. R. (1989) in *High Conductivity Solid Ionic Conductors*, ed. T. Takahashi, World Scientific, Singapore, p. 201.
Iwahara, H., Uchida, H. and Tanaka, S. (1986) *J. Appl. Electrochem.*, **16**, 663.
Jonscher, A. K. (1977) *Nature*, **267**, 673.

Jonscher, A. K. (1983) *Dielectric Relaxation in Solids*, Chelsea Dielectrics Press, London.
Kanno, R., Takeda, T. and Yamamoto, O. (1982) *Mat. Res. Bull.*, **16**, 999.
Kennedy, J. H. (1977) in *Solid Electrolytes*, ed. S. Geller, Springer-Verlag, Berlin, p. 105.
Kreuer, K.-D., Kohler, H. and Maier, J. (1989) in *High Conductivity Solid Ionic Conductors*, ed. T. Takahashi, World Scientific, Singapore, p. 242.
Kudo, T. and Fueki, K. (1990) *Solid State Ionics*, VCH, Tokyo, p. 77.
Kudo, T. and Obayashi, H. (1976) *J. Electrochem. Soc.*, **123**, 417.
Kummer, J. T. (1972) *Progr. Solid State Chem.*, **7**, 141.
Kuwano, J. and West, A. R. (1980) *Mat. Res. Bull.*, **15**, 1661.
Lapp, T., Skaarup, S. and Hooper, A. (1983) *Solid State Ionics*, **11**, 97.
Liang, C. (1973) *J. Electrochem. Soc.*, **120**, 1289.
Lunden, A. (1987) *Solid State Ionics*, **25**, 231.
Lunden, A. and Thomas, J. O. (1989) in *High Conductivity Solid Ionic Conductors*, ed. T. Takahashi, World Scientific, Singapore, p. 45.
Lutz, H. D., Schmidt, W. and Haeuseler, H. (1981) *J. Phys. Chem. Solids*, **42**, 287.
Moseley, P. T. (1985) in *Sodium Sulfur Battery* eds. J. L. Sudworth and A. R. Tilley, Chapman & Hall, London, p. 19.
Nakamura, O., Kodama, T., Ogino, I. and Miyake, Y. (1979) *Chem. Lett.*, **1**, 17.
O'Keeffe, M. and Hyde, B. G. (1976) *Phil. Mag.*, **33**, 219.
Onsager, L. (1927) *Phys. Z.*, **27**, 388; (1926) **28**, 277.
Owens, B. B. and Argue, G. R. (1967) *Science*, **157**, 308.
Poulsen, F. W. (1989) in *High Conductivity Solid Ionic Conductors*, ed. T. Takahashi, World Scientific, Singapore, p. 166.
Roos, A., Aalders, A. F., Schoonman, J., Arts, A. F. M. and de Wijn, H. W. (1983) *Solid State Ionics*, **9/10**, 571.
Seevers, R., DeNuzzio, J., Farrington, G. C. and Dunn, B. (1983) *J. Solid State Chem.*, **50**, 146.
Shahi, K., Wagner, J. B. and Owens, B. (1983) in *Lithium batteries*, ed. J. P. Gabano, Academic Press, London, p. 407.
Smith, R. I. and West, A. R. (1991) *J. Mater. Chem.*, **1**, 91.
Steele, B. C. H. (1989) in *High Conductivity Solid Ionic Conductors*, ed. T. Takahashi, World Scientific, Singapore, p. 402.
Strock, L. W. (1934) *Z. Phys. Chem.*, **B25**, 441.
Strom, U. and Ngai, K. L. (1981) *Solid State Ionics*, **5**, 167.
Takahashi, T. and Iwahara, H. (1978) *Mat. Res. Bull.*, **13**, 1450.
Takahashi, T., Yamamoto, O., Yamada, S. and Hayashi, S. (1979) *J. Electrochem. Soc.*, **126**, 1654.
von Alpen, U., Fenner, J., Marcoll, J. D. and Rabenau, A. (1977) *Electrochim. Acta*, **22**, 801.
Wang, J. C. (1982) *Phys. Rev.*, **B26**, 5911.
Whittingham, M. S. and Huggins, R. A. (1972) in NBS Spec. Publ. 364, *Solid State Chemistry*, p. 139.
Wright, A. F. and Fender, B. E. F. (1977) *J. Phys. C.*, **10**, 2261.
Yoshiasa, A., Maeda, H., Ishii, T. and Koto, K. (1990) *Solid State Ionics*, **40/41**, 341.

3 Crystalline solid electrolytes II: Material design

JOHN B. GOODENOUGH

Center for Materials Science & Engineering, University of Texas at Austin

3.1 Quality criteria

A solid electrolyte is an ionic conductor and an electronic insulator. Ideally, it conducts only one ionic species. Aside from a few specialty applications in the electronics industry, solid electrolytes are used almost exclusively in electrochemical cells. They are particularly useful where the reactants of the electrochemical cell are either gaseous or liquid; however, they may be used as separators where one or both of the reactants are solids. Used as a separator, a solid electrolyte permits selection of two liquid or elastomer electrolytes each of which is matched to only the solid reactant with which it makes contact.

Electrochemical cells are of two types: power cells and sensors. In an ideal power cell, the ionic current through the electrolyte inside the cell matches an electronic current through an external load. The solid electrolyte is in the form of a membrane of thickness L and area A that separates electronically the two electrodes of the cell. Any internal electronic current across the electrolyte reduces the power output. The internal resistance to the ionic current is

$$R_i = L/\sigma_i A, \tag{3.1}$$

where σ_i is the ionic conductivity of the electrolyte. For a current I through the cell, the voltage IR_i represents a potential drop that is to be minimized. Even in potentiometric sensors, the resistance R_i must be maintained below a certain level to obtain satisfactory sensitivity and speed of response, and any electronic contribution to the internal cell current must be factored into the cell calibration.

These simple considerations lead to the following general criteria for the quality of a solid-electrolyte material to be used in an electrochemical cell.

(i) To minimize R_i, ease of fabrication into a mechanically strong membrane of small L and large A. Optimisation of cell design may also require the fabrication of membranes of complex shape.

(ii) Unless an exceptionally small L/A ratio is feasible, a satisfactory R_i generally requires a $\sigma_i > 10^{-2}\ \mathrm{S\,cm^{-1}}$ at the cell operating temperature T_{op}. *Note*: In general, the conductivity is a tensor, but for polycrystalline electrolytes a scalar σ_i is used. However, for tunnel or layered structures having ionic conductivity in one or two dimensions only, it is necessary to appreciate that $\sigma_{\parallel} \neq \sigma_{\perp}$ so the scalar σ_i measured may be considerably reduced from the ionic conductivity in the optimum direction within an individual grain.

(iii) A transport number t_i near unity:

$$t_i \equiv \sigma_i/\sigma \approx 1 \text{ where } \sigma = \textstyle\sum_j \sigma_j + \sigma_e, \tag{3.2}$$

where σ_e is the electronic conductivity. In general there is only one mobile ion in a solid electrolyte so that $\sum_j \sigma_j = \sigma_i$. Because any electron mobility is much greater than the ionic mobility ($u_e \gg u_i$), only a relatively small number of electronic carriers can degrade t_i to an unacceptable level.

(iv) A low resistance to ion transfer across the electrode/electrolyte interfaces.

(v) Chemical stability in the working environment requires that the electrolyte is neither reduced by the reductant at the negative electrode nor oxidised by the oxidant at the positive electrode. Thermodynamic stability is only achieved by placing the bottom of the electrolyte conduction band above the highest occupied molecular orbital (HOMO) of the reductant and the top of the electrolyte valence band below the lowest unoccupied molecular orbital (LUMO) of the oxidant, as illustrated in Fig. 3.1. *Note*: In metallic electrodes, the energies of the HOMO and LUMO are given by the Fermi energy of the anode and cathode, respectively.

(vi) The mobile ion of the electrolyte must be the working ion of the cell.

(vii) Cost of material and fabrication is always a consideration.

Given these quality criteria, the challenge is to design a material that meets all these demands at a useful operating temperature T_{op} and competitive cost. In order to meet this challenge for any given engineering cell, it is necessary first to understand the factors that limit and control these quality criteria. We address here the electrolyte 'window' needed for chemical stability and the conduction mechanisms that control σ_i and t_i at any given T_{op}.

3.2 Electronic energies

Ionic conductors are ionic compounds. Therefore it is appropriate to start with ions rather than atoms to construct the electronic energy level diagrams. Fig. 3.2 illustrates such a construction for the electronic and ionic insulator MgO. The energy levels $Mg^{2+/+}$ and $O^{-/2-}$ correspond respectively to the ionisation potential of Mg^+ and the electron affinity of O^-. E_I represents the energy required to remove an electron from the Mg^+ ion and place it on an O^- ion at infinite separation, thereby creating free Mg^{2+} and O^{2-} ions. Note that the O^- ion has a negative electron affinity – the O^-/O^{2-} redox energy lies above the bottom of the vacuum energies designated Vac in Fig. 3.2. The energy E_M is the electrostatic Madelung energy gained by arranging point charges 2+ and 2− at the

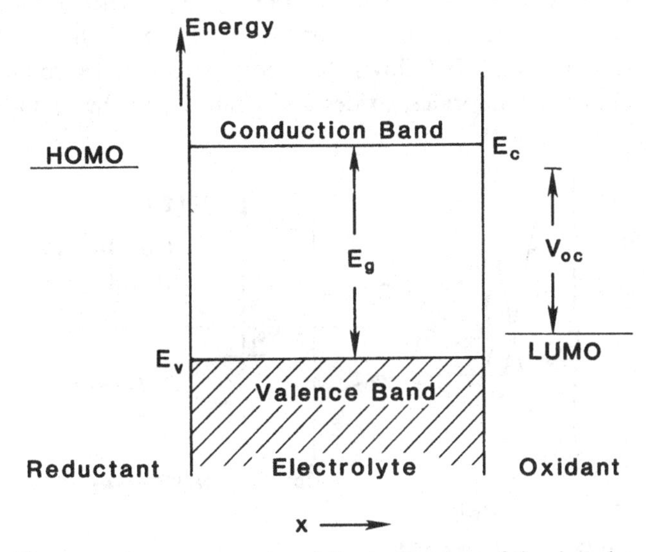

Fig. 3.1 Placement of reactant energies relative to the edges of the electrolyte conduction and valence bands in a thermodynamically stable electrochemical cell at flat-band potential.

Mg and O positions of the MgO rocksalt lattice. The ionic charges serve to invert the electron energies of Mg and O compared with those in the free ion state. Conservation of energy causes the anion and cation energies to be shifted in opposite directions by E_M. An $E_M > E_I$ is needed to stabilise an 'ionic' solid. Overlap of the 3s orbitals of Mg^{2+} with each other broadens the levels into a band. A similar process occurs for the 2p levels of O. However, back transfer of electronic charge from the O^{2-} to the Mg^{2+} ions gives a 'polarisation correction' that reduces the effective ionic charges and hence the energy E_M. This reduction in E_M is largely compensated by the quantum-mechanical repulsion between the Mg–O antibonding states of the conduction band. Therefore the introduction of a covalent contribution to the bonding introduces Mg 3s and 3p character into the O 2p band, but it alters little the mean energy of the band, which is primarily broadened by O–O interactions. The O 2p admixture into the Mg 3s and 3p bands is not identical; the bottom of the conduction band has Mg 3s parentage. In the case of MgO, an $E_g \approx 7.5$ eV is found.

Fig. 3.3 shows a similar ionic-model construction of the electronic energies for the isostructural transition-metal oxide MnO. In this case, the Madelung energy E_M also lifts the $Mn{:}3d^5$ level above the top of the bonding O 2p bands, and MnO is easily oxidised.

From the constructions of Figs. 3.2 and 3.3, it is clear that a large electrolyte stability 'window' E_g requires not only a large energy difference $E_M - E_I$, but also the absence of any cationic states above the top of the bonding 'anion-p' band. It follows that most practical electrolytes are generally confined to fluorides, oxides and chlorides of the main-group

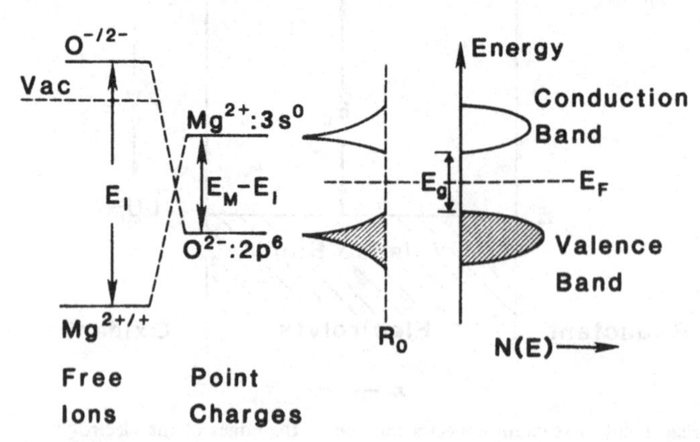

Fig. 3.2 Construction of conduction and valence bands of MgO with an ionic model.

elements. Exceptions on the one hand are the Group IV transition-metal cations Zr(IV) and Hf(IV), which have conduction 4d and 5d bands at high energies in the oxides, and trivalent rare-earth ions such as Gd(III) that have filled $4f^n$ configurations with energies below the top of the anion-p bands and empty $4f^{n+1}$ configurations above the bottom of the 5d conduction band. Exceptions in the other direction are the heavier Group-B main-group elements such as Sn, Pb, Sb or Bi that have energetically accessible 5s or 6s band states; these states are easily oxidised if occupied and easily reduced if empty. In Li_3N, for example, the N^{3-}: $2p^6$ valence band is too high in energy to make this Li^+-ion conductor a practical contender as a Li^+ ion electrolyte in a lithium storage battery; the empty Zr 4d band in the stabilised zirconias, on the other hand, is high enough in energy to make these O^{2-} ion electrolytes attractive for practical applications whereas electronic conductivity plagues the isostructural CeO_2-based electrolytes, which have a Ce:$4f^1$ energy level in the gap between the Ce 5d and O 2p bands. The Bi:$6s^2$ core is normally too easily oxidised to make practical O^{2-}-ion conductors based on Bi_2O_3.

In the fluorides, chlorides and oxides of the Group-A main-group metals and the transition metals zirconium and hafnium, aliovalent cation substitutions are generally charge-compensated by the introduction of native defects (e.g. an oxygen vacancy in $Zr_{1-x}Ca_xO_{2-x}$) because the intrinsic E_g is large; however, in some oxides neutral oxygen or water may

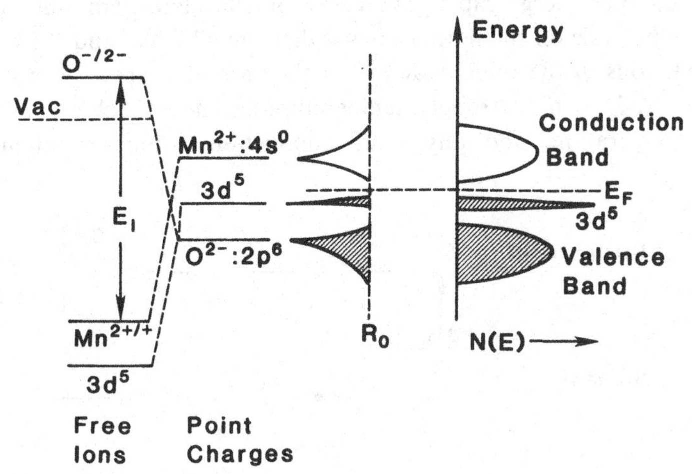

Fig. 3.3 Construction of Mn:$3d^5$ localised-electron configuration relative to conduction and valence bands in MnO with an ionic model.

be inserted at lower temperatures into oxygen vacancies. The neutral oxygen atoms react with neighbouring oxide ions to form the peroxide ion $(O_2)^{2-}$ or some other complex, thus oxidising the O 2p band by creating deep acceptor states associated with the O–O bonding. The insertion of water introduces protons that may be mobile in the structure. These complications together with the higher charge at an oxide ion have made difficult the search for a fully satisfactory O^{2-} ion electrolyte. The commercially available O^{2-} ion electrolytes now operate at a $T_{op} \approx 1000\,°C$ with a strongly reducing atmosphere on one side and/or a strongly oxidising atmosphere on the other. Measurement of the oxygen transport number t_O as a function of temperature and the partial pressure of oxygen is a necessary calibration and evaluation procedure for any oxide ion electrolyte.

3.3 Ionic energies

3.3.1 Intrinsic energy gap ΔH_g

At low temperatures, stoichiometric ionic solids have arrays of crystallographically equivalent sites completely filled with ions. We designate these sites as *normal* in order to distinguish them from *interstitial* sites having a higher ionic potential energy. The highest normal-site ionic energies are separated from the lowest interstitial-site energies by an energy gap ΔH_g as illustrated in Fig. 3.4; this ionic gap is analogous to the electronic energy gap E_g of an electronic insulator. No ionic conduction is possible on the normal sites if they are all filled, and there are no mobile ions in the interstitial sites if they are all empty. In most ionic solids, ΔH_g is not strongly temperature-dependent below the sintering temperature, and any ionic conduction requires near-sintering

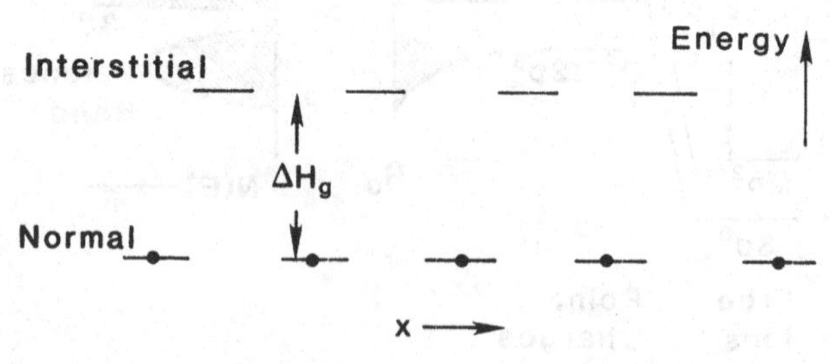

Fig. 3.4 Energies of a working ion in a stoichiometric ionic compound: the dots indicate an occupied site.

temperatures to excite ions from normal to interstitial positions, i.e. to make Frenkel defects, so as to create arrays of equivalent sites that are only partially occupied.

On the other hand, there are some stoichiometric ionic solids that undergo an order–disorder transition for only one ionic species at a transition temperature T_t that is below the melting point T_m. Where the order–disorder transition for one ionic species leaves the other atoms in their original positions, as occurs in PbF_2 and LnF_3, this phenomenon has been referred to as *sublattice melting* (O'Keeffe and Hyde, 1976). On the other hand, the order–disorder transition may induce a structural transformation of the other atoms, as occurs in AgI where disordering of the Ag^+ ions induces a transformation of the iodide ion array from close-packed to body-centred cubic. In either case, the energy gap ΔH_g for the mobile-ion species becomes strongly temperature-dependent as T approaches T_t, vanishing for $T > T_t$. An order parameter for the transition, which may be either smooth or first-order as illustrated in Fig. 3.5, is the ratio $\Delta H_g(T)/\Delta H_g(0)$ of the gap at temperature T relative to its value at $T = 0$ K. At temperatures $T > T_t$, a set of energetically equivalent sites of concentration C is partially occupied, and the site occupancy $n_c \equiv c/C$ of mobile ions of concentration c is a constant independent of temperature. At temperatures $T < T_t$, the site occupancy of interstitial ions n_c and normal-site vacancies $(1 - n_c)$ varies, according to the law of mass action, as $\exp(-\Delta H_g/2kT)$ in a manner analogous to the concentration of electrons and holes in an intrinsic semiconductor, but with a $\Delta H_g = \Delta H_g(T)$ as T approaches T_t – see Fig. 3.5.

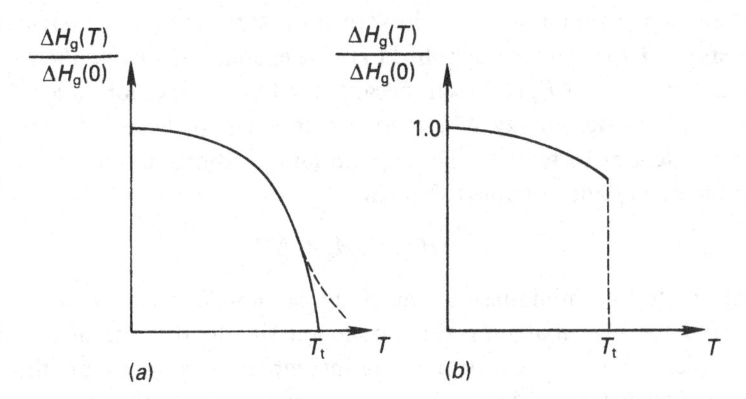

Fig. 3.5 The order parameter $\Delta H_g(T)/\Delta H_g(0)$ for an order–disorder transition that is (*a*) smooth, (*b*) first-order.

3.3.2 Motional enthalpy ΔH_m

Where ions are disordered over a partially occupied array of energetically equivalent sites, their motion is diffusive. Fig. 3.6 illustrates the variation in ionic potential with intersite position for three possible situations in which a set of energetically equivalent sites are partially occupied by mobile ions. The partially occupied sites may (*a*) share common faces in a continuously connected network through the structure, (*b*) be separated from one another by an array of empty sites, or (*c*) be separated from one another by an array of sites that are occupied by the mobile ions.

In the simplest situation (*a*), the common site faces act as barriers for the ionic motion. In order for an ion to pass through such a face, or 'bottleneck,' the shortest distance R_b from the centre of the face to a peripheral ion must be large enough to allow it to pass. Where R_b is less than the sum of the radii of the mobile and peripheral ions, thermal energy must be supplied to open up the face; where R_b is larger than the sum of the ionic radii, thermal energy may need to be supplied to push the mobile ion into the centre of the diffusion pathway. Even where R_b is roughly equal to the sum of the ionic radii, thermal energy must be supplied to the mobile ion to push it over the maximum in the potential energy curve at the saddle-point position; ions – with the exception of protons within a double-well hydrogen bond – are too heavy to tunnel from one site to another as do itinerant electrons in a partially filled band of electronic states.

The thermal energy needed for an ionic jump is the motional enthalpy ΔH_m. It has two components, a barrier energy ΔH_h for the ion to hop when the receptor and donor sites are at the same energy and a relaxation energy ΔH_r that must be supplied to make energetically equal the receptor and donor sites. ΔH_r is present because the time it takes for an ion to hop over the barrier energy ΔH_h is long compared to the time it takes the immobile ions to relax to their equilibrium positions, which are different at the empty and occupied sites. In

$$\Delta H_m = \Delta H_h + \Delta H_r \tag{3.3}$$

the first term is minimized by matching the mobile-ion size of R_b and the second term by matching the mobile-ion size to the size of the site it occupies. These two constraints are incompatible with one another, and the lowest values of ΔH_m are found where the effective sum of the ionic radii can be modified by a polarisation (or quadrupolarisation) of the

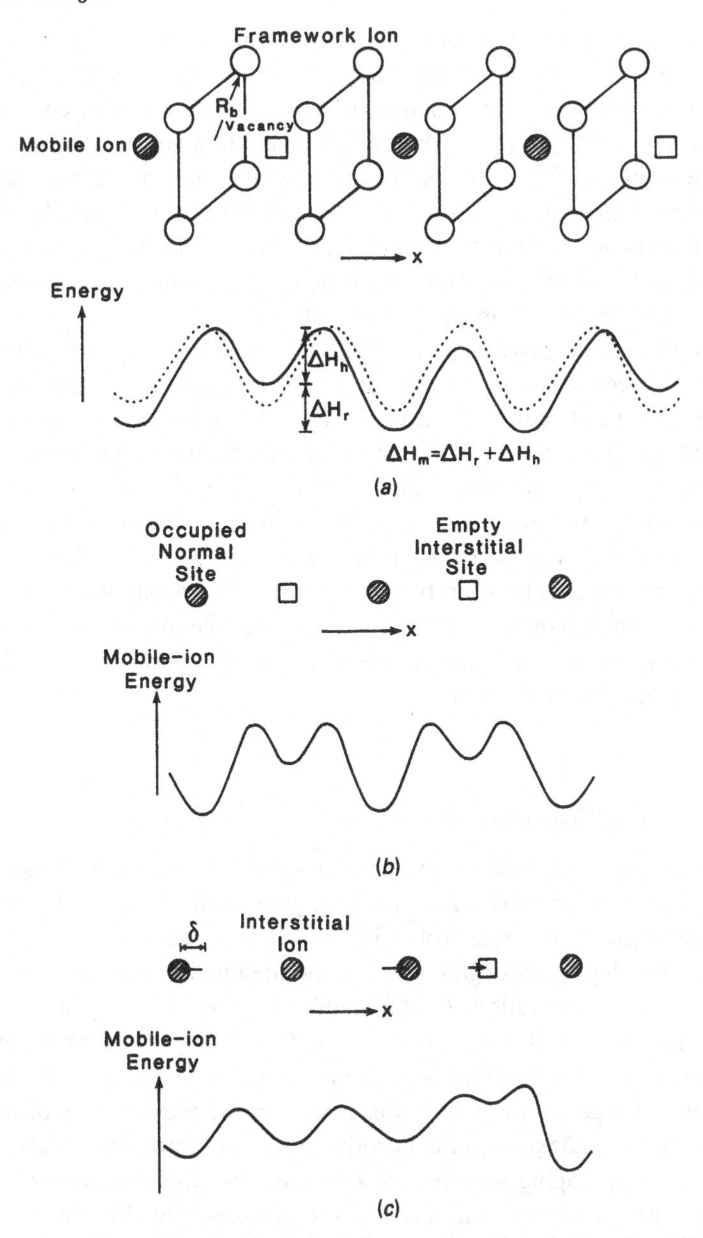

Fig. 3.6 (*a*) Definition of bottleneck shortest distance R_b and variation of mobile-ion potential with position with (solid line) and without (dotted line) relaxation of host structure. (*b*) Variation of mobile-ion potential for occupied sites separated by an array of inequivalent, empty sites. (*c*) Smoothing of the potential of (*b*) by introduction of mobile ions into second array of sites.

peripheral and/or the mobile ions. In AgI, for example, the large I^- ions are readily polarised and the Ag^+ ion can hybridise its $4d^{10}$ core with empty 5s states so as to change the shape of its repulsive core from a sphere to a prolate ellipsoid; the hybridisation energies involved in the polarisation of the I^- ion and quadrupolarisation of the Ag^+ ion are largely compensated by the gain in the covalent component of the Ag–I bond. Such an accommodation of the Ag–I bond length permits the size constraints of both ΔH_h and ΔH_r to be nearly optimised simultaneously, and a $\Delta H_m \approx 0.03$ eV in AgI is satisfactorily small.

In the more complex situations, (*b*) and (*c*) of Fig. 3.6, where the partially occupied sites are separated by inequivalent sites that are either empty or filled with mobile ions, fast ionic motion requires that the mobile-ion potential at the intervening sites be nearly the same as that at the partially occupied sites. (Of course, if the intervening sites are occupied by stationary ions, the ions in the partially occupied sites are immobilised.) From the constructions in Fig. 3.6, it is clear that the electrostatic forces between the mobile ions tend to smooth the potential where the inequivalent sites are filled with mobile ions whereas the local relaxation energy ΔH_r enhances the potential difference of the inequivalent sites where they are empty.

3.3.3 *Trapping energy* ΔH_t

Where there is no order–disorder transition temperature T_t above which ΔH_g collapses or where T_t is much higher than the desired operating temperature T_{op} of the electrolyte in a practical application, it is customary to employ doping strategies to create mobile-ion vacancies or interstitials as charge compensation. In the stabilised zirconia $Zr_{1-x}Ca_xO_{2-x}$, for example, the substitution of Ca^{2+} for Zr^{4+} not only stabilises the fluorite structure; it also introduces oxygen vacancies to compensate for the smaller charge on the Ca^{2+} ion. Superficially, the creation of oxygen vacancies is analogous to the introduction of holes into the valence band of silicon by doping with boron. However, the diffusive motion of ions as against the itinerant character of the valence-band electrons in silicon changes the magnitude of the energy ΔH_t by which a mobile ionic species is trapped at the dopant that creates it. In the case of an itinerant electron, both the kinetic and potential energies must be included. Consequently a hydrogenic model is appropriate; but screening of the coulombic attraction by the dielectric constant $\kappa > 10$ reduces the binding energy by a factor κ^{-2}, and the shallow trapping energies at the dopants in silicon are less than 0.1 eV. In an ionic conductor, on the other hand, only the potential energy need be considered, so the dielectric screening reduces the coulombic contribution ΔH_c to the binding energy by only a factor κ^{-1}, which makes $0.1 < \Delta H_c < 1$ eV. In addition, an ionic dopant of different size also creates a local elastic strain that attracts the mobile species with a relaxation energy ΔH_r, so the total trapping energy at a point dopant is

$$\Delta H_t = \Delta H_c + \Delta H_r. \tag{3.4}$$

Substitution of a larger Ca^{2+} ion for Zr^{4+} in stabilised zirconia, for example, creates a local elastic strain field that also attracts the oxygen vacancy to the dopant cation (Kilner and Factor, 1983; Steele, 1989). In order for the oxygen vacancy to contribute to a dc conductivity, it must either free itself from the dopant or find a tortuous percolation pathway through the solid via dopant near neighbours. However, doping heavily enough to provide such percolation pathways tends to make the material metastable with respect to phase segregation, and operation at higher temperatures can lead to the phenomenon of 'ageing' in which the mobile-ion vacancies become trapped in larger clusters. Ageing is a problem in the stabilised zirconias where relatively heavy doping is required to stabilise the fluorite structures and an $E_a \simeq 1$ eV makes $T_{op} > 800\,°C$.

3.4 Ionic conductivity

3.4.1 Phenomenology

In a crystal, the electronic and ionic conductivities are generally tensor quantities relating the current density i_0 to the applied electric field $\mathbf{E}$ in accordance with Ohm's law. The scalar expression for the mobile-ion current density in the different principal crystallographic directions has the form

$$i_0 = \sigma_i \mathbf{E} = c_i q v, \tag{3.5}$$

where v is the mean velocity of the mobile ions of charge q and c_i is their concentration. From the definition of the charge-carrier mobility $u \equiv v/\mathbf{E}$,

$$\sigma_i = c_i q u_i. \tag{3.6}$$

If we place an ionic conductor between parallel-plate blocking electrodes that produce an electric field **E** parallel to the x-axis, the electrostatic potential varies as $-x\mathbf{E}$ on passing from one electrode at $x = 0$ to the other. At equilibrium, the mobile-ion concentration $c_i(x)$ is proportional to $\exp(q\mathbf{E}x/kT)$, and the ionic drift-current density $\sigma_i\mathbf{E}$ in the field is balanced by a diffusion current due to the concentration gradient (Fick's law):

$$\sigma_i\mathbf{E} = Dq\,\partial c_i(x)/\partial x = Dq^2\mathbf{E}c_i(x)/kT, \tag{3.7}$$

from which it follows that

$$\sigma_i = c_i q(qD/kT) = c_i q u_i \tag{3.8}$$

which gives the Nernst–Einstein relationship between the ionic diffusion coefficient the ionic mobility

$$u_i = qD/kT. \tag{3.9}$$

The diffusion coefficient

$$D = D_0 \exp(-\Delta G_m/kT) \tag{3.10}$$

contains the motional free energy

$$\Delta G_m = \Delta H_m - T\Delta S_m. \tag{3.11}$$

Substitution of Eqns (3.9) and (3.10) into (3.8) gives finally

$$\sigma_i = (A_T/T)\exp(-E_a/kT) \tag{3.12}$$

where, for a temperature-independent mobile-ion site occupancy, $n_c = c_i/C$ of mobile ions on an array of energetically equivalent sites of concentration C

$$E_a = \Delta H_m \tag{3.13}$$

$$A_T = (Cq^2/k)n_c D_0 \exp(\Delta S_m/k). \tag{3.14}$$

To obtain a more complete description, we need to find an analytic expression for the pre-exponential factor D_0 of the diffusion coefficient by considering the microscopic mechanism of diffusion. The most straightforward approach, which neglects correlated motion between the ions, is given by the *random-walk* theory. In this model, an individual ion of charge q reacts to a uniform electric field along the x-axis supplied, in this case, by reversible nonblocking electrodes such that $\partial c_i(x)/\partial x = 0$. Since two atoms cannot occupy the same site and we have $0 < n_c < 1$, the transition probability for an ion to jump to a neighbouring site is $z(1 - n_c)f\nu(E)$, where $(1 - n_c)$ is the probability that an equivalent near-neighbour site is empty, z is the number of these sites, and f is a geometrical factor of order unity that depends on the jump path. The jump frequency

$$\nu(E) = \nu_0 \exp(-\Delta G_m/kT) \tag{3.15}$$

contains the optical-mode vibrational frequency ν_0 (10^{12}–10^{13} Hz) of the mobile ion within its site; ν_0 is called the *attempt frequency*. The presence of an electric field $\mathbf{E}$ makes the enthalpy $\Delta H_m - (q\mathbf{E}l_x/2)$ for jumps in the direction of the field lower than that in the opposite direction, $\Delta H_m + (q\mathbf{E}l_x/2)$, where l_x is the x-component of the jump distance l between neighbouring equivalent sites; the maximum in the potential-energy barrier occurs at $l_x/2$. The net drift velocity v_x in the direction of the field is the product of $l_x/2$ and the differences in the jump probabilities in the forward and backward directions:

$$v_x = (l_x/2)z(1 - n_c)f\nu[\exp(-q\mathbf{E}l_x/2kT) - \exp(q\mathbf{E}l_x/2kT)], \tag{3.16}$$

where $\nu = \nu(\mathbf{E} = 0)$. In general, a $q\mathbf{E}l_x/2 \ll kT$ reduces this expression to

$$v_x \approx (q\mathbf{E}/kT)(l_x^2/2)z(1 - n_c)f\nu. \tag{3.17}$$

For a cubic crystal, $l_x^2 = l_y^2 = l_z^2$ and $l_x^2/2$ may be replaced by $l^2/6$; hence from $u_i = v_x/\mathbf{E} = qD/kT$ we obtain

$$D \approx (l^2/6)z(1 - n_c)f\nu. \tag{3.18}$$

Eqn (3.14) then becomes

$$A_T = \gamma(Cq^2/k)n_c(1 - n_c)l^2\nu_0 \tag{3.19}$$

$$\gamma \approx f(z/6)\exp(\Delta S_m/k) \tag{3.20}$$

for the case $0 < n_c < 1$ and three-dimensional conduction. For two- or one-dimensional conduction, $\sigma_\perp$ and $\sigma_\parallel$ have $z/6 \to z/4$ and $z/2$, respectively, in Eqn (3.20).

In a stoichiometric material at $T < T_t$, an $n_c = 0.5\exp(-\Delta H_g/2kT)$ introduces an additional term into the activation energy E_a of Eqn. (3.12); Eqn (3.13) becomes

$$E_a = \Delta H_m + \tfrac{1}{2}\Delta H_g \tag{3.21}$$

providing ΔH_g is temperature-independent. As T approaches a T_t, ΔH_g

becomes temperature-dependent, which increases both the pre-exponential term A_T and the apparent value of E_a – see the discussion of Eqn (3.23). A $\Delta H_g(T)$ may even induce a first-order phase change. A measurement of $\log(\sigma_i T)$ vs $1/T$ over a finite temperature interval that gives an empirical value of A significantly greater than the value obtained from Eqn (3.19) signals a $\Delta H_g = \Delta H_g(T)$.

In a doped material, the trapping energy ΔH_t adds to the activation energy E_a of Eqn (3.12); Eqn (3.13) becomes

$$E_a = \Delta H_m + \tfrac{1}{2}\Delta H_t. \tag{3.22}$$

From these simple phenomenological considerations, it is clear that the two dominant considerations in the design of a solid electrolyte are the creation of a temperature-independent $n_c(1 - n_c)$ with n_c approaching its optimum value of 0.5 and a minimisation of the motional enthalpy ΔH_m; and minimisation of the energy E_a appearing in an exponent must demand priority.

3.4.2 *Correlated ionic movements*

In the random-walk model, the individual ions are assumed to move independently of one another. However, long-range electrostatic interactions between the mobile ions make such an assumption unrealistic unless n_c is quite small. Although corrections to account for correlated motions of the mobile ions at higher values of n_c may be expected to alter only the factor γ of the pre-exponential factor A_T, there are at least two situations where correlated ionic motions must be considered explicitly. The first occurs in stoichiometric compounds having an $n_c = 1$, but a low ΔH_m for a cluster rotation; the second occurs for the situation illustrated in Fig. 3.6(*c*).

A low ΔH_m for a cooperative cluster rotation allows excitation of a cluster of atoms from normal to saddle-point positions. Such an excitation may, in turn, lower the energy of the saddle-point sites relative to the normal sites, thus effectively introducing a $\Delta H_g(T)$ that collapses in a smooth transition. At temperatures $T > T_t$, the mobile ions become disordered over the normal and saddle-point sites. Such a situation appears to be illustrated by stoichiometric Li_3N and PbF_2 (Goodenough, 1984).

The situation illustrated in Fig. 3.6(*c*) may give rise to domains or clusters rich in mobile ions within which the atomic potential is smoothed by the coulombic repulsions between mobile ions and domains poor in mobile ions in which the lattice relaxation energy ΔH_r destabilises the empty sites. A cooperative motion of the cluster rich in mobile ions can, in such a case, give a lower average ΔH_m than is possible with isolated ionic jumps.

3.4.3 Proton movements

Among the ions, the proton is unique in the character of its bonding and hence in the variety of movements available to it in an ionic compound. In an ionic material, the hydrogen atom forms molecular orbitals with anions having acceptor electronic states more stable than the H:1s energy level. As the smallest cation, the proton tends to coordinate at most two nearest anion neighbours. Lone-pair electrons of a neighbouring anion X, or from two neighbouring anions, are stabilised by a virtual charge transfer from X:p^6 anion states back to the empty H^+:1s state, which polarises the outer electrons of the neighbouring anion(s) toward the proton to reduce its effective positive charge. This charge transfer increases the X–H bonding and so reduces the X–H bond length. Where two like atoms coordinating a proton are inequivalent, the proton is shifted toward the more polarisable anion.

An asymmetric hydrogen bond is common even where a proton coordinates two equivalent anions. The π-bond repulsive forces between two coordinated anions tend to prohibit a close X–H–X separation, so competition between the two equivalent anions for the shorter X–H bond may set up a double-well potential for the equilibrium proton position between the two coordinated anions. With oxide anions, an O–H–O separation greater than 2.4 Å sets up a double-well potential and creates an asymmetric hydrogen bond, which we represent as O–H···O. Although displacement toward one anion may be energetically equivalent to a displacement toward the other, one well is made deeper than the other by an amount ΔH_r as a result of the motion of the proton from the centre of the bond.

Where a smaller π-bond repulsion between filled anion p orbitals permits a closer X–H–X separation, a proton may bond two anions equally strongly on opposite sides of itself. Such a symmetric hydrogen bond may have an angle slightly bent from 180°; the bending results from a weaker charge transfer from the anion p states to the higher energy H^+

2p states (Potier, 1983). Formation of symmetric hydrogen bonds tends to occur with F^- ions or where two equivalent oxide ions are strongly polarised to the opposite side by neighbouring cations. The $O_2H_5^+$ dioxonium ion, for example, consists of two water molecules bonded by a symmetric hydrogen bond, but the O–H–O bond angle may be bent by as much as 6°.

Cooperative displacements of protons within a network of asymmetric hydrogen bonds may give rise to a spontaneous polarisation P_s of the solid. Ordering of the displacements occurs below a critical temperature in such ferroelectric materials. Reversal of P_s to $-P_s$ in a dc electric field gives rise to a transient current; it is a giant displacement current. This motion does not give rise to a diffusive, continuous transport of protons. However, where the proton forms a single strong bond with one anion neighbour, its association with the second neighbour in an asymmetric bond may be broken easily. For example, the axis of the shorter X–H bond may librate at lower temperatures and become reoriented from one bond direction to another either by an external electric field or, randomly, by thermal energy. With more heat, the molecule may tumble, thus freely rotating its attached proton. At still higher temperatures, the anion to which the proton is attached may diffuse through the solid over a long range. In all these movements – libration, reorientation, tumbling rotation or translation – the proton moves in association with its anion partner; it rides piggyback on the anion.

Translational piggyback motion is commonly referred to as 'vehicular' motion. In vehicular motion, the mobile molecule may carry a positive charge (e.g. NH_4^+ or OH_3^+), a negative charge (e.g. NH^{2-} or OH^-), or be neutral (e.g. NH_3 or OH_2). Translation of a charged molecule over a long distance gives rise to a dc current provided there is a source and a sink for the mobile species, respectively, at the two electrodes. Translation of neutral species only produces a flux of mass. Neutral species are generally volatile at modest temperatures.

A bare proton may diffuse through a hydrogen-bond system by the Grotthus mechanism, which consists of a combination of piggyback reorientations and proton displacements within hydrogen bonds. The cooperative displacements give rise to a transient displacement current as in a ferroelectric. Once displaced, the protons are blocked from further translation unless the molecular dipoles are free to rotate in the electric field to a new bond position, which resets the system for another proton displacement. Thus long-range translation of a base proton requires two distinct steps, an intrabond displacement and a piggyback rotation. Even if the displacement step occurs via a tunnelling through the potential barrier, thermal energy is required to overcome ΔH_r. Therefore the translational motion of a bare proton, like piggyback translation, is diffusive and requires a motional enthalpy ΔH_m.

3.5 Examples

3.5.1 Stoichiometric compounds

The compounds $Ba_2In_2O_5$, AgI and PbF_2 illustrate ionic conductivity in stoichiometric compounds. The first is a fast oxide-ion conductor above a first-order order–disorder transition at 930 °C that leaves the BaIn array unchanged; the second is a fast Ag^+-ion conductor above a first-order transition at which the I^--ion array changes from close-packed to body-centred cubic; and the third exhibits a smooth transition to a fast F^- ion conductor without changing the face-centred-cubic array of Pb^{2+} ions.

(a) $Ba_2In_2O_5$

The ideal cubic ABO_3 perovskite structure has an ordered CsAu array for the cations with 180° B–O–B bonding. At room temperature, $Ba_2In_2O_5$ has the brownmillerite structure of Fig. 3.7(*a*) (Mader and Müller-Buschbaum, 1990); it is an anion-deficient perovskite $BaInO_{2.5}$ with an ordering of the oxygen vacancies within alternate In–O planes. An In^{3+} ion is stable in either octahedral or tetrahedral coordination, and the vacancy ordering creates corner-shared planes of InO_6 octahedra sharing corners along the perpendicular *c*-axis with corner-shared planes of InO_4 tetrahedra. Ordering of the anion vacancies makes the vacancy and occupied sites both crystallographically and energetically inequivalent; the energy of the vacancy sites is separated from that of the occupied sites by an energy gap ΔH_g. If ΔH_g is not too large, thermal excitations of oxygen into the vacancy sites can be expected at moderate temperatures. These excitations would lower the ordering energy

$$\Delta H_g = \Delta H_g(0) - n_c\varepsilon, \qquad (3.23)$$

where the occupancy, n_c, of excited atoms on the vacancy sites is proportional to $\exp(-\Delta H_g/2kT)$ and the energy ε is the same order of magnitude as $\Delta H_g(0)$. Since ΔH_g decreases with $n_c \sim \exp(-\Delta H_g/2kT)$, there is a strong positive feedback to the disordering process that may induce a first-order phase change at the order–disorder transition temperature T_t. The plot of log σ vs $1/T$ shown in Fig. 3.7(*b*) confirms that such a first-order transition occurs at a $T_t \approx 930\,°C$ (Goodenough, Ruiz-Diaz and Zhen, 1990). The slope of the line for $T > T_t$ provides a measure of $E_a = \Delta H_m$ in the disordered phase; the slope of the line below 600 °C gives a measure of

$$E_a = \Delta H_m + \tfrac{1}{2}\Delta H_g \tag{3.24}$$

where $\Delta H_g \approx \Delta H_g(0)$. (A plot of $\log(\sigma T)$ vs $1/T$ is similar; it should strictly be used to obtain E_a.) In the interval $600\,°C < T < T_t$, the slope of the log σ vs $1/T$ plot reflects a $\Delta H_g(T)$ that is decreasing with increasing T due to the onset of increasing oxygen disorder as T approaches T_t. In this case, a first-order transition at T_t indicates that the order parameter $\Delta H_g(T)/\Delta H_g(0)$ varies as illustrated in Fig. 3.5(*b*). Extrapolation of the O^{2-}-ion conductivity σ_O for $T > T_t$ to temperatures $T < T_t$ shows that

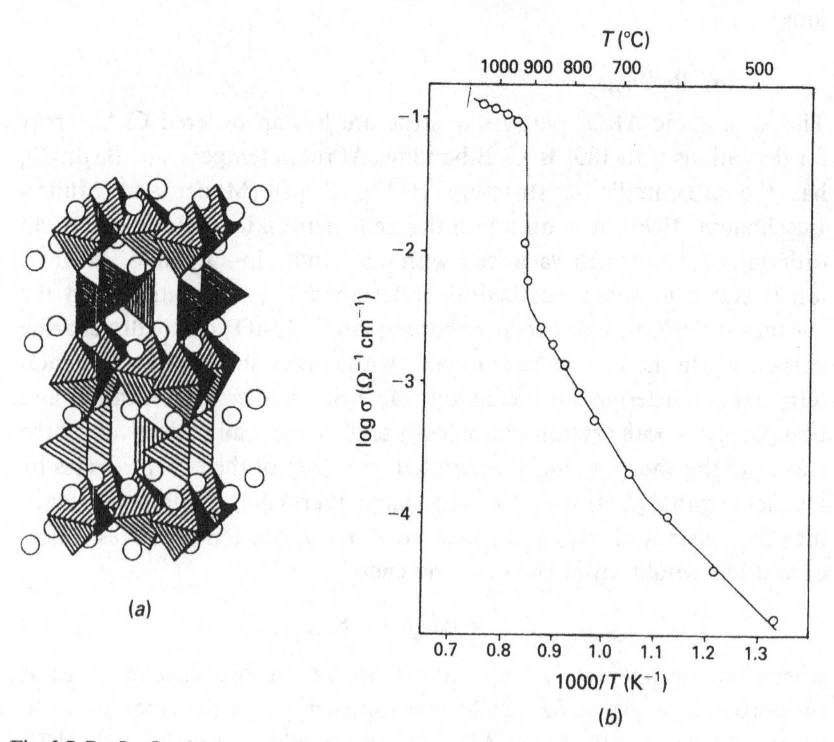

Fig. 3.7 $Bn_2In_2O_5$: (*a*) brownmillerite structure $T < T_t$ and (*b*) conductivity vs reciprocal temperature.

this perovskite structure would support a $\sigma_O \geq 10^{-2}$ S cm^{-1} at an operating temperature $T_{op} = 500$ °C were it possible to lower T_t to below 500 °C.

A similar situation is illustrated by Bi_2O_3. The high-temperature δ-Bi_2O_3 phase has a face-centred-cubic Bi-atom array with O atoms occupying in a disordered manner three-quarters of the tetrahedral interstices. At lower temperatures, the anion vacancies become ordered on these sites, and the energy splitting ΔH_g of the vacant and occupied sites varies with the degree of order as described by equation (3.23).

(b) *AgI*

A similar, but somewhat different situation is present in AgI. In this case the large I^- ions form a close-packed array with a mixture of cubic and hexagonal stacking at room temperature. At the α–β transition temperature $T_t = 147$ °C, the iodide-ion array undergoes a first-order transition from close-packed to body-centred cubic (Strock, 1934; Hoshino, 1957; Geller, 1967; Wright and Fender, 1977). Ag^+-ion transport in the β phase is described in Chapter 2. As already discussed, the polarisability of the large I^- ions and the quadrupolarisability of the Ag^+-ion core results in a $\Delta H_m = 0.03$ eV, which is the smallest activation energy E_a for any known solid electrolyte.

Substitution of Rb for Ag has led to an ordered RbI_5 array within which Ag^+ ions are disordered at room temperature over an array of face-shared tetrahedral sites; although the $E_a = \Delta H_m$ in this compound is a little larger, stabilisation of fast ionic conduction to room temperature with the elimination of a first-order phase change between room temperature and the melting point was a major technical accomplishment (Bradley and Greene, 1966; Owens and Argue, 1967). Unfortunately there are few technical applications other than in electronic components that can use Ag^+ as the working ion.

(c) *PbF_2*

The F^--ion conductor first discovered by Faraday represents a more complex order–disorder transition to fast ionic conduction. At all temperatures, PbF_2 is reported to have the fluorite structure in which the F^- ions occupy all the tetrahedral sites of a face-centred-cubic Pb^{2+}-ion array; however, the site potential of the Pb^{2+} ions is asymmetric, and a measurement of the charge density with increasing temperature indicates that the F^- ions spend an increasing percentage of the time at the saddle-point positions between tetrahedral sites as the temperature rises (Schultz, 1982). In order to understand this rather surprising result, it is instructive to consider the room-temperature structure of KY_3F_{10} shown in Fig. 3.8. In this structure, the K^+ and Y^{3+} ions are ordered within a face-centred-cubic array as are Au and Cu atoms in the alloy $AuCu_3$. Surprisingly, the F^- ions are not evenly distributed within this array, but are segregated into F^--poor quadrants containing eight F^- ions on all the tetrahedral sites and F^--rich quadrants containing twelve F^- ions on all the saddle-point positions between the tetrahedral sites. This structure demonstrates two interesting points: first, there is little difference in the potential energy for F^- ions on tetrahedral sites and saddle-point positions between these sites in a face-centred-cubic array of cations, in confirmation of the beautiful X-ray studies of Schultz on PbF_2; and second, the occupancy of saddle-point positions in one quadrant can coexist stably with the occupancy of tetrahedral sites in the neighbouring quadrants. This latter observation demonstrates that in the fluorite structure of PbF_2 *cluster excitation* of F^- ions within a quadrant from tetrahedral to saddle-point positions is energetically accessible (Goodenough, 1984). It is more accessible in PbF_2 than in isostructural SrF_2 because of the polarisability of the $Pb^{2+}:6s^2$ core, which reduces ΔH_m. In $PbSnF_4$, a lower melting point gives an even lower T_t. The energetic accessibility of a cluster excitation to saddle-point sites of larger density allows a disordering of the F^- ions over tetrahedral and

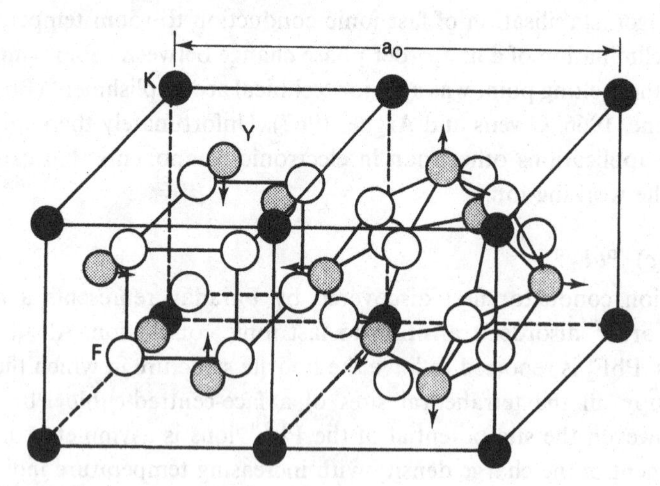

Fig. 3.8 Two quadrants of the KY_3F_{10} structure.

saddle-point sites, which in turn permits long-range F^--ion diffusion with a relatively small $E_a = \Delta H_m$.

In support of this model, it is noted that LnF_3 – which has F^- ions on both octahedral and tetrahedral sites of a face-centred-cubic Ln^{3+}-ion array – becomes a fast F^--ion conductor below its melting point without any change in the cation array (O'Keeffe and Hyde, 1975). This observation shows that some low-energy excitation other than the displacement of F^- ions into octahedral sites is operative, as is postulated with the cluster-rotation model for PbF_2.

A similar situation applies to stoichiometric Li_3N, which shows fast Li^+-ion conduction within Li_2N planes (Von Alpen, Rabenau and Talat, 1977; Schulz and Thiemann, 1978; Schulz, 1982; Goodenough, 1984).

3.5.2 Doping strategies

The number of stoichiometric compounds exhibiting technically useful order–disorder transitions is extremely limited, and no practical strategy exists for identifying where to look for them. Consequently most attempts at the design of solid electrolytes have relied on chemical doping. This approach generally introduces a trapping energy ΔH_t, which makes

$$E_a = \Delta H_m + \tfrac{1}{2}\Delta H_t \qquad (3.25)$$

but there appear to be conditions where the effective ΔH_t becomes acceptably small. However, strategies to reduce ΔH_t have not been explored systematically. As a preliminary step in this direction, three approaches are briefly reviewed.

(a) Point substitutions

The earliest approach to the introduction of mobile ionic species is represented by the stabilised zirconias. For example, the substitution of Ca^{2+} for Zr^{4+} in $Zr_{1-x}Ca_xO_{2-x}$ is charge compensated by the introduction of oxygen vacancies rather than by holes in the valence band. In this case, the oxygen vacancies introduced by the doping can be nearest neighbours of the dopant, so ΔH_t is relatively large. Heavier doping so as to create percolation pathways that have Ca^{2+}-ion neighbours at every site lead to problems with chemical inhomogeneities that form regions within which the vacancies become even more strongly trapped. Although the stabilised zirconias are the best O^{2-} ion electrolytes commercially available, they are inadequate for several important potential applications.

The compound $Ca_3Fe_2TiO_8$ is reported (Rodriguez-Carvajal, Vallet-Regi and Gonzalez-Calbet, 1989) to have a defect perovskite structure in which the oxygen vacancies order so as to give a layer of corner-shared tetrahedra alternating with a layer containing two planes of corner-shared octahedra – rather than one as in brownmillerite, i.e. $\cdots$OOTOOT$\cdots$ vs the $\cdots$OTOT$\cdots$ stacking of Fig. 3.7(*a*). On the assumption that $Ba_3In_2ZrO_8$ would form the same structure as $Ca_3Fe_2TiO_8$, fast extrinsic ionic conduction below 400 °C was interpreted to reflect the trapping by Zr^{4+} ions in the tetrahedral-site planes of oxygen from the octahedral-site layers (Goodenough, Ruiz-Diaz and Zhou, 1990). In fact, long-range vacancy ordering does not occur, and the low-temperature extrinsic ionic conduction is due to the insertion below 400 °C of neutral oxygen and/or water. Oxygen is extracted even from dry N_2 contaminated with O_2; in moist air, water is also inserted (Manthiram, Kuo and Goodenough, 1993). The inserted oxygen atoms apparently oxidise the oxide-ion array by forming oxygen-atom clusters; the simplest such cluster would be a peroxide ion $(O_2)^{2-}$. Inserted water introduces mobile protons. This type of complication can be expected where an oxide of large, basic cations such as Ba^{2+}, La^{3+}, or Zr^{4+} is forced by stoichiometry to have too low an oxygen coordination.

An interesting variation of this approach is found in the O^{2-}-ion conductors $Bi_4V_{2-x}M_xO_{11-y}$. $Bi_4V_2O_{11}$ itself is a stoichiometric layered compound containing oxygen vacancies. The ideal structure is that of Bi_2MoO_6 illustrated in Fig. 3.9; it has an intergrowth of $(Bi_2O_2)^{2+}$ layers alternating with $(MoO_4)^{2-}$ layers of corner-shared MoO_6 octahedra. In $Bi_4V_2O_{11}$, one-quarter of the apical oxygen sites coordinating the V^{5+} ions are vacant (Abraham, Debreuille-Gresse, Mairesse and Nowogrocki, 1988). Ordering of the oxygen vacancies and cooperative displacements of the V^{5+} ions from their ideal positions to form some shorter V–O bonds appears to be responsible for a distortion to orthorhombic symmetry at room temperature. On heating, the room-temperature α-$Bi_4V_2O_{11}$ phase transforms through a β phase $450 < T < 570$ °C to tetragonal γ-$Bi_4V_2O_{11}$ in which the V^{5+}-ion displacements are dynamic and oxygen vacancies are disordered on the apical oxygen sites. These phase changes are accompanied by a marked decrease in the apparent activation energy E_a; in the γ phase an $E_a = \Delta H_m$ for two-dimensional conduction is achieved, Fig. 3.10. The γ phase was stabilised to room temperature by a partial substitution of V by Cu (Abraham, Bovin, Mairesse and Nowogrocki, 1990); $Bi_4V_{1.8}Cu_{0.2}O_{11-\delta}$ exhibits a remarkably high O^{2-}-ion conductivity above 200 °C without any indication of a T_t. Although the transport number t_O is reported to be high in air at temperatures $T < 500$ °C, the presence of transition-metal cations and of Bi^{3+} ions makes too restricted the range of oxygen partial pressures over which t_O is high. Nevertheless, the origin of the low value of ΔH_m and a lowering of T_t without appreciably increasing ΔH_m is instructive.

A subsequent study (Goodenough, Manthiram, Parenthamen and Zhen, 1992) compared the influence of several substitutional atoms M in $Bi_4V_{2-x}M_xO_{11-y}$. In all cases, the room-temperature orthorhombic distortion of the α phase was suppressed for $x \geq 0.2$, but only M = Nb or Ti gave conductivities σ_O comparable to or slightly better than for M = Cu. Like the V^{5+} ion, the Nb^{5+} and Ti^{4+} ions exhibit displacements from the centre of symmetry of an octahedral interstice toward one, two, or three neighbouring oxygen. Cooperative displacements of this type give rise, for example, to the ferroelectric phases in the perovskite $BaTiO_3$. In V_2O_5, the V^{5+}-ion displacements stabilise a layered structure. Moreover,

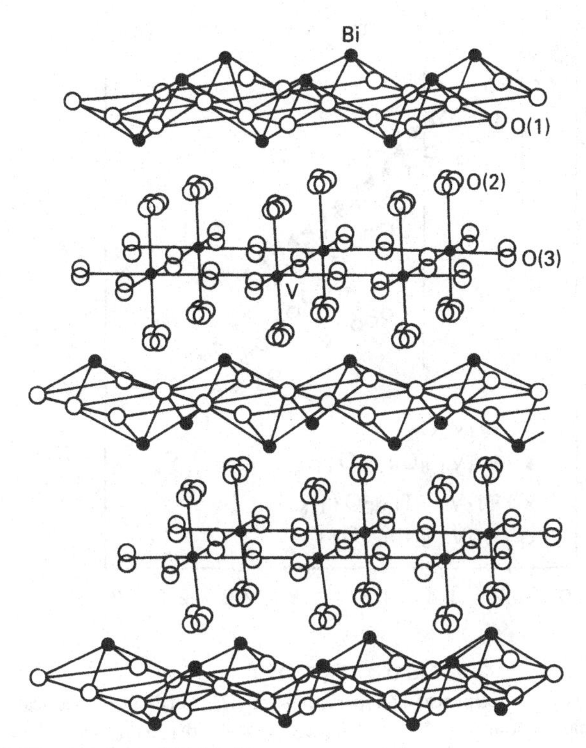

Fig. 3.9 The structure of Bi_2MoO_6.

work on the copper-oxide superconductors has shown that a Cu^{2+} ion in a CuO_2 sheet can be stabilised in square-coplanar, square-pyramidal, or octahedral coordination; addition or removal of an apical oxygen at Cu^{2+} ions in a CuO_2 sheet takes little energy. The lowering of T_t in $Bi_4V_{2-x}M_xO_{11}$ by the substitution of $x \geq 0.2$ foreign atoms is caused by a suppression of the cooperative, static V^{5+}-ion displacements; in the γ phase, displacements of a V^{5+} ion toward an apical oxygen can be dynamic along the *c*-axis where there is octahedral coordination and randomly static where there is square-pyramidal coordination. The relative energies of the apical-oxygen site potentials are modulated by the V^{5+}-ion displacements. The ions Nb^{5+} and Ti^{4+} are uniquely suited to participate in the modulation of the site potential, and the binding of an

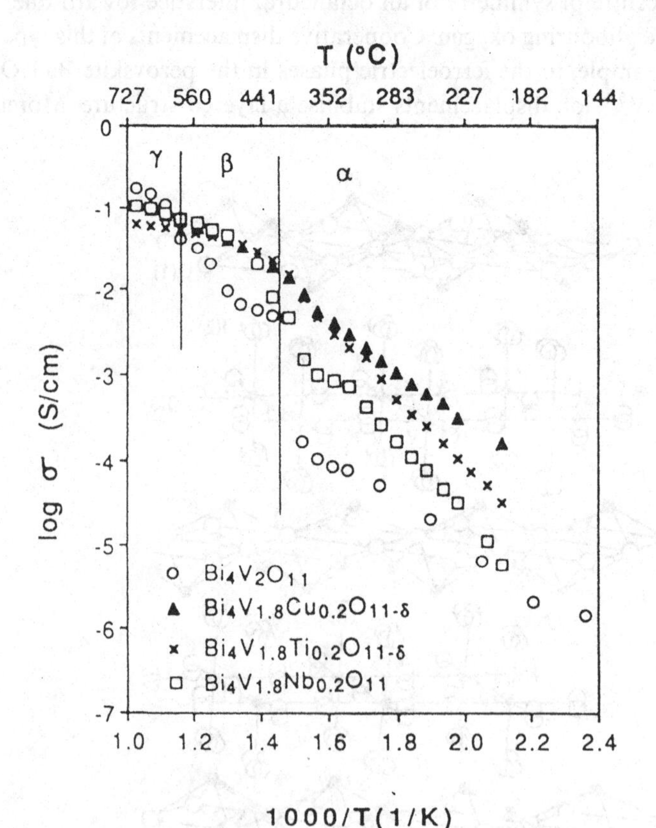

Fig. 3.10 Conductivity vs reciprocal temperature: circles, $Bi_4V_2O_{11}$; vertical lines separate α, β, γ phases; triangles, $V_{1.8}Cu_{0.2}O_{11-\delta}$; crosses $Bi_4V_{1.8}T_{0.2}O_{11-\delta}$; squares, $Bi_4V_{1.8}Nb_{0.2}O_{11-\delta}$.

apical oxygen to the Cu^{2+} ion is modulated by changes in the relative stabilities of the x^2–y^2 and z^2 orbitals at a Cu^{2+} ion. Systematic exploration of the influence of Jahn–Teller or ferroic ions on the ionic mobility of a solid electrolyte has not been made because these are all transition-metal ions.

(b) Framework structures

The use of framework structures to minimize ΔH_m for alkali-ion electrolytes has been demonstrated to provide a means of opening up the 'bottlenecks' to cation motion in a number of oxides (Goodenough, Hong and Kafalas, 1976). Framework structures may provide one-dimensional tunnels as in hollandite, two-dimensional transport in planes as in the β-aluminas, or three-dimensional transport as in NASICON and LISICON. Since one-dimensional tunnels are readily blocked, the two- and three-dimensional conductors are the more interesting.

The β-aluminas are described in some detail in Chapter 2, only a few specific features are noted here. In the β''-aluminas, the spinel blocks are stacked in such a way that the energetically equivalent sites occupied by Na^+ ions are ideally just half-filled; in the β-aluminas the spinel blocks are stacked so as to distinguish two types of Na^+-ion sites of different potential energy, the Beevers–Ross (BR) and anti-Beevers–Ross (aBR) sites. In the Na–O planes, the shortest bottleneck distance $R_b \approx 2.7$ Å is just a little greater than the sum of the ionic radii, 2.4 Å, at room temperature, so a small value of ΔH_m can be anticipated. The discovery of fast Na^+-ion conductivity in the Na β-aluminas (Yao and Kummer, 1967; Kummer and Weber, 1967) led to the invention of the Na/S battery that triggered extensive interest in the solid-electrolyte problem.

The Na β- and β''-aluminas are generally not stoichiometric, but stoichiometric β and β'' $Na_2O.11Ga_2O_3$ have been prepared (Chandrashekhar and Foster, 1977); ordering of the Na^+ ions into the BR sites in the β phase suppresses Na^+ ion conductivity whereas the β'' phase is an excellent Na^+ ion conductor. In the more practical β- and β''-aluminas, excess Na and/or Na_2O is present in the Na–O planes, which smoothes the Na^+-ion potential in the phase by changing from the situation depicted in Fig. 3.6(*b*) to that of Fig. 3.6(*c*). The β'' phase is stabilised by substituting some Mg^{2+} for Al^{3+} in the spinel blocks, and excess Na^+ charge compensates. Because the Mg^{2+} is within the spinel blocks, ΔH_m is minimized. The situation is a bit more complex in the β-aluminas where excess Na_2O is introduced into the Na–O planes

(Roth, Reidinger and La Placa, 1976). The excess Na_2O strengthens the ceramic, and practical ceramic membranes contain a mixture of the β and β'' phases.

In the NASICON family, Li^+ or Na^+ ions move in a $B_2(XO_4)_3$ framework, which contains an array of M_I interstitial sites separated by an array of M_{II} interstitial sites; there are three times as many M_{II} as M_I sites. In the initial study (Goodenough, Hong and Kafalas, 1976), the mobile ions were Na^+, B was Zr, and X was P and/or Si, i.e. $Na_{1+3x}Zr_2(P_{1-x}Si_xO_4)_3$. In the end-member $NaZr_2(PO_4)_3$, all the M_I sites are filled with Na^+ ions, and a large ΔH_r accentuates the energy difference between the Na^+-ion potentials at M_I and the M_{II} sites; the situation is illustrated by Fig. 3.6(*b*). Doping that removes Na^+ ions from the M_I sites does not give excellent Na^+-ion conductivity; but doping that adds Na^+ ions to the M_{II} sites smoothes the potential as illustrated in Fig. 3.6(*c*). In the end-member $Na_4Zr_2(SiO_4)_3$, both M_I and M_{II} sites are filled, and the compound is not an ionic conductor. The best Na^+-ion conductivity occurs for compositions near $x = 2$ where the potential energy at the M_{II} sites is stabilised to a value close to that at the M_I sites. Moreover, at $x = 2$ the populations of both Si and P atoms are large enough to make inapplicable any model of ΔH_t based on point-dopant trapping. In addition, the Si and P atoms are removed from the interstitial pathway for Na^+-ion diffusion in this framework structure, which has a nearly optimal R_b for Na^+-ion conduction as required for a small ΔH_h in ΔH_m.

The γ phase Li^+-ion electrolytes $Li_{2+2x}Zn_{1-x}GeO_4$ (LISICON) and $Li_{3+x}(Ge_xV_{1-x})O_4$ have been shown to illustrate a motion of Li-rich clusters that smoothes the Li^+ ion potentials (Bruce and Abrahams, 1991). The γ phases have an essentially hexagonal-close-packed array of anions with cations in half the tetrahedral sites. In γ-Li_3VO_4, the VO_4 tetrahedra are isolated and share only corners with LiO_4 tetrahedra. In γ-Li_2ZnGeO_4, Ge^{4+} ions replace V^{5+} ions and Zn^{2+} ions replace one-third of the Li^+ ions. In the doped $Li_{3+x}(Ge_xV_{1-x})O_4$ system, the charge-compensating Li^+ ions occupy interstitial octahedral sites; in the doped $Li_{2+2x}Zn_{1-x}GeO_4$ system, a Li^+ ion replaces a Zn^{2+} ion and the additional charge-compensating Li^+ ion occupies interstitial octahedral sites. The octahedral sites share common faces with the tetrahedral sites, and coulomb repulsions between the Li^+ ions of neighbouring tetrahedral and octahedral sites displace the Li^+ ions towards opposite faces to smooth the Li^+-ion potential as illustrated in Fig. 3.6(*c*). A similar situation occurs in the interstitial space of the spinel $[B_2]X_4$ framework as illustrated by the insertion system $Li_{1-x}[Ti_2]O_4$ (Goodenough, Manthiram, and Wnetrzewski, 1992). A doping of the γ phases that introduces Li^+-ion vacancies on the tetrahedral sites has little interest for Li^+-ion electrolytes; but the introduction of octahedral-site Li gives excellent Li^+-ion conduction. Smoothing of the Li^+-ion potential apparently leads to a clustering into Li-rich and Li-poor domains that occupy a volume too large to be trapped at individual dopants, and the ΔH_m for cooperative movements of the clusters is small. In this cooperative movement, octahedral-site Li^+ ions displace tetrahedral-site Li^+-ions, which jump into octahedral sites. It is a cooperative motion suitable for fast Li^+-ion conduction in a close-packed anion structure; the Li^+ ion is small enough to pass through a triangular site face with a relatively small ΔH_m whereas the larger alkali ions such as Na^+ and K^+ require more open frameworks.

(*c*) *Composites*

Since the initial observation (Liang, 1973; Liang, Joshi, and Hamilton, 1978) that the dispersion of small Al_2O_3 particles in LiI gives a dramatic enhancement of the Li^+-ion conductivity, a number of studies have been devoted to the phenomenon. The enhancement is greater the smaller the dispersoid particle size, and it exhibits a maximum between 10 and 40 mole % dispersoid. The phenomenology of the enhancement mechanism can be modelled – with an adjustable parameter – in terms of an increase in the defect concentration responsible for extrinsic conduction in a narrow space-charge layer in the host at the particle–host interfaces (Shukla, Manoharan and Goodenough, 1988). In this case, doping of the stoichiometric salt is accomplished by the higher affinity of the Li^+ ions for an oxide than an iodide ion. Attachment of the Li^+ ions to the surface of the oxide charges the Al_2O_3 particle positively; the Li^+-ion vacancies in the iodide give a negatively charged space-charge layer. The vacancies thus created are trapped in the surface space-charge layer, but such a layer is connected in two dimensions and can percolate from particle to particle through the solid. The oxide particles thus provide an alternative way of doping a stoichiometric solid.

3.5.3 Proton conductors

Because proton movements are distinct, the general principles developed for the conduction of other ions in electrolytes need to be modified for protonic conduction.

Most inorganic protonic conductors are hydrated oxides. The intercalation of interstitial water is facile in many framework and layered compounds; it can even occur in oxygen-vacancy perovskites since a water molecule is smaller than an O^{2-} ion. Since interstitial water tends to be lost above 100 °C and bulk water above about 300 °C, proton conductors are generally confined to near-room-temperature applications. Measurement of the protonic transport requires the making of pellets. Since sintering would involve a loss of the interstitial water, the pellets are made by cold-pressing wet samples. This procedure creates a particle hydrate, and the protonic conductivity measured commonly reflects the conductivity of an immobilised, intergranular aqueous matrix rather than the bulk conductivity of interest (England, Cross, Hamnett, Wiseman and Goodenough, 1980).

Layered oxides that may contain variable amounts of water between the layers include, for example, various clays. An important family of layered hydrates are the lamellar acid salts $[M(IV)(XO_4)_2]_n^{2n-}$, where M(IV) = Ti, Zr, Mf, Ge, Sn, Pb, Ce or Th and X = P or As (Alberti and Constantino, 1982). In order to obtain interesting protonic conductivity at room temperature (10^{-3} S cm^{-1}), it is necessary to have more than one layer of water between the oxide layers (Casciola, Constantino and D'Amico, 1986). The excess water provides hydrogenated oxygen species with a rotational degree of freedom, which is a requirement in the Grotthus mechanism. The hydrated uranyl phosphate $H_3OVO_2PO_4 . 3H_2O$ (HUP) represents another group of layered hydrates in which good H^+-ion conductivity at room temperature has been reported (Shilton and Howe, 1977), but here also extra water is needed. A Grotthus diffusion mechanism is operative in the interlayer water (Fitch, 1982). Antimonic acid is a framework hydrate with an $(Sb_2O_6)^{2-}$ framework of the cubic pyrochlore structure $A_2B_2O_6O'$; the interstitial A_2O' sites are occupied by $2H_3O^+ + H_2O$ (England *et al.*, 1980). In such a strongly acidic framework, the protons may move in the interstitial space by either a vehicular or a Grotthus mechanism. However, this hydrate is also prepared as a wet powder, and interparticle diffusion dominates. All these examples, when present with extra water, may be categorised as particle hydrates. In a particle hydrate, the oxide particles dope the excess water, which carries most of the protonic current, in a manner analogous to the doping of LiI by Al_2O_3 in a composite.

Particle hydrates consist of small particles, commonly oxides, imbedded in a hydrogen-bonded aqueous matrix. Because hydrogen bonds can be broken by application of pressure, these composite materials can be easily fabricated into dense sheets at room temperature by cold-pressing. (For practical applications, a binder is needed.) Metal ions at the surface of an oxide particle bind water in order to complete their normal oxygen coordination. The protons associated with this water distribute themselves over the particle to create surface hydroxyl anions. The concentration of surface protons also comes into equilibrium with that of the aqueous matrix. 'Acidic' particles push protons from the surface into a pH 7 aqueous matrix; 'basic' particles attract protons from a pH 7 matrix. Colloidal particles have a large surface-to-volume ratio, so they are more effective than larger particles at changing the pH of the aqueous matrix.

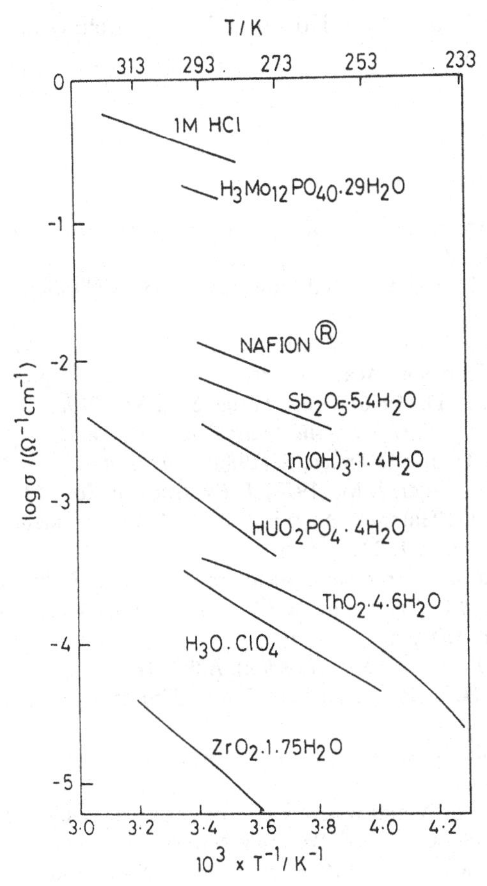

Fig. 3.11 Proton conductivity vs reciprocal temperature for several particle hydrates compared with that of 1 M HCl.

The more acidic and smaller an oxide particle, the higher the mobile-ion concentration $[H^+]$ in the matrix for a given particle fraction. The protons move in the aqueous matrix by a Grotthus mechanism, and the mobilities of the protons (or proton vacancies) are greater the less structured is the water in which they move; the structure of the aqueous matrix decreases with distance from the surface of the colloidal particle. Therefore the protonic mobility is higher the greater the fraction of water in the composite, and the protonic conductivity of a particle hydrate drops off with time if water is being lost. Technical applications are restricted to an operating temperature $T_{op} < 60\,°C$ unless operated under a high pressure of water vapor. Fig. 3.11 compares the protonic conductivities of several particle hydrates with that of 1 M HCl.

The Robert A. Welch Foundation, Houston, TX is gratefully acknowledged for financial support.

References

Abraham, F., Debreuille-Gresse, M. F., Mairesse, G. and Nowogrocki, G. (1988) *Solid State Ionics*, **28–30**, 529.

Abraham, F., Bovin, J. C., Mairesse, G. and Nowogrocki, G. (1990) *Solid State Ionics*, **40–41**, 934.

Alberti, G. and Constantino, U. (1982) in *Intercalation Chemistry*, Ed. M. S. Whittingham and A. J. Jacobson, Academic Press, New York, Chapter 5.

Bradley, J. N. and Greene, P. D. (1966) *Trans. Farad. Soc.*, **62**, 2069.

Bruce, P. G. and Abrahams, I. (1991) *J. Solid State Chem.*, **95**, 74.

Casciola, M., Constantino, U. and D'Amico, S. (1986) *Solid State Ionics*, **22**, 17.

Chandrashekhar, G. V. and Foster, J. M. (1977) *J. Electrochem. Soc.*, **124**, 329.

England, W. A., Cross, M. G., Hamnett, A., Wiseman, P. J. and Goodenough, J. B. (1980) *Solid States Ionics*, **1**, 231.

Fitch, A. N. (1982) in *Solid State Protonic Conductors (I) for Fuel Cells and Sensors*, Eds. J. Jensen & M. Kleitz, Odense University Press, Odense, p. 235.

Geller, S. A. (1967) *Science*, **159**, 310.

Goodenough, J. B. (1984) *Proc. Roy. Soc. (London)*, **A393**, 215.

Goodenough, J. B., Hong, H. Y.-P. and Kafalas, J. A. (1976) *Mat. Res. Bull.*, **11**, 203.

Goodenough, J. B., Manthiram, A., Parenthamen, P. and Zhen, Y. S. (1992) *Mat. Sci. Eng.*, B, **12**, 357.

Goodenough, J. B., Manthiram, A. and Wnetzewski, B. (1992) *Proceedings of the Sixth International Meeting on Lithium Batteries*, Münster, in press.

Goodenough, J. B., Ruiz-Diaz, J. E. and Zhen, Y. S. (1990) *Solid State Ionics*, **44**, 21.

Hoshino, S. (1957) *J. Phys. Soc. Japan*, **12**, 315.

References

Kilner, J. A. and Factor, J. D. (1983) in *Progress in Solid Electrolytes*, Eds. T. A. Wheat, A. Ahmed and A. K. Kuriakose, Energy, Mines, and Resources Erp/MSL 83-94 (TR), Ottawa, p. 347.

Kummer, J. T. and Weber, N. (1967) in *Proceedings of the 21st Annual Power Sources Conference*, Red Bank, NJ, PSC Publications Committee, NS, p. 37.

Kuo, J.-F., Manthiram, A. and Goodenough, J. B. (1992) unpublished.

Liang, C. C. (1973) *J. Electrochem. Soc.*, **120**, 1289.

Liang, C. C., Joshi, A. V. and Hamilton, N. E. (1978) *J. Appl. Electrochem.*, **8**, 445.

Mader, K. and Müller-Buschbaum, H. K. (1990) *J. Less Common Metals*, **157**, 71.

Manthiram, A., Kuo, J.-F. and Goodenough, J. B. (1993) *Solid State Ionics*, **65**, 225.

O'Keeffe, M. and Hyde, B. G. (1975) *J. Solid State Chem.*, **13**, 172.

O'Keeffe, M. and Hyde, B. G. (1976) *Phil. Mag.*, **33**, 219.

Owens, B. B. and Argue, G. R. (1967) *Science*, **157**, 308.

Potier, A. (1983) in *Solid State Protonic Conductors (II) for Fuel Cells and Sensors*, Eds. J. B. Goodenough, J. Jensen, and M. Kleitz, Odense University Press, Odense, p. 173.

Rodriguez-Carvajal, J., Vallet-Regi, M. and Gonzalez-Calbet, J. M. (1989) *Mat. Res. Bull.*, **24**, 423.

Roth, W. L., Reidinger, F. and La Placa, S. (1976) in *Superionic Conductors*, Eds. G. D. Mahan and W. L. Roth, Plenum Press, New York, p. 223.

Schultz, H. (1982) in *Proceedings of the 2nd European Conference on Solid State Chemistry*. Veldhoven, Netherlands, Eds. R. Metselaar, H. J. M. Heijligers and J. Schoonman, Elsevier Science, Amsterdam, p. 117.

Schultz, H. (1983) *Ann. Rev. Mat. Sci.*, **12**, 351.

Schultz, H. and Thiemann, K. H. (1978), *Acta Crystall.*, **A35**, 309.

Shilton, M. G. and Howe, A. T. (1977) *Mat. Res. Bull.*, **12**, 701.

Shukla, A. K., Manoharan, R. and Goodenough, J. B. (1988) *Solid State Ionics*, **26**, 5.

Steele, B. C. H. (1989) in *High Conductivity Solid Ionic Conductors, Recent Trends and Applications*, Ed. T. Takahashi, World Scientific, Singapore, p. 402.

Strock, L. W. (1934) *Z. Phys. Chem.*, **B25**, 441.

Von Alpen, V., Rabenau, A. and Talat, G. M. (1977) *Appl. Phys. Lett.*, **30**, 621.

Wright, A. F. and Fender, B. E. F. (1977) *J. Phys. C*, **10**, 2261.

Yao, Y. F. and Kummer, J. T. (1967) *J. Inorg. Nucl. Chem.*, **29**, 2453.

4 Ionic transport in glassy electrolytes

J. L. SOUQUET

Laboratoire d'Ionique et d'Electrochimie du Solide, URA D1213 CNRS, Institut National Polytechnique de Grenoble

4.1 Ionic transport: experimental facts

Inorganic glasses are among the oldest known solid electrolytes. As early as 1884, Warburg proved the existence of Na^+ ionic conductivity in Thuringer glass, verifying Faraday's laws for sodium transfer through a thin glass membrane separating two amalgams. Thereafter, for most oxide or sulphide based glasses, a purely cationic conductivity has been confirmed. Less common vitreous electrolytes, which are also less conductive, exhibit anionic transport. This is the case for amorphous silicates containing lead halides which conduct by the motion of the halide ions (Shulze and Mizzoni, 1973; Ravaine and Leroy, 1980).

Whatever the mobile ion is, all the vitreous electrolytes have a transport number of unity and below their vitreous transition temperature, ionic conductivity follows an Arrhenius law:

$$\sigma = \sigma_0 \exp(-E_a/RT). \tag{4.1}$$

Almost all vitreous electrolytes have similar values for the preexponential term σ_0. (that is the extrapolated value for conductivity when temperature tends to infinity) of between 10 and 10^3 S cm^{-1}. They differ among themselves mainly because of different values for the activation energy E_a, which is very sensitive to the concentration and the nature of the mobile cation being usually between 0.2 and 1 eV. Consequently, near room temperature a large variation in the conductivity is observed, between 10^{-2} S cm^{-1} and 10^{-11} S cm^{-1}. This general behaviour is illustrated in Fig. 4.1(*a*).

For most oxide based glasses, the best conductivities are only some 10^{-7} S cm^{-1} at ambient temperature and 10^{-3} S cm^{-1} at 300 °C and as a result their use is largely confined to high temperatures and the electrolytes must be prepared as thin films (Levine, 1983; Jourdaine *et al.*,

1988). In the last 20 years, new sulphide, sulphate, molybdate, halide, etc., based compositions have been obtained in the glassy state (Ingram, 1987). They have much higher ionic conductivity than most oxide glasses at ambient temperature, e.g. from 10^{-5} to 10^{-2} S cm^{-1} in the case of some lithium or silver conducting glasses (Fig. 4.1(*b*)).

The high sensitivity of ionic conductivity to chemical composition is illustrated in Fig. 4.2 for some silver and alkali-ion conducting glasses. In all cases a very small change in the concentration of the conducting species leads to a very large variation in conductivity. A typical example is found with the $(AgPO_3)_{1-x}(AgI)_x$ glass system for which the room-temperature conductivity increases by four orders of magnitude when the silver concentration varies by only 10%, namely from 2.3×10^{-2} mol cm^{-3} to 2.5×10^{-2} mol cm^{-3} (Malugani, 1976).

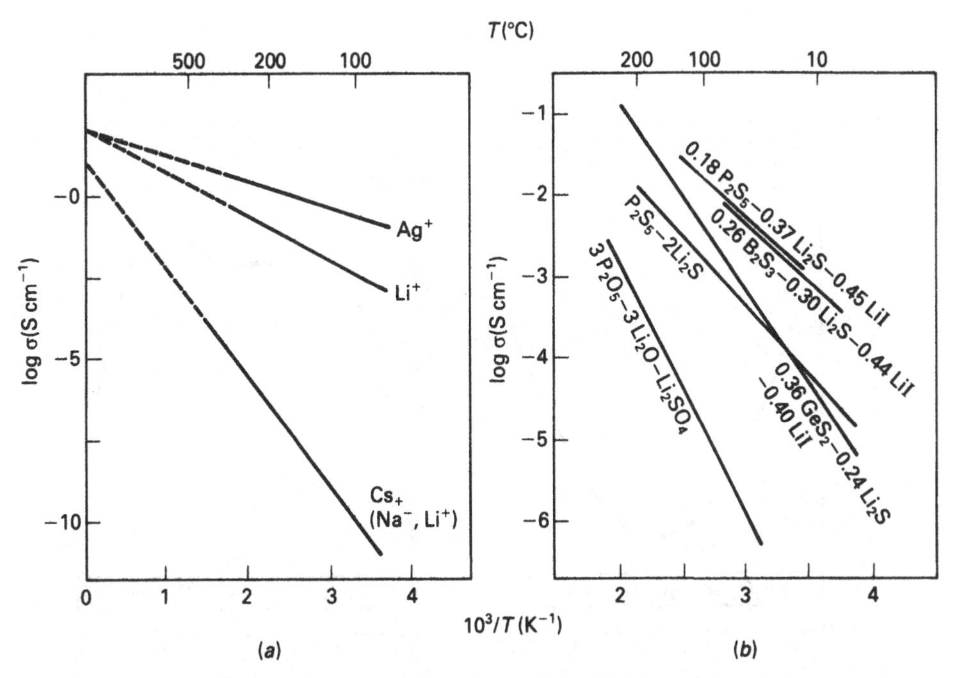

Fig. 4.1 Arrhenius plots for ionic conductivity. (*a*) General behaviour for ionically conducting glasses. At room temperature, the most conductive (Li^+ or Ag^+) have the lowest activation energy. For the less conductive glasses (Cs^+ or mixed alkali glasses) the activation energy is around 1 eV. (*b*) Experimental data for Li^+ conducting glasses (Souquet and Kone, 1986).

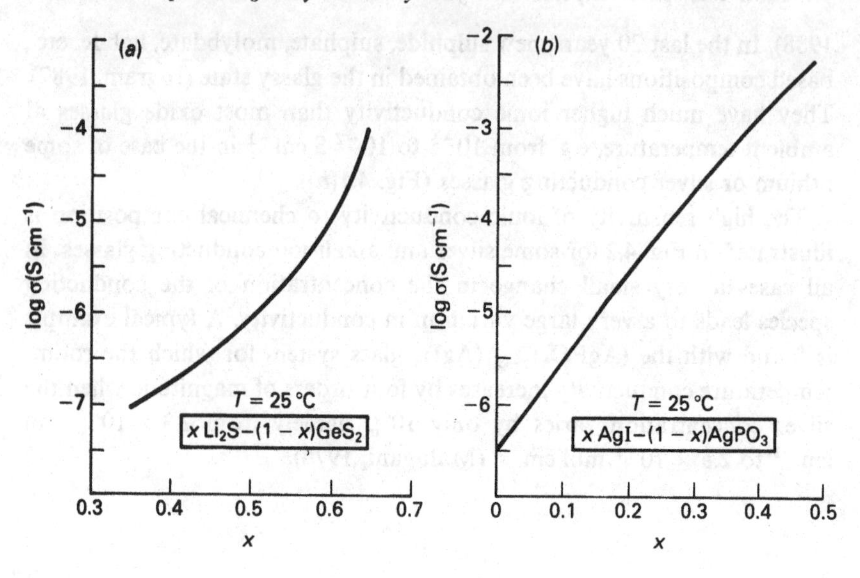

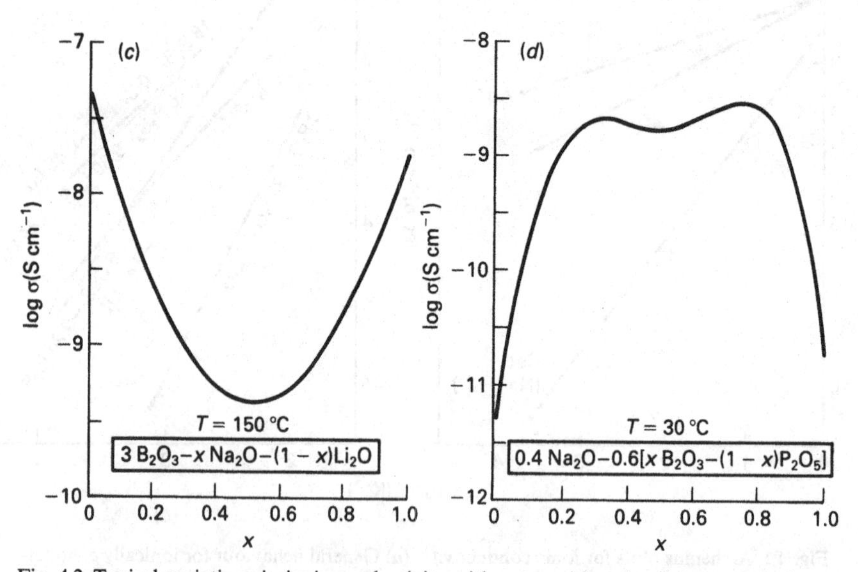

Fig. 4.2 Typical variations in ionic conductivity with composition. In all cases, variations in alkali or silver content are very low compared to the observed variation in log σ: (*a*) influence of the network modifier (Li_2S); (*b*) influence of a doping salt; (*c*) mixed alkali effect; (*d*) mixed anion effect. References for data are indicated in Souquet and Perera (1990).

4.2 *Chemical composition of ionically conductive glasses*

The glassy systems mentioned in Figs. 4.1(*b*) and 4.2 show that quite complex chemical compositions have been prepared in the glassy state. Up to three basic constituents are present in all ionically conducting glasses: network formers, network modifiers and ionic salts, in different proportions.

Network formers are compounds of a covalent nature such as SiO_2, P_2O_5, B_2O_3, GeS_2, P_2S_5, B_2S_3, etc. They form macromolecular chains which are strongly cross-linked by an assembly consisting of tetrahedra (SiO_4, PO_4, BO_4 ...) or triangles (BO_3) which combine to form macromolecular chains by sharing corners or edges. When pure, network formers readily form glasses by cooling from the liquid phase. A certain range of bond angles and lengths characterises the disorder existing in the vitreous state. The existence of local order which is associated with the stability of the tetrahedral or triangular entities is the result of the covalent character of the bonds. The possibility of deforming this local order is on the other hand the result of the partially ionic character of these same bonds. Usually network former oxides or sulphides are characterised by a difference in electronegativity between the anion and the network former cations amounting to 0.4–1.7 on the Pauling scale.

Network modifiers include oxides or sulphides (e.g. Ag_2O, Li_2O, Ag_2S, Li_2S, etc.) which interact strongly with the structure of network formers. A true chemical reaction is involved, leading to the breaking of the oxygen or sulphur bridge linking two network former cations. The addition of a modifier introduces two ionic bonds.

For instance, the reaction between silica and lithium oxide may be expressed schematically as:

$$\begin{array}{ccccccccc} & | & & | & & | & \mathrm{Li}^+ & | & \\ - & \mathrm{Si} & -\mathrm{O}- & \mathrm{Si} & - + \mathrm{Li_2O} \rightarrow - & \mathrm{Si} & -\mathrm{O}^- \quad {}^-\mathrm{O}- & \mathrm{Si} & - \\ & | & & | & & | & \mathrm{Li}^+ & | & \end{array} \qquad (4.2)$$

The increasing addition of a modifier to a given network former leads to the progressive breaking of all oxygen bridges as in (4.2).

As the number of non-bridging oxygen or sulphur atoms increases, the average length of the macromolecular chains decreases. The chemical reaction symbolised in (4.2) is strongly exothermic, and the mixing enthalpies are of the order of some hundreds of kilojoules. The magnitude of these values is difficult to account for on the basis of the energy balance of the bonds described in (4.2). The origin could be the stabilisation of the negative charge carried by the non-bridging oxygen atom by interaction of the oxygen p orbitals and the silicon d orbitals. The result is a reinforcement of the bond, representing the probable origin of the increase in the force constant observed by IR and Raman spectroscopy, and the shortening of this same bond observed by X-ray crystallography on the recrystallised glasses (Zarzycki and Naudin, 1960).

The case of boron as a network former cation is somewhat specific in that this element has no available d orbitals. However, a p orbital is available when the boron has a coordination number of 3, which allows stabilisation of an electronic doublet of the oxygen or sulphur introduced by the modifier. This oxygen or sulphur giving up a doublet to another boron atom increases the cross-linking by the formation of two BO_4 tetrahedra. In hybridisation terms, the boron is altered from the sp^2 configuration to the sp^3 configuration. The coordination change of boron has been especially well observed by NMR (Bray and O'Keefe, 1963; Muller-Warmuth and Eckert, 1982).

Although the environment of the network former cation is relatively well known, that of the modifier cation is much less so, due to the lack of appropriate spectroscopic techniques. The absence of direct experimental data has given rise to the coexistence in the literature of very different hypotheses ranging from models based on a totally random distribution of ionic bonds to those based on zones rich in modifier cations which alternate with less rich zones (Greaves, 1985).

Ionic salts are often added to a glassy matrix containing a network former and a network modifier. Such an addition significantly increases the ionic conductivity as shown in Fig. 4.2(*b*). For this reason these ionic salts are often referred to as doping salts. They are generally halide salts or in some cases sulphates. From a structural viewpoint, it is almost certain that the halide anions are not inserted in the macromolecular chain. Indeed, no modifications in the vibrations of the macromolecular chain have been revealed by spectroscopic analysis. In fact the only certainty to date is the absence of chemical reactions with the macromolecular chains. The arrangement of ions from the halide salt with respect to one another is still unknown. Hypotheses range from the formation of salt clusters (Malugani and Mercier, 1984) to a uniform distribution throughout the mass of the glass (Borjesson and McGreevy, 1992). From a thermodynamic viewpoint, the absence of chemical reactions is likely to lead to low mixing enthalpies of the order of a few kilojoules per mole (Reggiani, Malugani and Bernard, 1978).

It has been proposed that this mixing enthalpy is of purely electrostatic origin representing a slight modification of the environment near the ions. We can, for instance, envisage dissolution of silver iodide in silver phosphate schematically as follows

$$\begin{array}{ccccccccccc}
 & | & & \mathrm{Ag^+} & & | & & | & & \mathrm{Ag^+} & \\
\mathrm{O{=}} & \mathrm{P} & \mathrm{{-}O^-} & & \mathrm{{}^-O{-}} & \mathrm{P} & \mathrm{{=}O + 2AgI \rightarrow 2O{=}} & \mathrm{P} & \mathrm{{-}O^-} & \mathrm{I^-} & \qquad (4.3) \\
 & | & & \mathrm{Ag^+} & & | & & | & & \mathrm{Ag^+} &
\end{array}$$

Qualitatively, the dipole–dipole interactions between the macromolecular chains and the halide salt compensate for the lattice energy of the halide crystal and tend to decrease the interactions existing in the glass between the oxide macroanions. This decrease is probably the reason for the significant drop in the glass transition temperature resulting from the addition of a halide salt (Reggiani *et al.*, 1978). Furthermore this type of reaction is consistent with the fact that dissolution of a halide salt in a vitreous solvent requires the existence of ionic bonds provided by a network modifier.

Finally, mixtures of ionic salts may form glasses, which contain discrete anions (iodide, or molybdate...) without any macromolecular anions. This is the case for glasses in the $AgI–AgMoO_4$ system for which the pure limiting compositions AgI or $AgMoO_4$ do not form glasses.

4.3 Kinetic and thermodynamic characteristics of glassy electrolytes

Whatever their chemical composition most glasses are obtained by quenching from the liquid state. The quenching process which produces a glass from the liquid must be sufficiently fast to avoid crystallization kinetically and to leave a material that is not in thermodynamic equilibrium. Quenching rates of between $10\,°C\,s^{-1}$ and $10^7\,°C\,s^{-1}$ are used to produce a wide range of ionic conducting glasses. The resulting arrangement of atoms, obtained by X-ray and neutron scattering, is practically the same as in the original liquid.

The main difference between a glass and its liquid is not structural but kinetic and depends on a microscopic quantity called the structural relaxation time τ. This time is the mean life time for the movement of a structural unit over a distance equivalent to its size. Such a structural unit may consist of several SiO_4 units in the case of a silicate glass. The structural relaxation time is strongly dependent on the temperature. At the highest temperatures, τ is small and may reach a value of 10^{-13}–10^{-12} s which is the time of an elementary vibration in the potential well formed by the neighbouring units. As the temperature is lowered below the melting point, T_m, the reponse time in the supercooled liquid increases rapidly, eventually surpassing the observational time scale. When this happens, large scale flow processes cease and the material appears solid on a human time scale. The temperature T_g, at which the relaxational and observational time scales cross, depends on the observer and does not represent any intrinsic temperature of the system itself. Conventionally, the vitreous transition temperature T_g corresponds to $\tau = 10^2$ s, i.e. to viscosities in excess of 10^{13} dPa s (10^{14} poises) since structural relaxation time and viscosity are proportional quantities ($\eta = \tau G$, where η is the viscosity and G is the shear modulus).

Currently, the dependence of τ on temperature is deduced from viscosity–temperature measurements. At $T < T_g$, the temperature dependence of τ obeys an Arrhenius law, but this dependence is much more complex at $T > T_g$. In the latter case it is referred to an empirical Vogel–Tamman–Fulcher (VTF) law (Vogel, 1921; Tamman and Hesse, 1926; Fulcher, 1925).

$$\tau = \tau_0 \exp\left[\frac{B}{R(T - T_0)}\right] \tag{4.4}$$

with B constant or weakly temperature dependent and T_0 a characteristic temperature of the material. From Eqn (4.4) it may be seen that T_0 would be the glass transition temperature measured with an infinite observational time. For this reason, T_0 is called the ideal vitreous transition temperature. When ionic conductivity is measured above T_g, i.e. for 'molten' glasses or in the elastomeric domain for polymers, a VTF behaviour with temperature is also observed, and this will be discussed in Section 4.7.

The glass transition temperature is thus closely related to kinetic parameters and to the duration of the experiment conducted on the material. Thus, the glass transition temperature is an increasing function of the quenching rate. In practice a variation of about 10–20 K for T_g may be observed for the same glass (Menetrier, Hojjaji, Estournes and Levasseur, 1991). Note also that for a well defined compound which may be obtained in the form of a glass T_g and T_0 are linked by an empirical relationship to the melting temperature T_m of the crystalline form:

> $T_g \approx \frac{2}{3}T_m$ (Sakka and Mackenzie, 1971) and $T_0 \approx \frac{1}{2}T_m$ (Caillot, Duclot and Souquet, 1991).

As a consequence of the disorder existing in the vitreous state a glass possesses a higher enthalpy and a higher entropy than the corresponding crystalline compound. The excess enthalpy arises from the range of bond lengths and angles. It has been suggested that an excess enthalpy of about 5 kJ atom^{-1} is appropriate. The excess entropy is in fact the entropy which is frozen into the supercooled liquid at T_g and is usually around 4 J K^{-1} atom^{-1}. Such values mean that a glass is not at thermodynamic equilibrium. For the same chemical composition, the excess free energy contained in a glass compared to the crystalline phase depends on the preparation procedure, especially on the quenching rate, and physical characteristics such as density, refractive index and ionic conductivity may differ slightly. Nevertheless, thermodynamics can be applied to such metastable phases since the relaxation times of the molecular degrees of freedom are either very short (e.g. vibrational motion) or very long (e.g. structural relaxation time) compared with the observational time scale (Jäckle, 1981).

4.4 A microscopic approach to ionic transport in glasses

Regardless of the conduction mechanism, electrical transport can be expressed as the product of the charge carrier concentration c, the mobility u and the charge. Conductivity is expressed in S cm^{-1}, mobility in cm^2 V^{-1} s^{-1} and if the charge carrier concentration is expressed in mole cm^{-3} then the charge per mole of ions must be used. For singly charged ions this is F, Faraday's constant (96 500 C). Using the subscript + in the case of cationic transport we get the relationship

$$\sigma_+ = c_+ F u_+ \tag{4.5}$$

Note that this is equivalent to the expression $\sigma_i = c_i q u_i$ used in Chapters 2 and 3 except that c_i and q refer to the individual ions rather than moles of ions.

Since the relative dielectric constant ε_r of inorganic glasses is low, typically between 5 and 15, ionic species are strongly associated. For instance most of the Ag^+ cations will be associated with non-bridging oxygens in $AgPO_3$ glass. Nevertheless, thermal vibrations allow a partial dissociation and Fig. 4.3 suggests a possible two-step displacement mechanism: creation of a charged defect (*a*), then defect migration (*b*).

(*a*) For the usual concentration C of alkali or silver, i.e. 0.01 to 0.03 mole cm^{-3}, anionic sites are close enough (2–4 Å) for a cation to leave its normal site and move to a neighbouring site which is already occupied. This defect formation in the glass structure is formally analogous to the formation of a Frenkel defect in an ionic crystal (Chapter 2). The concentration of interstitial cationic pairs thus formed is obviously equal to the concentration of vacated cation sites, it represents the concentration c_+ of charge carriers. Conventional techniques do not allow the measurement of c_+. Nevertheless, it has been estimated that, for a pure silver phosphate glass, silver cations in interstitial positions represent only 10^{-7} of all the silver cations (Clement, Ravaine and Deportes, 1988).

From a thermodynamic point of view the formation of an interstitial pair obeys the chemical equilibrium (Kittel, 1968)

cation on a normal site	+	available interstitial site	⇌	interstitial cation	+	vacated cation site
$(C - c_+)$		$(C - c_+)$		c_+		c_+

Taking into account that every associated pair may accept an interstitial cation and that $c_+ \ll C$ it then follows that

$$c_+ = C \exp(-\Delta G_f/2RT) \tag{4.6}$$

where $\Delta G_f = \Delta H_f - T\Delta S_f$ is the free energy associated with the formation of a defect.

(*b*) The second step is the defect migration caused by the electric field. The suggested mechanism in Fig. 4.3 is an indirect interstitialcy mechanism.

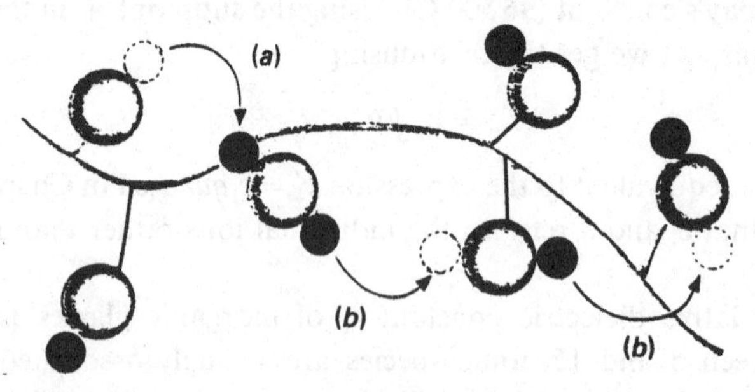

Fig. 4.3 Schematic representation of interstitial cationic pair formation (*a*) and migration from one non bridging oxygen to another in a cation conducting glass (*b*).

Such a mechanism is not incompatible with a Haven ratio between 0.3 and 0.6 which is usually found for mineral glasses (Haven and Verkerk, 1965; Terai and Hayami, 1975; Lim and Day, 1978). The Haven ratio, that is the ratio of the tracer diffusion coefficient D^* determined by radioactive tracer methods to D, the diffusion coefficient obtained from conductivity via the Nernst–Einstein relationship (defined in Chapter 3) can be measured with great accuracy. The simultaneous measurement of D^* and D by analysis of the diffusion profile obtained under an electrical field (Kant, Kaps and Offermann, 1988) allows the Haven ratio to be determined with an accuracy better than 5%. From random walk theory of ion hopping the conductivity diffusion coefficient $D = (e^2/\sigma)\nu_0$ in an isotropic medium. Hence for an indirect interstitial mechanism, the corresponding mobility is expressed by

$$u_+ = \frac{F}{RT}\frac{l^2}{6}\nu_0 \exp\left(-\frac{\Delta H_m}{RT}\right), \tag{4.7}$$

where ν_0 is the vibrational frequency for interstitial cations, l is the jump distance and ΔH_m is the enthalpy needed for an elementary jump. Obviously, this displacement in a disordered medium implies mean values for l and ΔH_m. Then

$$\sigma_+ = Fu_+c_+ = \frac{F^2 l^2 \nu_0 C}{6RT} \exp\left(\frac{-\Delta G_f/2 - \Delta H_m}{RT}\right). \tag{4.8}$$

By identification with the experimental law

$$\sigma_+ = \sigma_0 \exp(-E_a/RT)$$

we get the following formal expressions for the preexponential term and activation energy E_a

$$\sigma_0 = \frac{F^2 l^2 \nu_0 C}{6RT} \exp\left(\frac{\Delta S_f}{2R}\right) \tag{4.9}$$

$$E_a = \frac{\Delta H_f}{2} + \Delta H_m \tag{4.10}$$

For the usual values of the physical parameters l, ν_0 and C, calculated values for σ_0 are between 10 and 10^3 S cm^{-1} as found experimentally. This agreement means that there is little influence of the entropic term ΔS_f.

Since all glasses have comparable values for σ_0, isothermal conductivity variations represented in Fig. 4.2 are related to $\Delta H_f/2$ and ΔH_m variations with composition. At this point, without any additional assumptions, the relative influence of the two terms is unknown. In other words, when the chemical composition varies, the corresponding variation in charge carrier concentration and mobility are inseparable.

4.5 Thermodynamics of charge carriers: weak electrolyte theory

As suggested in Fig. 4.3 the charge carrier formation in step (*a*) may be compared to a dissociation leading to the following equilibrium

$$M_2O \rightleftharpoons OM^- + M^+, \tag{4.11}$$

where the M^+ species represents the alkali ion in an interstitial position. An equilibrium constant K_{diss}, which is a function of the dissociation free energy, links the thermodynamic activities of the species involved in equilibrium (4.11)

$$K_{diss} = \exp\left(-\frac{\Delta G^\circ_{diss}}{RT}\right) = \frac{a_{M^+}a_{OM^-}}{a_{M_2O}}. \tag{4.12}$$

In a medium with a relatively low dielectric constant like a glass, the dissociation constant is expected to be small, and the thermodynamic ionic activities proportional to their concentrations. An approximate expression for (4.12) is then

$$K'_{diss} = \frac{[M^+][OM^-]}{a_{M_2O}} \tag{4.13}$$

in which the K'_{diss} expression includes the ionic activity coefficients γ_{M^+} and γ_{OM^-} depending on the chosen concentration scale. In this case the charge carrier concentration c_+ is then

$$c_+ = K'^{1/2} a_{M_2O}{}^{1/2} \tag{4.14}$$

or

$$c_+ = K'^{1/2} \exp\left(\frac{\Delta G_{M_2O}}{2RT}\right), \tag{4.15}$$

where the M_2O thermodynamic activity is expressed as a function of the network modifier partial molar free energy.

Many fast ion conducting glasses contain several salts of the same alkali metal to optimise the conductivity. The expression for the charge carrier concentration in terms of the thermodynamic activities of all the components is difficult to establish since several dissociation equilibria are involved simultaneously. Nevertheless, if the dissociation of one of the salts MY is expected to dominate greatly over all the others, Eqn (4.14) may be used for this salt alone as a convenient first approximation. Experimental evidence supporting the predominant dissociation of one salt is provided by the large increase of the ionic conductivity with the salt content. This is clearly the case for silver iodide when added to silver phosphate, as is shown in Fig. 4.2. Generally ionic salts with a large anion should have a high dissociation constant. The Fuoss expression (1958) relates log K_{diss}, the relative dielectric constant ε_r and the ionic radii.

$$\log K_{diss} = \frac{N_A e^2}{(r_+ + r_-)RT} \frac{1}{4\pi\varepsilon_0\varepsilon_r}, \tag{4.16}$$

where N_A is Avogadro's number, e is the electronic charge, r_+ and r_- are the radii of the cation and anion and ε_0 is the static permittivity. From this equation, it is easily seen that a large ionic radius is expected to increase K_{diss} through larger values of r and ε_r.

Such a chemical approach which links ionic conductivity with thermodynamic characteristics of the dissociating species was initially proposed by Ravaine and Souquet (1977). Since it simply extends to glasses the theory of electrolytic dissociation proposed a century ago by Arrhenius for liquid ionic solutions, this approach is currently called the *weak electrolyte theory*. The weak electrolyte approach allows, for a glass in which the ionic conductivity is mainly dominated by an MY salt, a simple relationship between the cationic conductivity σ_+, the electrical mobility u_+ of the charge carrier, the dissociation constant K_{diss} and the thermodynamic activity of the salt with a partial molar free energy $\overline{\Delta G}_{MY}$ with respect to an arbitrary reference state:

$$\sigma_+ = Fu_+K'^{1/2}a_{MY}{}^{1/2} = Fu_+K'^{1/2}\exp\left(\frac{\overline{\Delta G}_{MY}}{2RT}\right). \tag{4.17a}$$

From this relationship we may expect σ_+ to be proportional to the salt thermodynamic characteristics, if u_+ and K_{diss} have constant values at constant temperature and pressure in a given glassy system. The square root dependency of ionic conductivity on a_{MY} has been experimentally verified over several orders of magnitude. The dissociating species is either a network modifier or a doping salt. Potentiometric (Ravaine and Souquet, 1977) or calorimetric (Reggiani, Malugani and Bernard, 1978) techniques have been used for thermodynamic activity measurements, Fig. 4.4. Such results suggest that the main contribution to the variations of ionic conductivity as a function of composition is related to the large variation in the number of charge carriers rather than to the variation in the u_+ or K_{diss}.

An interesting limiting case may be found at very low concentrations

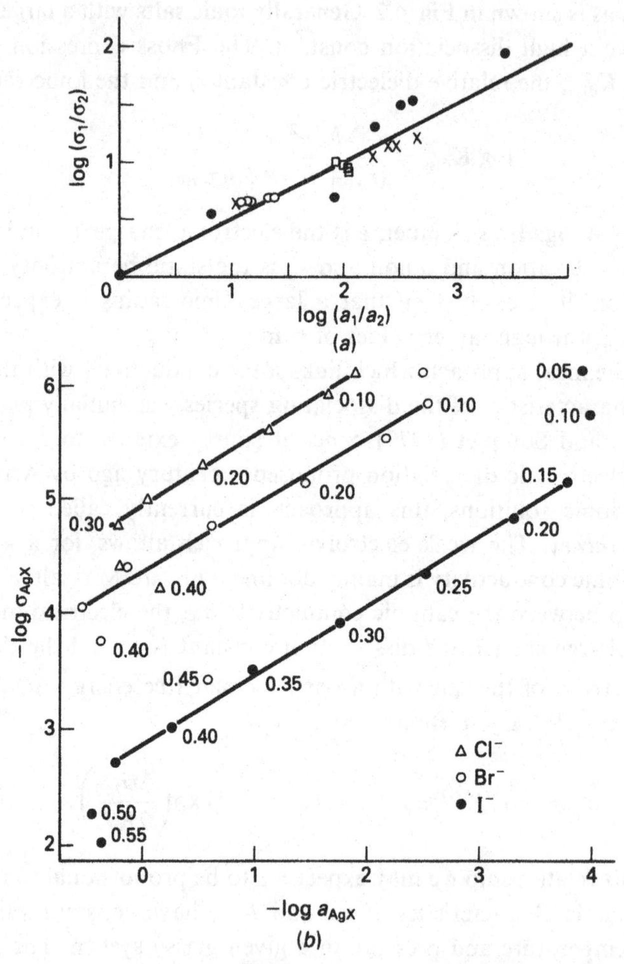

Fig. 4.4 (*a*) Comparison on a logarithmic scale of the conductivity ratio and the thermodynamic activity ratio of alkali oxide in several silica based glasses. Activity ratios are deduced from potentiometric measurements. (*b*) On the same scale, conductivity vs activity of AgX (X = Cl, Br, I) in phosphate glasses. Activity values are deduced from calorimetric measurements.

for the ionic salt, when its thermodynamic activity is proportional to its concentration, C. In this case, ionic conductivity varies as $C^{1/2}$ and equivalent conductivity $\Lambda = \sigma/C$ varies as $C^{-1/2}$. This behaviour has been shown for the GeO_2–Na_2O system for $C_{Na} < 10^{-4}$ mole cm^{-3} (Cordado and Tomozawa, 1980). The same $C^{1/2}$ dependence has also been suggested for organic polymers containing a small amount of ionic impurities (Blythe, 1980).

4.6 *Conductivity measurements and the thermodynamics of glass*

Experiment shows that the variation of ionic conductivity with the content of MY salt depends on its partial molar free energy through the term $\exp(\overline{\Delta G}_{MY}/2RT)$. If x is any concentration scale, for instance the molar ratio, which allows a measure of MY content, experimental results follow an Arrhenius law that can be expressed by:

$$\sigma(x) = \sigma_0(x)\exp[-E_a(x)/RT] \tag{4.17b}$$

Partial differentiation of Eqns (4.17a) and (4.17b) with respect to x allows us easily to relate the variation in the preexponential term with the variation in partial molar entropy of MY, and the variation in activation energy with the variation in partial molar enthalpy by the relationships (Fig. 4.5)

$$\frac{\partial \log \sigma_0}{\partial x} = -\frac{\partial \bar{S}_{MY}}{2R\,\partial x}, \tag{4.18}$$

$$\frac{\partial E_a}{\partial x} = -\frac{\partial \bar{H}_{MY}}{\partial x}. \tag{4.19}$$

So from electrical data, it is possible to get information on partial thermodynamic functions of the salt and then develop thermodynamic models for quantitative interpretation of the conductivity variation with composition. These models are not very different from those already developed for molten salt mixtures or metallic alloys.

For the four typical variations in ionic conductivity with composition illustrated in Fig. 4.2, thermodynamic interpretations and predictions have been advanced.

For a simple glass made of a network former and a network modifier, M_2X, a large increase in ionic conductivity with alkali content results in a large increase of $\overline{\Delta G}_{M_2X}$, mainly from entropic origin. In the case of sulphide or oxide glasses, a simple model, based on the existence of three oxygen or sulphur configurations in chemical equilibrium, has been developed (Pradel, Henn, Souquet and Ribes, 1989). This model is similar to that initially proposed by Fincham and Richardson for liquid silicates (1954).

The large increase in ionic conductivity obtained by halide dissolution in oxide or sulphide based glasses is easily explained by the increase of the halide partial free energy $\overline{\Delta G_{MY}}$ as a consequence of a regular solution formed between the halide salt (MY) and the glassy solvent (Kone and Souquet, 1986). The microscopic origin of this thermodynamic behaviour probably lies in the formation of electrostatic quadrupoles made by two alkali cations associated with a non-bridging oxygen or sulphur and a halide anion as shown in Eqn (4.3). The mixing enthalpies resulting from the formation of such an ionic association may be deduced from conductivity data. They are of the order of several tens of kilojoules per mole, comparable to the mixing enthalpies for molten salt mixtures involving one alkali cation and two different anions.

No simple thermodynamic model is yet able to explain the large decrease in conductivity when partially substituting one alkali metal for another, i.e. the so-called mixed alkali effect. Systematic analysis of the pre-

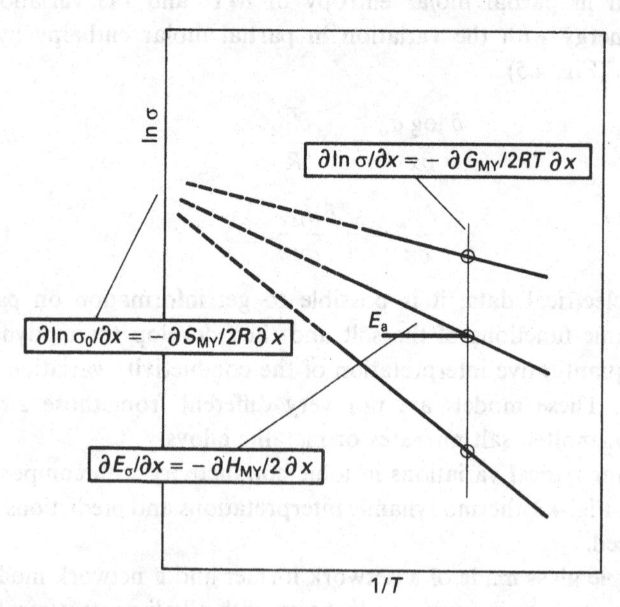

Fig. 4.5 Variations of σ_0 and E_a in an Arrhenius plot as a function of the molar ratio x of the dissociating salt MY dissolved in the glass.

exponential term and activation energy with composition suggests a strong decrease in partial molar entropy and partial molar enthalpy of alkali cations while mixing. The microscopic interpretation of such a behaviour remains unclear although recent X-ray spectroscopic studies have suggested that the origin may be structural. However, a significant (some kJ $mole^{-1}$) exothermic mixing enthalpy between the limiting compositions may be predicted and has been found experimentally by calorimetry (Terai and Sugita, 1978; Kone, Reggiani and Souquet, 1989).

A similar analysis may be developed for the mixed anion effect. In that case an endothermic mixing enthalpy between the limiting compositions may enhance the partial molar free energy of the network modifier (Souquet, 1988). The fact that all the glassy mixtures which show a mixed anion effect tend to separate into two phases is a qualitative verification of an endothermic mixture. Endothermic mixing enthalpies, as a consequence of the mixing of different anions, have been observed by calorimetry for halogenated salts and glasses (Janz, 1967; Hervig and Navrotsky, 1985). In the case of glasses, endothermic reactions may be interpreted as an increase of the coordination number of one of the network former cations which is very often boron. From an experimental point of view it is interesting to correlate an increase of conductivity with the variation of boron coordination from 3 to 4. Such a comparison may be done in the Li_2O–B_2O_3–TeO_2 and Li_2S–B_2S_3–P_2S_5 systems (Martins Rodrigues and Duclot, 1988; Zhang *et al.*, 1989) using conductivity and NMR data.

Eqns (4.18) and (4.19) relate the partial molar entropy and partial molar enthalpy to the preexponential term and activation energy. These partial thermodynamic functions may vary as a function of the alkali content but also, at constant composition, as a function of the quenching rate. The enthalpy and entropy stored in a glass depends on the temperature interval in which the material is a supercooled liquid, before the structure is frozen in at the vitreous transition temperature T_g. Their values will be higher with a faster quenching rate. If ΔT_g is the difference between the T_g values associated with two different quenching rates and $\overline{\Delta C^{P}_{MY}}$ the difference in partial molar thermal capacity of the dissociating salt in the liquid and vitreous state, we may estimate the excess in partial molar entropy and enthalpy stored in the glass obtained by fast quenching

$$d\bar{S}_{MY} \simeq \frac{\overline{\Delta C^{P}_{MY}} \Delta T_g}{T_g}, \tag{4.20}$$

$$\mathrm{d}\bar{H}_{\mathrm{MY}} \simeq \overline{C^{\mathrm{P}}_{\mathrm{MY}}} \Delta T_{\mathrm{g}}. \tag{4.21}$$

These relationships imply that $\Delta T_{\mathrm{g}} \ll T_{\mathrm{g}}$ which is the case since ΔT_{g} does not usually exceed 10–20 K. $\overline{\Delta C^{\mathrm{P}}_{\mathrm{MY}}}$ values are generally unknown but the ΔC^{P} for an atom is estimated to be 10 J K^{-1} atom^{-1} (Angell, 1968).

Using these values as a first approximation, the effect of quenching rate would lead to

$$\frac{\mathrm{d}\sigma_0}{\sigma_0} = -\frac{\mathrm{d}\bar{S}_{\mathrm{MY}}}{2R} \approx 3\%$$

and

$$\mathrm{d}E_{\mathrm{a}} = -\frac{\mathrm{d}\bar{H}_{\mathrm{MY}}}{2} \approx 1.5 \times 10^{-3}\ \mathrm{eV}.$$

These variations are low but cumulative and large enough to be observed since the glass obtained with a high quenching rate shows a conductivity about 10% higher at room temperature (Menetrier *et al.*, 1991; Boesch and Moynihan, 1975). This difference vanishes after annealing the glass near T_{g}.

4.7 *A microscopic model for ionic transport above the vitreous transition temperature*

Above the vitreous transition temperature T_{g}, ionic conductivity increases steeply as represented in Fig. 4.6 from data obtained in the AgI–$AgMoO_4$ mixture. Above T_{g}, ionic conductivity is no longer represented by an Arrhenius law (4.1) and experimental results are better represented by an empirical relationship

$$\sigma_{+} = \sigma_0 \exp\left[\frac{-B}{R(T - T_0)}\right]. \tag{4.22}$$

Such a relationship is also commonly observed for salt–polymer complexes (Armand, Chabagno and Duclot, 1979). As mentioned in the context of the structural relaxation time (Section 4.3) it is referred to as VTF behaviour.

Generally, the VTF behaviour of all transport properties may be understood from the free volume concept introduced by Doolittle (1951) and further developed by Cohen and Turnbull (1959). Essentially, any diffusing species is depicted as encaged by the nearest atoms in a cell of temperature dependent volume V. Above a critical value of the temperature T_0, and consequently of a volume V_0, the excess volume V_f ($V_f = V - V_0$) is considered as free, that is redistributable around its mean value $\langle V_f \rangle$ without any enthalpic contribution. The temperature dependence of this mean free volume is then simply expressed by $\langle V_f \rangle = \Delta\alpha V_0(T - T_0)$ where $\Delta\alpha$ is the difference in the volumetric dilatation coefficient of the liquid and crystalline phases.

Finally, the structural relaxation time for any diffusing species in the supercooled liquid is proportional to the probability of this species having access to a free volume over the minimum value V_f^* required for an elementary displacement.

$$\tau \sim \exp\left(\frac{-V_f^*}{\langle V_f \rangle}\right) = \exp\left(\frac{-V_f^*}{\Delta\alpha V_0(T - T_0)}\right). \tag{4.23}$$

Obviously, V_f^*, and therefore the B constant in the final VTF expression (4.22) will have different values depending on the ion and the chain segments involved in the conductivity process.

Fig. 4.7 is an attempt to illustrate an ionic displacement for an interstitial pair by a VTF mechanism along a macromolecular chain. The

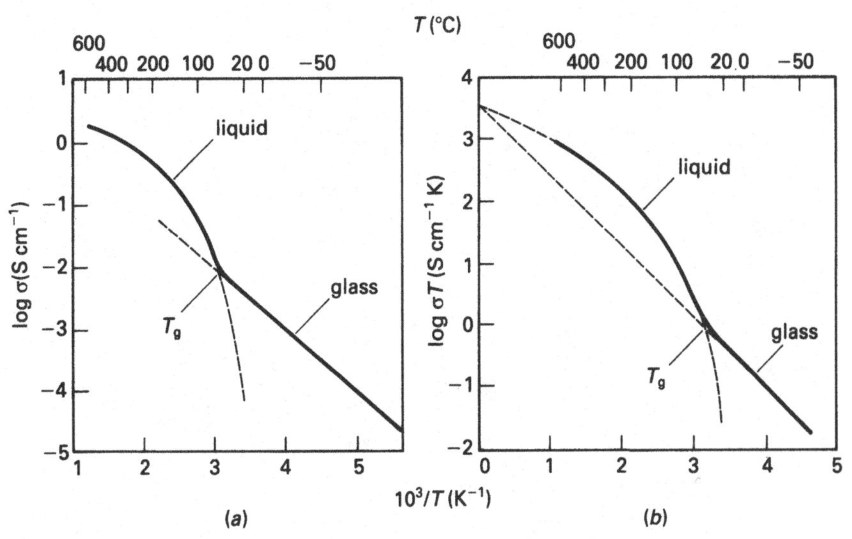

Fig. 4.6 Arrhenius plot for (*a*) σ and (*b*) σT for the $(AgI)_{0.7}$–$(AgMoO_4)_{0.3}$ mixtures. Data from Kawamura and Shimoji (1986). Note on the σT plot that extrapolated values from the glassy and liquid state are the same when $T \to \infty$.

first step (*b*) is similar to the dissociation represented in (*a*) of Fig. 4.3. The essential difference lies in the transfer mechanism for the interstitial cation represented in Fig. 4.7(*c*). The transfer needs a local deformation of the macromolecular chain involving a local free volume over a minimal value V_f^*. When such a transfer occurs, it means that the $\exp(-\Delta H_m/RT)$ probability term in the mobility expression (4.7) has to be replaced by the term $\exp[-V_f^*/\Delta\alpha V_0(T-T_0)]$. A complete expression for cationic conductivity as a function of temperature above T_g is then:

$$\sigma_+ = \frac{F^2}{RT} l^2 \nu_0 C \exp\frac{\Delta G_f}{2RT} \exp\left[\frac{-V_f^*}{\Delta\alpha V_0(T-T_0)}\right] \tag{4.24}$$

in which all terms have been previously defined. By splitting entropic and enthalpic terms Eqn. (4.24) becomes

$$\sigma_+ = \frac{F^2}{RT} l^2 \nu_0 C \exp\left(\frac{\Delta S_f}{2R}\right) \exp\left(\frac{\Delta H_f}{2RT}\right) \exp\left[\frac{-V_f^*}{\Delta\alpha V_0(T-T_0)}\right]. \tag{4.25}$$

For temperatures above T_g, the deviation from an Arrhenius behaviour

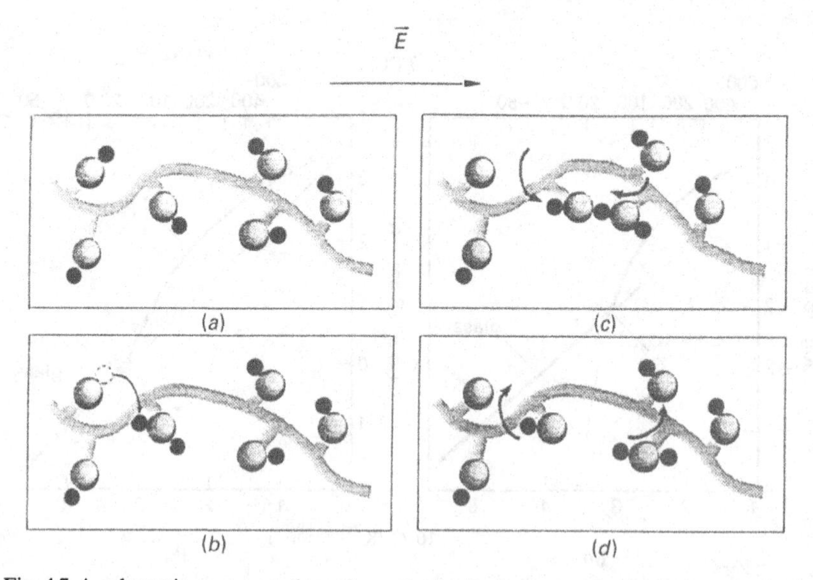

Fig. 4.7 A schematic representation of a cationic displacement along a polymeric chain above its T_g. (*a*) An initial activated step (*b*) allows the formation of an interstitial pair, the migration of which (*c*) and (*d*) is assisted by local free volume redistribution.

is imposed by the last exponential term leading to an experimental behaviour of Eqn (4.22). In fact Eqn (4.25) is similar to Eqn (6.11) for polymer electrolytes in which ion transport may also be described by a free volume approach and an Arrhenius term representing the creation of charge carriers has been included.

Note that, for a material with an ionic conductivity that can be measured above and below T_g, extrapolated data for the σT term in the two domains should give an identical value when T approaches infinity $(\sigma T)_{T \to \infty} = (F^2/R)l^2 v_0 C$. Such behaviour is illustrated in Fig. 4.6(*b*) for the $(AgI)_{0.7}$–$(AgMoO_4)_{0.3}$ mixture.

References

Angell, C. A. (1968) *J. Am. Ceramic Soc.*, **51**, 117–33.

Armand, M. B., Chabagno, J. M. and Duclot, M. J. (1979) in *Fast Ion Transport in Solids*, Eds. P. Vashishta, J. N. Mundy and G. K. Shenoy, North-Holland, Amsterdam, p. 131–6.

Blythe, A. R. (1980) in *Electrical Properties of Polymers*, Cambridge University Press, Cambridge.

Borjesson, L., McGreevy, R. L. and Wicks, A. (1992) *Journal de Physique III*, **2**, 107–16.

Boesch, L. P. and Moynihan, C. T. (1975) *J. Non-Cryst. Solids*, **17**, 44–60.

Bray, P. J. and O'Keefe, J. G. (1963) *Phys. and Chem. of Glasses*, **4**, 37.

Caillot, E., Duclot, M. J. and Souquet, J. L. (1991) *C. R. Acad. Sci. Paris*, **312**, II, 447–9.

Clement, V., Ravaine, D., Deportes, C. and Billat, R. (1988) *Solid State Ionics*, **28–30**, 1572–8.

Cohen, M. H. and Turnbull, D. (1959) *J. Chem. Phys.*, **31**, 1164–9.

Cordado, J. F. and Tomozawa, M. (1980) *Phys. and Chem. of Glasses*, **19**, 115–20.

Doolittle, A. K. (1951) *J. Appl. Phys.*, **22**, 1471–80.

Fincham, C. I. B. and Richardson, F. D. (1954) *Proc. Roy. Soc., A*, **223**, 40.

Fulcher, G. S. (1925) *J. Am. Ceramic Soc.*, **8**, 339–56.

Fuoss, J. (1958) *J. Am. Chem. Soc.*, **80**, 5059–61.

Greaves, G. N. (1985) *J. Non-Cryst. Solids*, **71**, 203–17.

Haven, Y. and Verkerk, B. (1965) *Phys. and Chem. of Glasses*, **6**, 38–45.

Hervig, R. I. and Navrotsky, A. (1985) *J. Am. Ceramic Soc.*, **68**, 314–19.

Ingram, M. D. (1987) *Phys. and Chem. of Glasses*, **28**, 215–34.

Jäckle, J. (1981) *Phil. Mag. B*, **44**, 533–45.

Janz, G. J. (1967) *Molten Salt Handbook*, Academic Press, New York.

Jourdaine, L., Souquet, J. L., Delord, V. and Ribes, M. (1988) *Solid State Ionics*, **28–30**, 1490–4.

Kant, H., Kaps, C. and Offermann, J. (1988) *Solid State Ionics*, **31**, 215–20.

Kawamura, J. and Shimoji, M. (1986) *J. Non-Cryst. Solids*, **88**, 281–94.

Kittel, C. (1968) *Introduction to Solid State Physics*, John Wiley & Sons, New York.
Kone, A., Reggiani, J. C. and Souquet, J. L. (1989) in *Proceedings of the 10th Risø International Symposium on Metallurgy and Material Science*, 435–40, Eds. J. B. Bide Sorensen *et al.*, Risø National Laboratory, Roskilde, Denmark.
Kone, A. and Souquet, J. L. (1986) *Solid State Ionics*, **18–19**, 454–60.
Levine, C. A. (1983) in *The Sulfur Electrode*, Ed. R. P. Tischer, Academic Press, New York.
Lim, C. and Day, D. E. (1978) *J. Am. Ceramic Soc.*, **61**, 99–102.
Malugani, J. P. (1976) PhD Thesis (University of Besançon).
Malugani, J. P. and Mercier, R. (1984) *Solid State Ionics*, **13**, 293–9.
Martins Rodrigues, A. C. and Duclot, M. J. (1988) *Solid State Ionics*, **28–30**, 729–31.
Menetrier, M., Hojjaji, A., Estournes, C. and Levasseur, A. (1991) *Solid State Ionics*, **48**, 325–30.
Muller-Warmuth, W. and Eckert, H. (1982) *Phys. Rep.*, **2**, 91–149.
Pradel, A., Henn, F., Souquet, J. L. and Ribes, M. (1989) *Phil. Mag. B*, **60**, 741–51.
Ravaine, D. and Souquet, J. L. (1977) *Phys. and Chem. of Glasses*, **18**, 27–31.
Ravaine, D. and Leroy, D. (1980) *J. Non-Cryst. Solids*, **38–39**, 575–9.
Reggiani, D., Malugani, J. P. and Bernard, J. (1978) *J. Chim. Physique*, **75**, 245–9.
Sakka, S. and Mackenzie, J. D. (1971) *J. Non-Cryst. Solids*, **25**, 145–62.
Shulze, P. C. and Mizzoni, M. S. (1973) *J. Am. Ceramic Soc.*, **56**, 65–8.
Souquet, J. L. (1988) *Solid State Ionics*, **28–30**, 693–703.
Souquet, J. L. and Kone, A. (1986) in *Materials for Solid State Batteries*, Eds. B. V. R. Chowdari and S. Radhakrishna, World Scientific Publ. Co., Singapore.
Souquet, J. L. and Perera, W. G. (1990) *Solid State Ionics*, **40–41**, 595–604.
Tammann, G. and Hesse, W. (1926) *Z. Anorg. Allgem. Chem.*, **156**, 245–57.
Terai, R. and Hayami, R. (1975) *J. Non-Cryst. Solids*, **18**, 217–64.
Terai, R. and Sugita, A. (1978) *Q. Rep. Govt. Ind. Res. Inst.*, **20**, 353–6.
Vogel, H. (1921) *Phys. Z.*, **22**, 645–6.
Warburg, E. (1884) *Ann. der Physik und Chemie*, **2**, 622–46.
Zarzycki, J. and Naudin, F. (1960) *Verres et Réfractaires*, 113.
Zhang, Z., Kennedy, J. H., Thompson, J., Anderson, S., Lathop, D. A. and Eckert, H. (1989) *Appl. Phys. A*, **49**, 41–54.

5 Polymer electrolytes I: General principles

D. F. SHRIVER
Department of Chemistry and Materials Research Center, Northwestern University
and
P. G. BRUCE
Department of Chemistry, University of St. Andrews

5.1 *Background*

Polymer electrolytes are the newest area of solid ionics to receive wide attention for applications in electrochemical devices such as batteries and electrochromic windows. Unlike inorganic glass or ceramic electrolytes, the polymer electrolytes are compliant and this property makes it possible to construct solid state batteries in which the polymer conforms to the volume changes of the electrodes that typically occur during discharge and charging cycles. The potential utility of polymer electrolytes has stimulated the synthesis of new polymer electrolytes, physical studies of their structure and charge transport, and theoretical modelling of the charge transport processes. The rapid progress in this field has led to many reviews (Armand, 1986; MacCallum and Vincent, 1987, 1989; Ratner and Shriver, 1988; Vincent, 1989; Tonge and Shriver, 1989; Cowie and Cree, 1989; Bruce and Vincent, 1993; Linford, 1987, 1990).

The structures and charge transport mechanisms for polymer electrolytes differ greatly from those of inorganic solid electrolytes, therefore the purpose of this chapter is to describe the general nature of polymer electrolytes. We shall see that most of the research on new polymer electrolytes has been guided by the principle that ion transport is strongly dependent on local motion of the polymer (segmental motion) in the vicinity of the ion.

In the broad use of the word polymer, ion-containing polymers are ubiquitous. They include inorganic substances such as silicate and borosilicate glasses discussed in Chapter 4, most biopolymers, solvent-swollen synthetic ion exchangers and some synthetic structural polymers. With few exceptions, these exhibit the characteristic feature of an electrolyte, ion mobility. In this chapter we consider the group of synthetic ion-containing polymers some of which are of interest for applications in electrochemical devices such as batteries and sensors.

Two general types of polymer electrolytes have been intensively investigated, polymer–salt complexes and polyelectrolytes. A typical polymer–salt complex consists of a coordinating polymer, usually a polyether, in which a salt, e.g. $LiClO_4$, is dissolved, Fig. 5.1(*a*). Both anions and cations can be mobile in these types of electrolytes. By contrast, polyelectrolytes contain charged groups, either cations or anions, covalently attached to the polymer, Fig. 5.1(*b*), so only the counterion is mobile.

5.2 Polymer–salt complexes

5.2.1 Early developments

The interaction of poly(ethylene oxide) and other polar polymers with metal salts has been known for many years (Bailey and Koleska, 1976). Fenton, Parker and Wright (1973) reported that alkali metal salts form crystalline complexes with poly(ethylene oxide) and a few years later, Wright (1975) reported that these materials exhibit significant ionic conductivity. Armand, Chabagno and Duclot (1978, 1979) recognised the potential of these materials in electro-chemical devices and this prompted them to perform more detailed electrical characterisation. These reports kindled research on the fundamentals of ion transport in polymers and detailed studies of the applications of polymer–salt complexes in a wide variety of devices.

Poly(ethylene oxide) (usually abbreviated to PEO) has been the most intensively studied host polymer for polymer electrolytes and it serves as a prototype for the structural features in most of the more advanced polymer electrolyte hosts. It consists of —O—CH_2—CH_2— repeat units and occurs as a semicrystalline solid. The relative orientations of groups

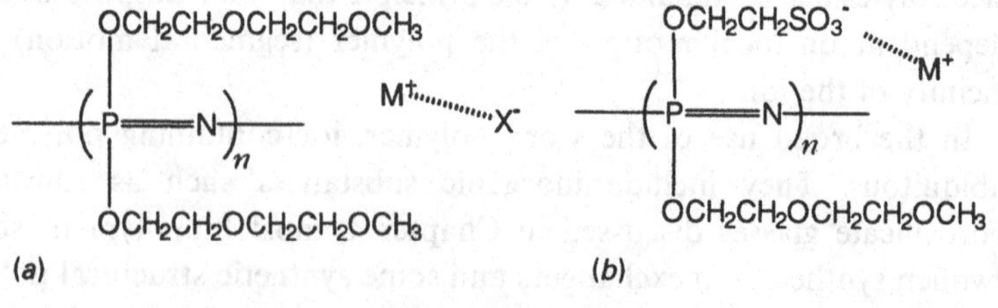

Fig. 5.1 Contrast between (*a*) a polymer electrolyte containing a salt MX and (*b*) a polyelectrolyte in which the anion is attached to the polymer.

along the chain (chain conformations) have been investigated using IR and Raman spectroscopy (Matsuura and Miyazawa, 1969; Maxfield and Shepherd, 1975), which give information on the local conformation of the CH_2 groups relative to each other, and the longer-range order has been determined from X-ray data on oriented fibres (Yoshihara, Tadokoro and Murahashi, 1964; Liu and Parsons, 1969). These studies indicate that crystalline poly(ethylene oxide) has an extended helical structure, Fig. 5.2, with a 19.3 Å repeat unit consisting of seven CH_2CH_2O units in two turns of the helix. The chain conformations become disordered in the molten state.

5.2.2 Polymer segment motion and ion transport

Many polymer–salt complexes based on PEO can be obtained as crystalline or amorphous phases depending on the composition, temperature and method of preparation. The crystalline polymer–salt complexes invariably exhibit inferior conductivity to the amorphous complexes above their glass transition temperatures, where segments of the polymer are in rapid motion. This indicates the importance of polymer segmental motion in ion transport. The high conductivity of the amorphous phase is vividly seen in the temperature-dependent conductivity of poly(ethylene oxide) complexes of metal salts, Fig. 5.3, for which a metastable amorphous phase can be prepared and compared with the corresponding crystalline material (Stainer, Hardy, Whitmore and Shriver, 1984). For systems where the amorphous and crystalline polymer–salt coexist, NMR also indicates that ion transport occurs predominantly in the amorphous phase. An early observation by Armand and later confirmed by others was that the

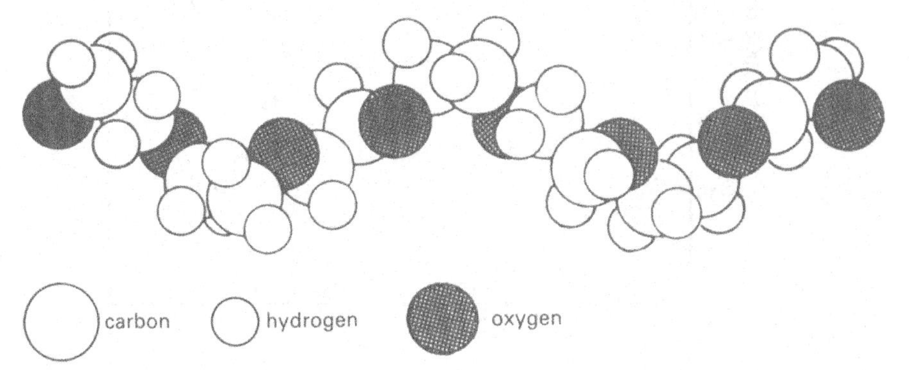

Fig. 5.2 The structure of crystalline poly(ethylene oxide) showing the contents of one unit cell.

conductivity, σ, collected over a wide temperature range on amorphous polymer–salt complexes is more accurately represented by a Vogel–Tamman–Fulcher (VTF) equation than a simple Arrhenius function, Fig. 5.4 (Armand *et al.*, 1979; Bruce, Gray, Shi and Vincent, 1991)

$$\sigma = \sigma_0 \exp[-B/(T - T_0)]$$

The parameters in this equation are discussed in Chapter 6. VTF behaviour also describes the diffusion of uncharged molecules through a variety of disordered media such as fluid solutions and polymers. The functional form of the VTF equation can be derived on the assumption that the ions are transported by the semirandom motion of short polymer segments. Typically, the polymer motions that are prominent above the glass transition temperature are crank-shaft torsional motion around C—C or C—O bonds. The onset of segmental motion occurs in the vicinity of the glass transition temperature, T_g, and becomes more rapid as the temperature of the sample is raised beyond that temperature. In a typical polymer electrolyte this segmental motion occurs at above 1 GHz, at room temperature (Shriver *et al.*, 1981; Ansari *et al.*, 1986) approxi-

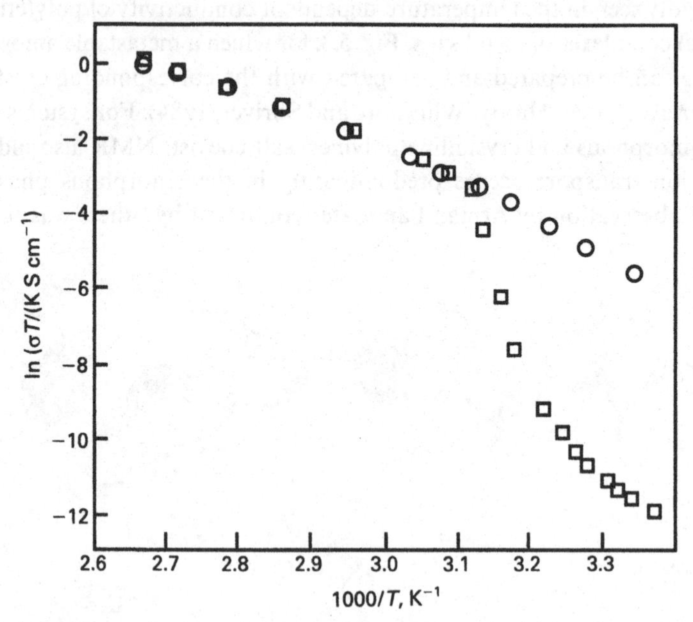

Fig. 5.3 Comparison of the temperature-dependent conductivities for amorphous PEO:NH_4SCN (circles) and crystalline PEO:NH_4SCN (squares).

mately 10^3–10^5 slower than the more localised C—C and C—H stretching vibrations that are detectable in the infrared. The segmental motions are thought to promote ion motion by making and breaking the coordination sphere of the solvated ion and by providing space (free volume) into which the ion may diffuse under the influence of the electrical field, Fig. 5.5. In Chapter 6 we shall see that this simple physical model will have to be modified somewhat to account for ion–ion interaction.

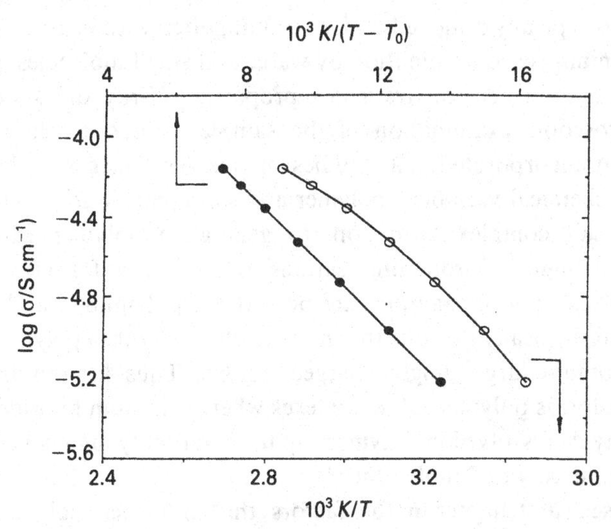

Fig. 5.4 Dependence of conductivity on reciprocal absolute temperature (open circles) and as plotted according to the VTF equation (filled circles) for an amorphous PEO–$LiClO_4$ electrolyte.

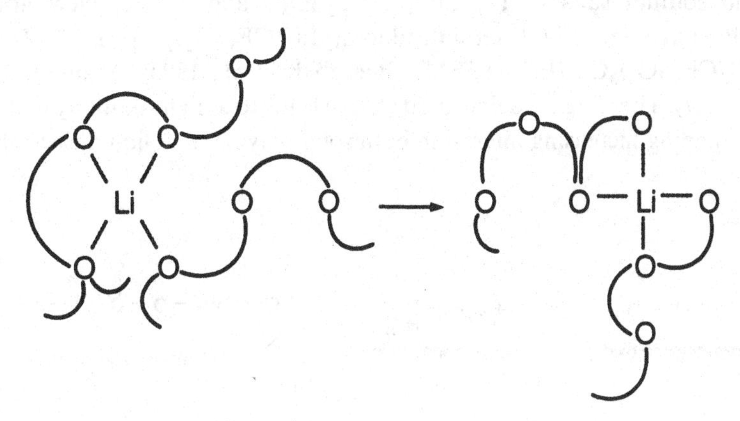

Fig. 5.5 Cartoon of ion motion in a polymer.

The indication that polymer segmental motion is necessary for ion transport has focused most of the current research and development on amorphous materials with low glass transition temperatures.

5.2.3 *Formation*

Salt crystals may be directly diffused in a polymer to produce a polymer–salt complex but these complexes are usually prepared in nonaqueous solutions of the dried polymer and dry salt in a dry nitrogen atmosphere. The solvent is typically removed under vacuum, often with heating. These procedures minimise contamination by water and small molecules, which are known to influence ion transport properties. X-ray diffraction or optical microscopic examination of the sample is used to detect the presence of unincorporated salt. Studies of this type have revealed the scope of the chemical variables (polymer and salt composition) that lead to polymer–salt complex formation. In general, a polymer which is capable of strongly coordinating cations is necessary for electrolyte formation. Some typical examples are polyethers, polyimines, and polyesters, Fig. 5.6. Typically the salts that most readily enter into polymer–salt formation contain large, singly charged anions. Thus lithium triflate, $Li[CF_3SO_3]$, forms polymer salt complexes whereas lithium fluoride does not. Salts may be dissolved in polymers up to remarkably high concentrations, often in excess of 2 moles dm^{-3}.

As discussed in Chapter 6 the factors that influence polymer salt formation can be understood by analysing polymer–salt interaction vs the lattice energy for the salt. Some of the most studied salts are lithium and sodium salts of I^-, ClO_4^-, $CF_3SO_3^-$ and AsF_6^-. New lithium salts have recently been introduced, $Li[(CF_3SO_2)_2N]$ (LITFSI) and $Li[(CF_3SO_2)_3C]$ (LiTriTFSM) (Benrabah *et al.*, 1993; Dominey, 1991: Fig. 5.7). Their highly delocalised charge leads to a high solubility and they are finding increasing interest in both solid polymer and liquid electrolytes.

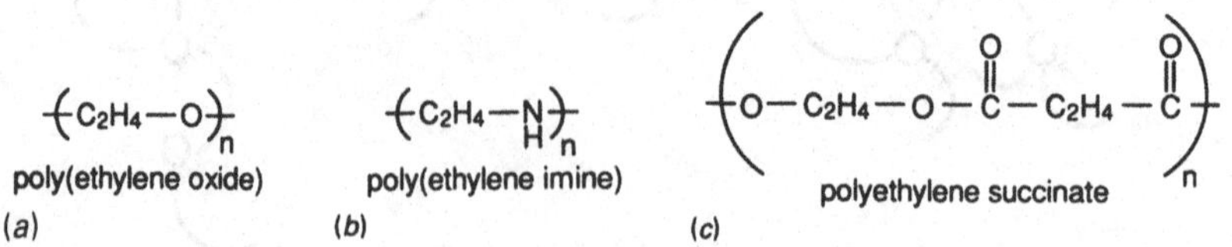

$$\left(C_2H_4-O \right)_n$$
poly(ethylene oxide)
(*a*)

$$\left(C_2H_4-NH \right)_n$$
poly(ethylene imine)
(*b*)

$$\left(O-C_2H_4-O-\overset{O}{\overset{\|}{C}}-C_2H_4-\overset{O}{\overset{\|}{C}} \right)_n$$
polyethylene succinate
(*c*)

Fig. 5.6 Coordinating polymers which can act as 'solid-solvents'.

However, salts of many other cations including mono-, di- and tripositive ions may be dissolved in coordinating polymers, Fig. 5.8. In contrast to the monovalent cations, some di- and trivalent cations are mobile while others are not. This leads to polymer electrolytes which are purely anionic conductors. Cation mobility depends on the lability of the cation–polymer interactions: if these are strong, cation transport is suppressed, in marked contrast to low molecular weight liquid electrolytes where the solvent is transported with the ion. Polymer salt complexes of poly(ethylene oxide) and some other polymer hosts can have a high degree of crystallinity and these crystalline phases represent well defined stoichiometric polymer–salt complexes. Depending on the composition and temperature these systems can also contain the pure crystalline polymer or an amorphous phase. The multiphase character of these polymer electrolytes means that they are best described by a phase diagram. Since the phase changes can be quite sluggish due to the slow diffusion of the entangled polymer chains, these diagrams are not as precise as those for simple inorganic substances; therefore the term pseudophase diagrams is appropriate for the polymer–salt systems. An early estimate of the pseudophase diagram for PEO–NH_4SCN, (Stainer, Hardy, Whitmore, and Shriver, 1984) indicated a

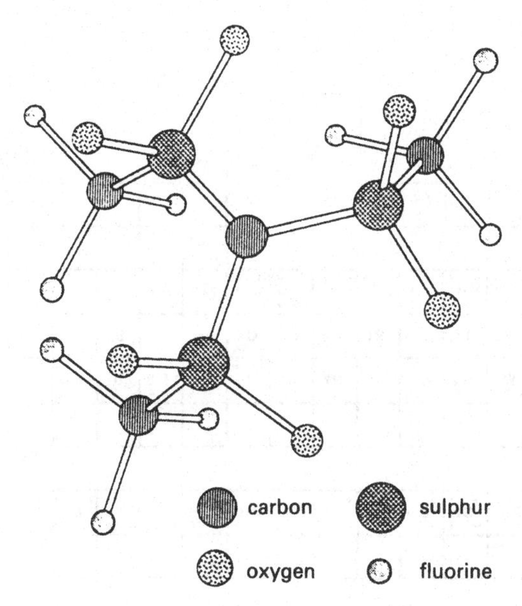

Fig. 5.7 Molecular structure of the $[(CF_2SO_2)_3C]^-$ anion. From Turowsky and Seppelt (1988).

single compound *ca.* $(NH_4SCN)_{0.2}(EO)_{0.8}$, where EO represents an ethylene oxide unit. Subsequently several more precise pseudophase diagrams have been published. Those for PEO–$LiCF_3SO_3$ (Zahurak, Kaplin, Rietman, Murray and Cava, 1988; Vallée, Besner and Prud'homme, 1992) (Fig. 5.9) and PEO–$LiClO_4$ (Jacobs, Lorimer, Russer and Wasiucionek, 1989; Vallée *et al.*, 1992) (Fig. 5.10) are of interest because these complexes are some of the best PEO-based electrolytes, and have been widely studied. Note in Fig. 5.9 that three discrete polymer–salt complexes are indicated with stoichiometries $(PEO)_3$:$LiCF_3SO_3$, PEO:$LiCF_3SO_3$ and $(PEO)_{0.5}$:$LiCF_3SO_3$. In the PEO–$LiClO_4$ system complexes of composition PEO:$LiClO_4$, $(PEO)_2$:$LiClO_4$, $(PEO)_3$:$LiClO_4$ and $(PEO)_6$:$LiClO_4$ are indicated by the most recent work.

The rate of growth of polymer–salt complexes can provide fundamentally important information that is difficult to determine otherwise. The rate of crystal growth of $(PEO)_3$:NaSCN from its undercooled liquid was measured and used to determine values for the diffusion coefficients of Na^+ and SCN^- (Lee, Sudarsana and Crist, 1991). Also it was shown that the rate of the salt diffusion is independent of the molecular weight of the polymer for PEO molecular weights above 10^6. This result is fully consistent with the concept that ion motion is due to local segmental motion of the polymer.

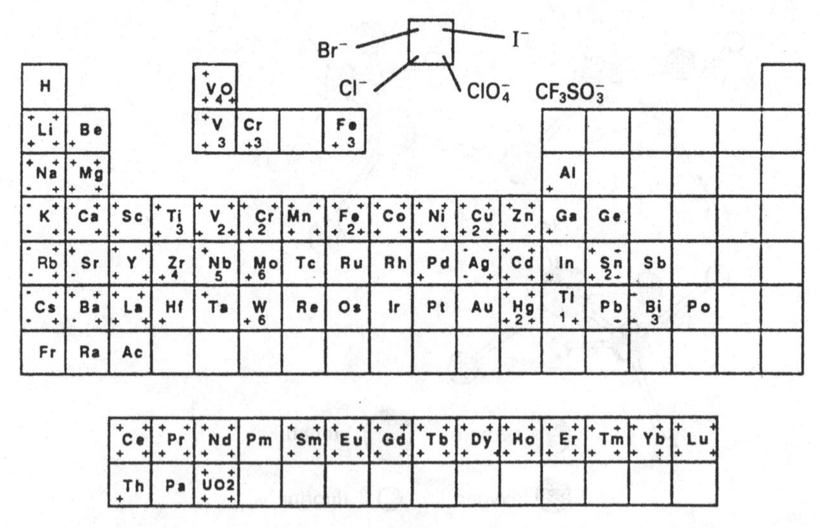

Fig. 5.8 Complex formation between PEO and various metal salts: + complex formed; – no evidence of complex. From Armand and Gauthier (1989).

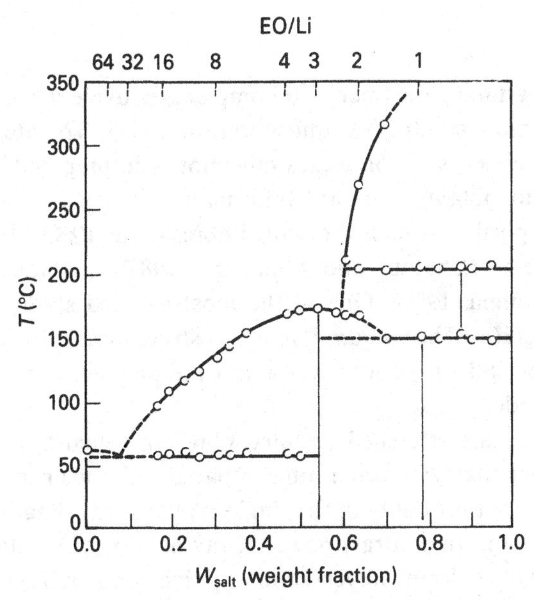

Fig. 5.9 Phase diagram of the PEO–$LiCF_3SO_3$ system. The 0.5/1 compound forms upon the incongruent melting of the 1/1 compound at 150 °C. From Vallée *et al.* (1992).

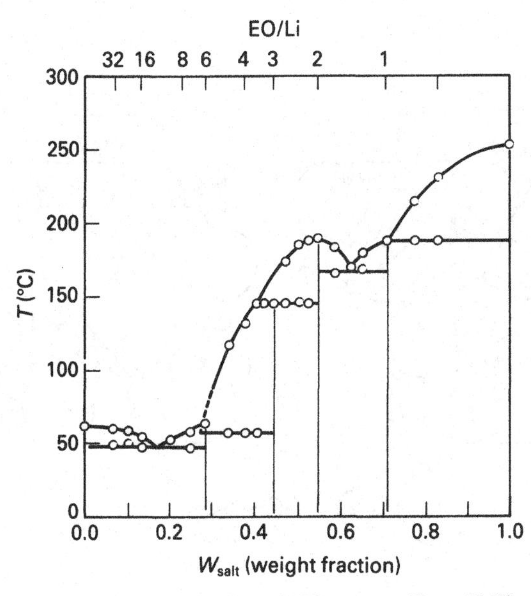

Fig. 5.10 Phase diagram of the PEO–$LiClO_4$ system. From Vallée *et al.* (1992).

5.2.4 *Structure*

The structures of crystalline polymer–salt complexes provide insight into the structure of the more conducting amorphous materials. To date, large single crystals of polymer–salt complexes have not been prepared, but it has been possible to obtain structural information from single crystal X-ray diffraction applied to stretched oriented fibres in the PEO:NaI and PEO:NaSCN systems (Chatani and Okamura, 1987; Chatani, Fujii, Takayanagi and Honma, 1990). One of the most detailed studies is of $(PEO)_3$:NaI, Fig. 5.11(*a*). The sodium ion in this structure is coordinated to both the polymer and to the iodide ion and the polymer is coiled in the form of an extended helix.

It is difficult to prepare stretched oriented fibres and such fibres may differ in their polymer chain conformation compared with the unstretched materials. Furthermore the quality of the single crystal X-ray data is poor and difficult to interpret. In contrast power X-ray data of relatively high quality may be obtained from polycrystalline polymer films. Lightfoot, Mehta and Bruce (1992) have obtained the first crystal structure of a polymer–salt complex, $PEO_3:NaClO_4$ from powder X-ray data, Fig. 5.11(*b*). The structure is similar to the corresponding PEO_3:NaI structure, the PEO chains are wrapped around the Na^+ ions with each Na^+

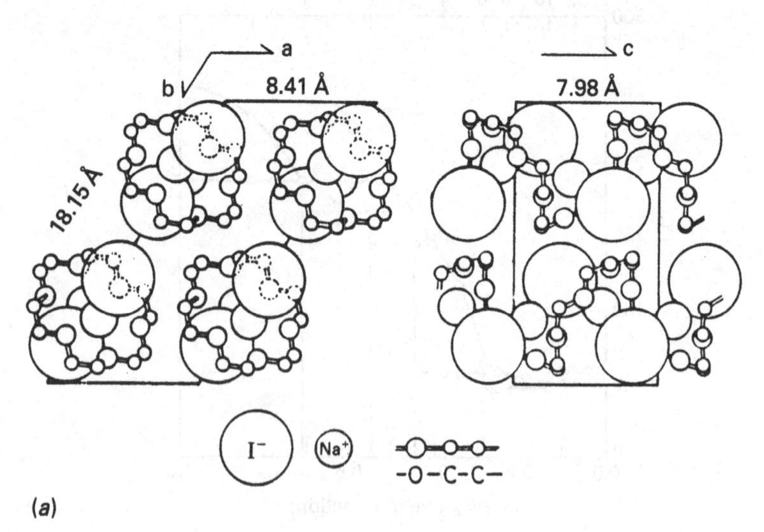

Fig. 5.11 (*a*) Crystal structure of the $(PEO)_3$:NaI complex.

ion being coordinated by four ether oxygens from the polymer chain and two oxygens each from a different ClO_4^- group. Each ClO_4^- ion bridges two neighbouring Na^+ ions by donating one oxygen to the coordination sphere of each Na^+. The powder X-ray results indicate that the ClO_4^- anions are either statically disordered or undergo some rotational motion at room temperature. Recently Bruce *et al* have extracted the crystal structure of the archetypal polymer electrolyte PEO_3:$LiCF_3SO_3$ as well as that of PEO_4:KSCN, PEO_4:NH_4SCN and PEO_4:RbSCN. Much can be learnt from these studies about the basic principles governing the structures of polymer electrolytes in general (Lightfoot, Mehta and Bruce, 1993; Bruce, 1992; Lightfoot, Nowinski and Bruce, 1994; Bruce 1995). The crystal structures reveal that for all ion sizes from Li^+ to Rb^+ the cations are accommodated within the PEO helix contrary to previously held views which placed ions larger than Na^+ outside the helix. In the case of PEO_3:$LiCF_3SO_3$, Li^+ is five coordinate composed of three ether oxygens and one oxygen from each of two triflate groups. There is a Li^+ ion in each turn of the helix. In the case of PEO_4:MSCN where M = K^+, NH_4^+ or Rb^+ the cations are seven coordinate (five ether oxygens and two nitrogens from the thiocyanate anions).

Infrared and Raman spectroscopy have been very useful in the study of polymer electrolytes because they provide indications of cation–anion

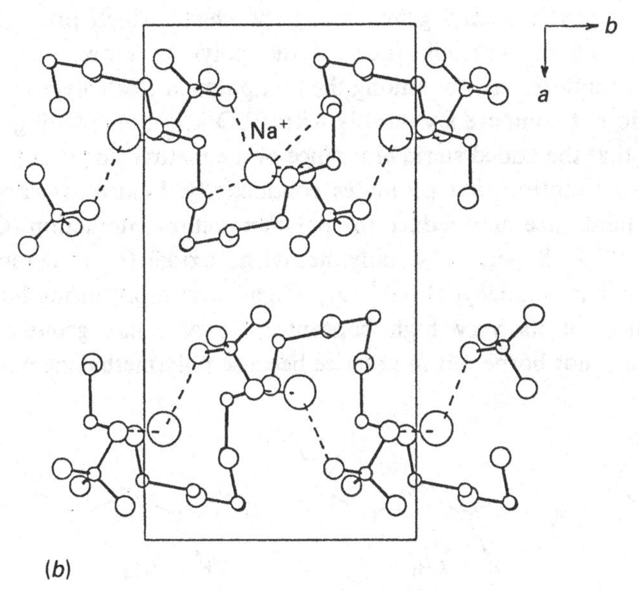

Fig. 5.11 (*b*) Crystal structure of the $(PEO)_3$:$NaClO_4$ complex.

contact in both crystalline and amorphous salt complexes over a wide concentration range (Papke, Ratner and Shriver, 1982; Dupon, Papke, Ratner, Whitmore and Shriver, 1982; Tetters and Frech, 1986; Kakihana, Sanchez and Torell, 1990). Vibrational spectra for polymer–salt complexes include low frequency modes characteristic of the cation vibrating in its solvent–anion cage, and vibrational frequencies characteristic of the polymer (Papke, Ratner and Shriver, 1981). Further details on the information which can be obtained from spectroscopic measurements including EXAFS is presented in Chapter 6.

5.2.5 *Host polymers*

Following the discovery of ion transport in PEO, other polymers have been investigated and these studies give insight into the nature of the ion–polymer interactions and ion transport. Since the presence of crystalline PEO below 60 °C reduces the room temperature conductivity of PEO:salt systems, much of the research has focused on materials which are amorphous at room temperature because of their higher conductivity. In general, amorphous polymer hosts lead to amorphous polymer–salt complexes and one widely-studied noncrystalline polymer host is atactic poly(propylene oxide), Fig. 5.12. The term atactic refers to the random arrangement of the methyl group along the chain, which prevents the order necessary for crystallisation of the polymer. However, atactic PPO–salt complexes are not among the best polymer electrolytes and, for example, do not compare favourably with PEO above its melting point. It is likely that the added steric hindrance of the methyl side group limits the segmental motion that promotes conductivity. It also appears that this steric hindrance may reduce the polymer–cation interaction (Cowie and Cree, 1989). Superficially, poly(methylene oxide) (more commonly known as polyformaldehyde) $(CH_2O)_n$, might seem a promising host for salts because of its very high concentration of polar groups. This expectation is not borne out in practice because poly(methylene oxide) is

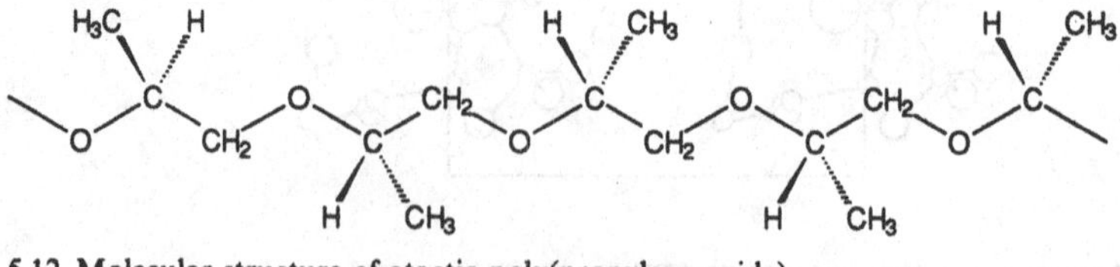

Fig. 5.12 Molecular structure of atactic poly(propylene oxide).

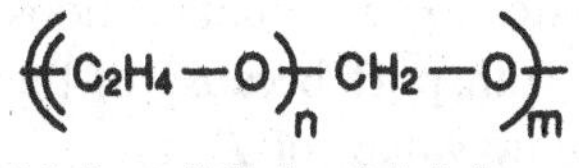

$$\left(\left(C_2H_4-O\right)_n CH_2-O\right)_m$$

Fig. 5.13 Methoxy linked poly(ethylene oxide).

a hard polymer with high cohesive energy that is not offset by its interaction with monopositive cations.

An amorphous material sometimes referred to as amorphous poly(ethylene oxide), aPEO, consists of medium but randomly-variable length segments of poly(ethylene oxide) joined by methyleneoxide units, Fig. 5.13 (Wilson, Nicholas, Mobbs, Booth and Giles, 1990). These methyleneoxide units break up the regular helical pattern of poly(ethylene oxide) and in doing so suppress crystallisation. The aPEO host polymer and its salt complexes can crystallise below room temperature, but this is not detrimental to the properties of the polymer–salt complexes at or above room temperature. Similarly, dimethyl siloxy units have been introduced between medium length poly(ethylene oxide) units to produce an amorphous polymer, Fig. 5.14 (Nagoka, Naruse, Shinohara and Watanabe, 1984).

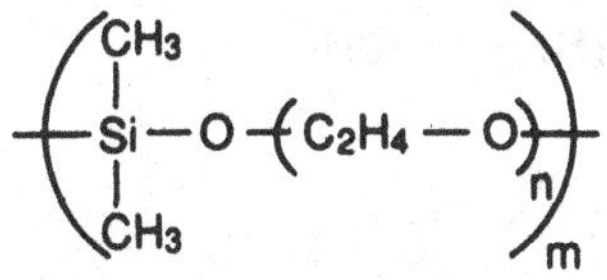

$$\left(Si(CH_3)_2-O\left(C_2H_4-O\right)_n\right)_m$$

Fig. 5.14 Dimethyl siloxy linked poly(ethylene oxide).

Amorphous comb polymers, with short-chain polyethers attached to a polyphosphazene backbone (Blonsky, Shriver, Austin and Allcock, 1984), Fig. 5.15, or a polysiloxane backbone (Xia, Soltz and Smid, 1984), Fig. 5.16, have been found to be excellent hosts for alkali metal salts. The PN

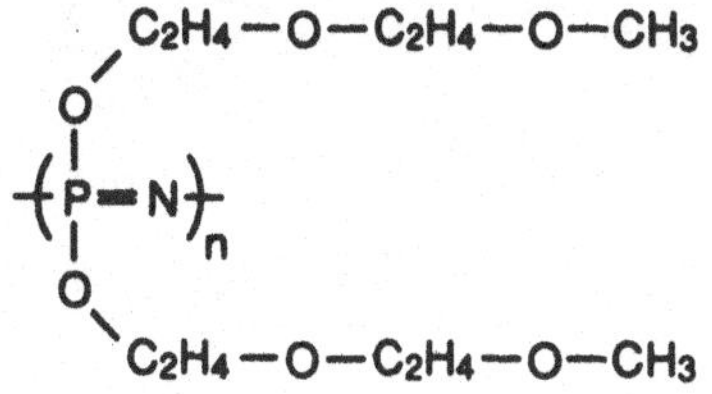

$$\left(P(O-C_2H_4-O-C_2H_4-O-CH_3)_2=N\right)_n$$

Fig. 5.15 MEEP (poly(bis-(methoxyethoxyethoxy)phosphazene)).

backbones have low polarity and appear not to coordinate cations. These backbones are thought to promote ion transport because of their high flexibility, as indicated by characteristic low values of T_g.

In addition to the linear and comb polymer topologies discussed so far, branched and network polymer ionics have been investigated (Killis, LeNest, Cheradame and Gandini, 1982). The degree of crosslinking in these polymer network materials provides control over the flow and deformation of the polymer. At a macroscopic level the polymer electrolyte must be sufficiently rigid to separate the anode and cathode materials, but also have sufficient macroscopic flexibility to accommodate changes in volume of the anode and cathode materials during operation of the device. At a microscopic level the degree of crosslinking must not be so great as to make the local polymer segments rigid, thereby increasing T_g and reducing ion transport.

One example of the chemical crosslinking strategy is the formation of urethane linkages between polyoxyethylene chains, Fig. 5.17 (Killis *et al.*, 1982). The urethane crosslinked polyether systems were exploited by Killis *et al.* in their extensive studies of the relation between conductivity and bulk mechanical properties of network electrolytes. Other chemically

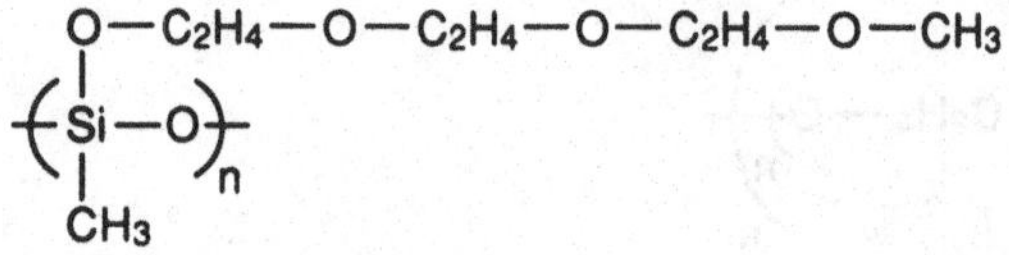

Fig. 5.16 Siloxane.

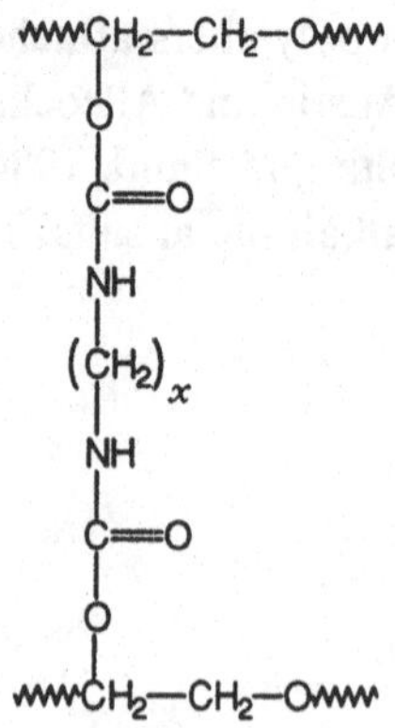

Fig. 5.17 Urethane linked polyoxyethylene chains.

crosslinked network materials that have been investigated for use in polymer electrolytes include polyether linked polyphosphazenes, Fig. 5.18 (Tonge and Shriver, 1987), and siloxy bridged poly(ethylene oxide) (Fish, Khan and Smid, 1986; Spindler and Shriver, 1987).

Alternatively, crosslinks can be introduced in a cast film of polymer or polymer–salt complex by chemical or irradiation crosslinking. *In situ*, crosslinking has the practical advantage that a device can be readily fashioned with the desired film thickness and then crosslinked to maintain this shape. In radiation crosslinking (MacCallum, Smith and Vincent, 1984) the sample is typically exposed to intense gamma radiation, which cleaves C—H bonds preferentially and the resulting carbon-based radicals in adjacent chains link to form the network structure. The general objective is to achieve sufficient crosslinking to avoid flow of the polymer, but without inducing a highly rigid material that would have low ion mobility and be incapable of conforming to changes in the electrode volume during charging or discharging of a battery.

The introduction of small polar molecules into polymer electrolytes can lead to significant improvement in conductivity (Abraham and Alamgir, 1990). Network polymers tend to retain their physical shape even upon the addition of polar cosolvents and systems of this type are being actively investigated for practical applications. In high energy density battery applications, this more volatile component may not be a serious problem because these cells are generally sealed to avoid contact of highly active electrode materials, e.g. lithium, with air. The most thoroughly studied systems employ a highly polar small molecule, such as propylene carbonate. For example, poly(acrylonitrile), and poly(vinyl pyrrolidone) have been used as host polymers for propylene carbonate when combined with

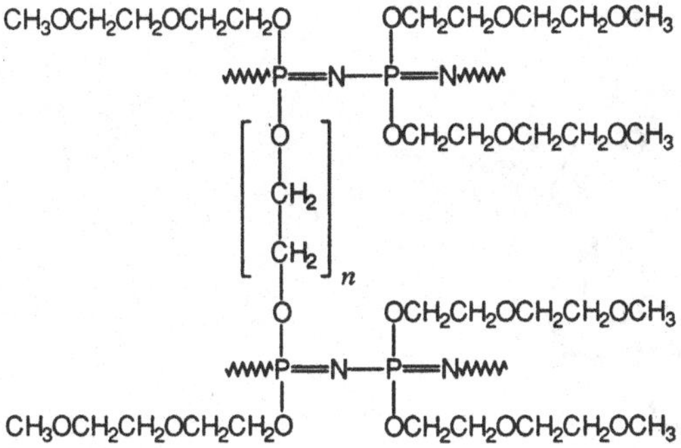

$CH_3OCH_2CH_2OCH_2CH_2O$ $OCH_2CH_2OCH_2CH_2OCH_3$
P=N—P=N
O $OCH_2CH_2OCH_2CH_2OCH_3$
$[CH_2$
$CH_2]_n$
O $OCH_2CH_2OCH_2CH_2OCH_3$
P=N—P=N
$CH_3OCH_2CH_2OCH_2CH_2O$ $OCH_2CH_2OCH_2CH_2OCH_3$

Fig. 5.18 Polyoxyethylene linked MEEP.

$LiClO_4$; this produces materials with remarkably high conductivities: 1.7×10^{-3} at 20 °C and 1.1×10^{-3} at −10 °C (Abraham and Alamgir, 1990). The small molecule appears to serve two functions: it may plasticise the host polymer and thereby induce flexibility and segmental motion in the host polymer chains, and it solvates the cation, or more rarely an anion, (thereby reducing ion–ion interactions).

Small molecule chelating agents have been added to polymer electrolytes with similar objectives in mind. Short-chain poly(ethylene glycol) is one such example. Cyclic polyethers are even more effective complexing agents for alkali metal cations. Two such complexing agents are the monocyclic crown ethers, Fig. 5.19, and the bicyclic cryptand ligands, Fig. 5.20 (Lindoy, 1989). The crown ethers often do not shield the cation from an anion as evidenced by single crystal X-ray data, Fig. 5.21(*a*). Structural data on simple cryptand–salt complexes indicates that the cation is generally shielded from the anion, Fig. 5.21(*b*). The influence of the cryptand complexing agents on polymer–salt complexes is often complicated by the precipitation of salt (Doan, Heyen, Ratner and Shriver, 1990). When the cation is encapsulated by a cryptand, the favourable interaction with the host polymer, which is largely responsible for salt

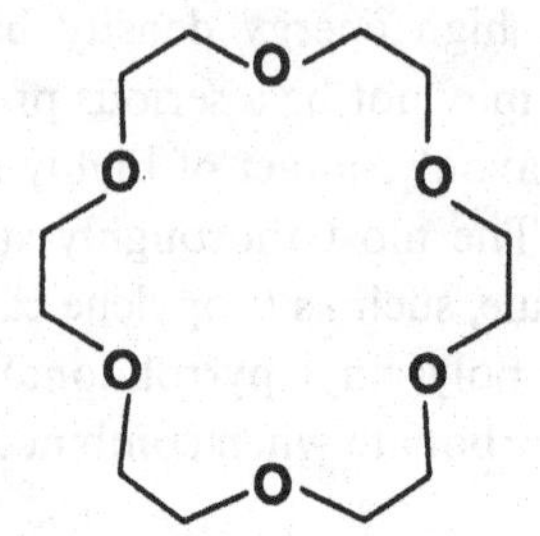

Fig. 5.19 Monocyclic crown ether.

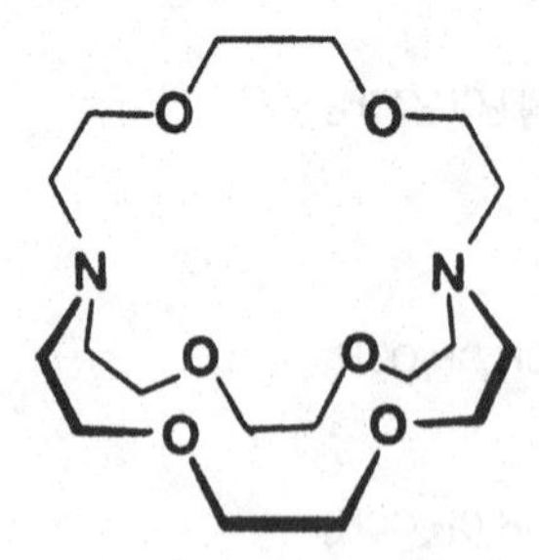

Fig. 5.20 Bicyclic cryptand.

dissolution, is lost, with a corresponding decrease in solubility of the salt complex.

The emphasis on ether polar groups in this chapter reflects the dominance of these polymer hosts in research on polymer electrolytes. This central role results from their combination of good ion solvation, high chain flexibility and high chemical stability of the polymeric ethers. Chemical stability is particularly important for high energy density batteries where highly oxidising cathode materials and highly reducing anode materials contact the polymer. Table 5.1 provides a range of other polar polymers which have been investigated as hosts for polymer electrolytes. Included in this table are esters, sulfides, and imines. Some of these have already been discussed in this chapter.

5.2.6 Proton conductors

As in other electrolytes, both liquid and solid, proton conduction is quite distinct from that of the other ions. Due to the highly polarisable nature of the H^+ it is invariably covalently bonded. Several classes of the H^+ conducting polymer electrolytes are now known. Poly(ethylene oxide) forms complexes with H_3PO_4 and exhibits a room temperature conductivity of 4×10^{-5} S cm^{-1} (Donoso, Gorecki, Berthier, Defendini, Poinsignon and Armand, 1988). Proton conductors may also be formed by complexing H_3PO_4 or H_2SO_4 with linear or branched poly(ethylene imines) (Daniel, Desbat and Lassegues, 1988). Perhaps the most interesting proton conducting complexes to date are the ormosils (organically modified silane electrolytes). These organically modified ceramics are prepared by

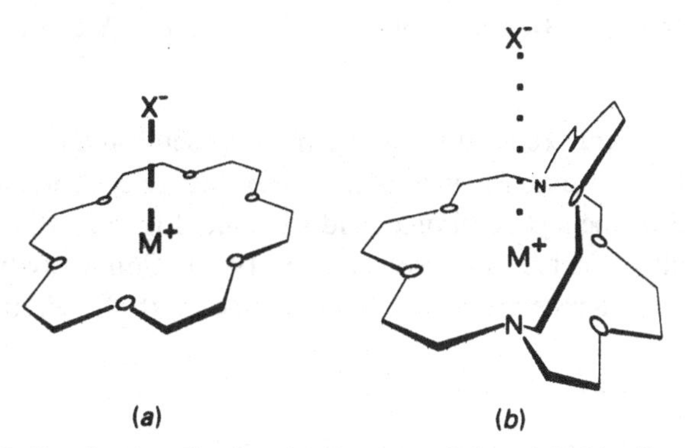

Fig. 5.21 Coordination of cations by (*a*) crown ethers and (*b*) bicyclic cryptands indicating that the cation is better shielded from the anion in (*b*).

Table 5.1. *Some coordinating polymers which have been used as solid solvents for polymer electrolytes*

Name	Monomer unit
Poly(propylene oxide)[a]	$[CH_2CH(CH_3)O]_n$
Poly(ethylenimine)[b]	$(CH_2CH_2NH)_n$
Poly(alkylene sulfides)[c]	$[(CH_2)_pS]_n$
Poly(ethylene succinate)[d]	$[OCH_2CH_2OC(O)CH_2CH_2C(O)]_n$
Poly(*N*-methylaziridine)[e]	$[CH_2CH_2N(CH_3)]_n$
Poly(epichlorohydrin)[f]	$[OCH_2CH(CH_2Cl)]_n$
Poly(vinyl acetate)[g]	$\{CH_2CH[OC(O)CH_3]\}_n$
Poly[bis(methoxyethoxyethoxy) phosphazene][h]	$\{NP[O(CH_2CH_2O)_2CH_3]_2\}_n$
Oxymethylene-linked poly(oxyethylene)[i]	$[(CH_2CH_2O)_mCH_2O]_n$

[a] Armand, M. B., Chabagno, J. M. and Duclot, M. J. (1979) in *Fast Ion Transport in Solids: Electrodes and Electrolytes*, Eds. P. Vashishta, J. N. Mundy, and G. K. Shenoy, North-Holland, New York, 131.
[b] Chiang, C. K., Davis, G. T. Harding, C. A. and Takahashi, T. (1985) *Macromolecules*, **18**, 825.
[c] Clancy, S., Shriver, D. F. and Ochrymowycz, L. A. (1986) *Macromolecules*, **19**, 606.
[d] Dupon, R., Papke, B. L., Ratner, M. A. and Shriver, D. F. (1984) *J. Electrochem. Soc.*, **131**, 586.
[e] Armand, M. B. (1983) *Solid State Ionics*, **9/10**, 745.
[f] Shriver, D. F., Papke, B. L., Ratner, M. A., Dupon, R., Wong, T. and Brodwin, M. (1981) *Solid State Ionics*, **5**, 83.
[g] Wintersgill, M. C., Fontanella, J. J., Calame, J. P., Greenbaum, S. G. and Andeen, C. G. (1984) *J. Electrochem. Soc.*, **121**, 2208.
[h] Blonsky, P. M., Shriver, D. F., Austin, P. and Allcock, H. R. (1986) *J. Am. Chem. Soc.*, **106**, 6854.
[i] Craven, J. R., Mobbs, R. H. and Booth, C. (1986) *Makromol. Chem. Rapid Commun.*, **7**, 81.

sol–gel routes. They are expected to show superior chemical and thermal stability particularly in devices such as methanol fuel cells. One specific example is based on benzyl sulphonic acid siloxane, Fig. 5.22 (Goutier-Lineau, Denoyelle, Sanchez and Poinsignon, 1992). Conductivities of 10^{-2} S cm^{-1} at room temperature have been obtained with this electrolyte.

5.3 Polyelectrolytes

This class of materials has either positively or negatively charged ions covalently attached to the polymer backbone and therefore only the unattached counterion has long range mobility. Some examples of polyelectrolytes that have been investigated in the context of solid ionics are collected in Table 5.2. Some well known polyelectrolytes such as sodium polystyrene sulphonate are highly rigid materials and these lack significant ion mobility. The introduction of a plasticiser such as a short-chain poly(ethylene glycol) converts the rigid, poorly conducting polyelectrolyte to a more compliant material with significant ionic conductivity (Hardy and Shriver, 1985). Spectroscopic data indicate that the ether oxygens coordinate the cations and thus reduce association of the cations with the immobile anions on the chains while also providing a locally mobile coordination environment which is conducive to ion motion. Another indication from IR spectra is that the poly(ethylene glycol) is retained in the poly(styrene sulphonate) by hydrogen bonding between terminal —OH groups, on the glycol, and the sulphonate ions, which are covalently attached to the polystyrene backbone. One good set of ion conductors are polyphosphazenes with attached short-chain polyethers and tetraalkylammonium groups (Chen, Ganapathiappan and Shriver, 1989). Good cation conductors are composed of polysiloxanes with pendant short-chain polyethers and a sterically hindered phenolate ion, where the bulky groups adjacent to the phenolate oxygen ion reduce ion association (Liu, Okamata, Skotheim, Pak and Greenbaum, 1985).

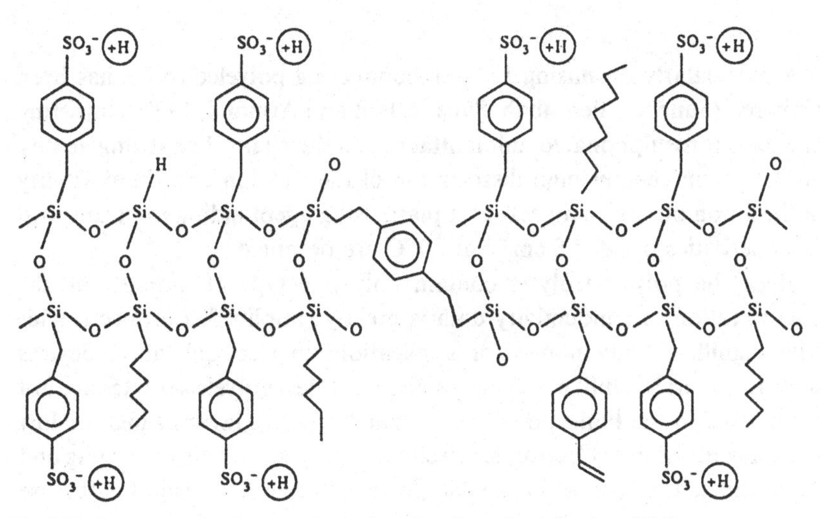

Fig. 5.22 Proton-conducting membrane based on benzyl sulphonic acid siloxane.

Table 5.2. *Some polyelectrolytes*

Name	Monomer Unit
Lithium poly(2-sulphoethyl methacrylate)[a]	$-(CH_2-C(CH_3)(CO_2CH_2CH_2SO_3^- Li^+))-$
Sodium poly(phosphazene sulphonates)[b]	$-(N{=}P(O(CH_2CH_2O)_nCH_3)_2)_x-(N{=}P(O(CH_2CH_2O)_nCH_3)(OCH_2CH_2SO_3^- Na^+))_y-$
Poly(diallyldimethylammonium chloride)[c]	dimethylpyrrolidinium ring, N^+ Cl^-
Sodium poly(styrene sulphonate)[d]	$-(CH_2-CH(C_6H_4SO_3^- Na^+))-$
Nafion derivative 117[e]	$(CF_2CF_2)_x-(CFCF_2)_y$, $OCF_2CF_2OCF_2CF_2SO_3^- Li^+$

[a] Bannister, D. J., Davies, G. R., Ward, I. M. and McIntyre, J. E. (1984) *Polymer*, **25**, 1291.
[b] Ganapathiappan, S., Chen, K. and Shriver, D. F. (1989) *J. Am. Chem. Soc.*, **111**, 4091.
[c] Hardy, L. C. and Shriver, D. F. (1984) *Macromolecules*, **17**, 975.
[d] Hardy, L. C. and Shriver, D. F. (1985) *J. Am. Chem. Soc.*, **107**, 3823.
[e] Tsuchida, E. and Shigehara, K. (1984) *Mol. Cryst. Liquid Cryst.*, **106**, 361.

A particularly promising Li^+ ion conducting polyelectrolyte has been prepared (Sanchez, Benrabah, Sylla, Alloin and Armand, 1993) consisting of a perfluorosulphonated anion attached to the chain. The strong acidity of this group ensures high dissociation of the Li^+ ion and the flexibility of the anion on the chain acts as a plasticising agent aiding ion transport. Conductivities of 10^{-5} S cm^{-1} at 30 °C are obtained.

Since the polyelectrolytes contain only one type of mobile ion, the interpretation of conductivity data is greatly simplified. Polyelectrolytes have significant advantages for applications in electrochemical devices such as batteries. Unlike polymer–salt complexes, polyelectrolytes are not susceptible to the build up of a potentially resistive layer of high or low salt concentration at electrolyte–electrolyte interfaces during charging and discharging. Unfortunately flexible polyelectrolyte films suitable for use in devices have not yet been prepared.

5.3.1 Influence of complexing agents and solvents

In the example above, a short-chain poly(ethylene glycol) was added to a rigid polyelectrolyte to plasticise the material and thereby increase polymer-solvent motion in the vicinity of mobile ions. This strategy has been widely explored as a means of improving ion transport in electrolytes.

The addition of a cryptand to some polyelectrolytes leads to significant increases in conductivity and in some cases IR and Raman spectroscopy demonstrate that the cryptand breaks up the ion–ion interactions (Chen, Doan, Ganapathiappan, Ratner and Shriver, 1991; Doan, Ratner and Shriver, 1991). Apparently the reduction of ion association more than offsets the reduction in mobility of the cation–crypt complex, which has a larger effective radius than the simple cation. It is also possible that the cryptand–ion complex is rendered more mobile by the reduction of polymer–cation complex formation, but this point has not been investigated in any detail.

Since the chelating ligand forms a strong complex with cations it is conceivable that this complex formation would lead to slow transfer of the ion into a cathode material:

$$[M(crypt)]^{+}(polymer) + (cathode) \rightarrow crypt(polymer) + M^{+}(cathode).$$

This issue has been investigated using amalgam cathodes, and the indications are that the cryptand has negligible influence on the rate of incorporation of a sodium ion into the amalgam (Chen *et al.*, 1991; Doan *et al*, 1991). One of the most interesting applications of the cryptands is that they increase the mobility of dipositive ions in polymer electrolytes (Chen and Shriver, 1991).

Mg^{2+} has been studied in a polyelectrolyte system where the counterion is a phosphazene-bound sulphonate group. In this case the finite but low conductivity is due to Mg^{2+} migration. A large increase in conductivity is noted when a bicyclic ligand, crypt [2.1.1.], is introduced (Hancock and Martell, 1988). Apparently, the cation–anion and cation–polymer interactions are reduced by complex formation, resulting in a more mobile cation, despite its larger effective radius.

Polymer–salt complexes involving the tripositive lanthanides have been investigated from the standpoint of conductivity, which is observed to be very low. In addition, the neutral complex $Nd(DMP)_3$ (DMP = 2,2,6,6-tetraethyl,-3,5-heptane dionate) will dissolve in PEO; although not electrically conductive this polymer may have utility as a laser material. To date the best known and most commercially promising polyelectrolyte is NAFION® and related fluorinated sulphonates, Fig. 5.23. These H^+ conducting materials exhibit high ionic conductivity in the presence of a suitable plasticiser, and are being extensively used in prototype solid polymer fuel cells in which H_2 is the fuel.

5.4 Summary

The investigation of ion transport in solvent-free polymers is now roughly 20 years old. Substantial progress has been made in that time in preparation, characterisation and understanding the charge transport process in these materials. The importance of polymer segmental motion to ion transport is well established and this, along with strategies for the reduction of cation–anion interaction appear to be the most promising avenues for future development. Nevertheless, our understanding of transport processes in these materials is still rudimentary and the issues of polymer electrolyte structure and ion transport across electrode–electrolyte interfaces is even less satisfactory. Clearly there is considerable scope for more incisive studies of ion transport mechanisms. Undoubtedly, there are many opportunities for new synthetic developments which optimise ion transport and improve the stability of the electrolyte. It is conceivable that new classes of polymer electrolytes will be discovered that transport charge by new mechanisms. For example, it may be possible to synthesise rigid polymers with channel size and composition appropriate for ion transport. These materials would be analogues of inorganic ceramic electrolytes such as β-alumina. This active area should continue to provide many scientific and technological challenges.

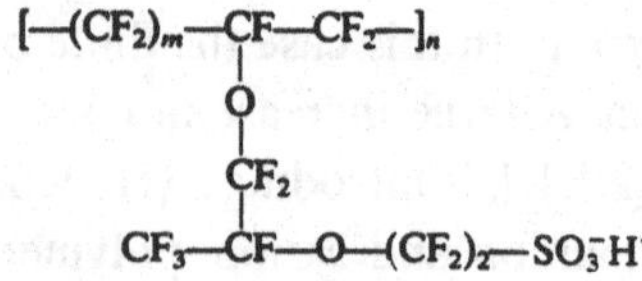

$[\text{—}(CF_2)_m\text{—}CF\text{—}CF_2\text{—}]_n$
O
CF_2
$CF_3\text{—}CF\text{—}O\text{—}(CF_2)_2\text{—}SO_3^-H^+$

Fig. 5.23 NAFION, *m* is usually around 13.

References

Abraham, K. M. and Alamgir, M. (1990) *J. Electrochem. Soc.*, **137**, 1657.
Ansari, S. M., Brodwin, M., Druger, S., Stainer, M., Ratner, M. A. and Shriver, D. F. (1986) *Solid State Ionics*, **17**, 101.
Armand, M. B. (1986) *Ann. Rev. Mater. Sci.*, **16**, 245.

References

Armand, M. B., Chabagno, J. M. and Duclot, M. (1978) *Extended Abstracts Second International Conference On Solid Electrolytes*, St Andrews, Scotland.
Armand, M. B., Chabagno, J. M. and Duclot, J. M. (1979) in *Fast Ion Transport in Solids*, eds Vahista, P., Mundy, J. N. and Shenoy, G. K., North-Holland, Amsterdam, p. 131.
Armand, M. and Gauthier, M. (1989) in *High Conductivity Solid Ionic Conductors*, Ed. T. Takahashi, World Scientific Press, Singapore, p. 117.
Bailey, F. E. and Koleska, J. V. (1976) *Poly(ethylene oxide)*, Academic Press, New York, pp. 94–7.
Benrabah, D., Bard, D., Sanchez, J.-Y., Armand, M. and Gard, G. G. (1993) *J. Chem. Soc. Faraday Trans.*, **89**, 355.
Blonsky, P. M., Shriver, D. F., Austin, P. and Allcock, H. R. (1984) *J. Am. Chem. Soc.*, **106**, 6854.
Bruce, P. G. (1992) in *Fast Ion Transport in Solids*, eds Scrosoti, B., Magistris, A., Mavi, C. M. and Maviotto, G., NATO ASI Series, Vol. 250, p. 87.
Bruce, P. G. (1995) *Electrochimica Acta*, in press.
Bruce, P. G., Gray, F. M., Shi, J. and Vincent, C. A. (1991) *Phil. Mag.*, **64**, 1091.
Bruce, P. G. and Vincent, C. A. (1993) *J. Chem. Soc. Faraday Transactions*, **89**, 3187.
Chatani, Y., Fujii, Y., Takayanagi, T. and Honma, A. (1990) *Polymer*, **31**, 2238.
Chatani, Y. and Okamura, S. (1987) *Polymer*, **28**, 1815.
Chen, K., Doan, K. E., Ganapathiappan, S., Ratner, M. and Shriver, D. F. (1991) *Proc. Mat. Res. Soc. Symp., Solid State Ionics II*, 201, 215.
Chen, K., Ganapathiappan, S. and Shriver, D. F. (1989) *Chem. Mater.*, **1**, 483.
Chen, K. and Shriver, D. F. (1991) *Chem. Mater.*, **3**, 771.
Cowie, J. M. G. and Cree, S. H. (1989) *Ann. Rev. Phys. Chem.*, **40**, 85.
Daniel, M. F., Desbat, B. and Lassegues, J. C. (1988) *Solid State Ionics*, **28–30**, 632.
Doan, K. E., Heyen, B. J., Ratner, M. A. and Shriver, D. F. (1990) *Chem. Mater.*, 2.
Doan, K. E., Ratner, M. A. and Shriver, D. F. (1991) *Chem. Mater.*, **3**, 418.
Dominey, L. (1991) *Third International Symposium on Polymer Electrolytes*, Annecy, France.
Donoso, P., Gorecki, W., Berthier, C., Defendini, F., Poinsignon, C. and Armand, M. (1988) *Solid State Ionics*, **28–30**, 969.
Dupon, R., Papke, B. L., Ratner, M. A., Whitmore, D. H. and Shriver, D. F. (1982) *J. Am. Chem. Soc.*, **104**, 6247.
Fenton, D. E., Parker, J. M. and Wright, P. V. (1973) *Polymer*, **14**, 589.
Fish, D., Khan, I. M. and Smid, J. (1986) *Polymer Preprints*, **27**, 325.
Fujita, M. and Honda, K. (1989) *Polymer Comm.*, **30**, 200.
Goutier-Lineau, I., Denoyelle, A., Sanchez, J.-Y. and Poinsignon, C. (1992) *Electrochimica Acta*, **37**, 1615.
Hancock, R. D. and Martell, A. E. (1988) *Comments on Organic Chemistry*, **6**, 237.
Hardy, L. C. and Shriver, D. F. (1985) *J. Am. Chem. Soc.*, **107**, 3823.
Jacobs, P. W. M., Lorimer, J. W., Russer, A. and Wasiucionek, M. (1989) *J. Power Sources*, **26**, 503.

Kakihana, M., Sanchez, S. and Torell, L. M. (1990) *J. Chem. Phys.*, **92**, 6271.
Killis, A., LeNest, J. F., Cheradame, H. and Gandini, A. (1982) *Macromol. Chem.*, **183**, 2835.
Lee, Y. L., Sudarsana, B. and Crist, B. (1991) *Solid State Ionics*, **45**, 215.
Lightfoot, P., Mehta, M. A. and Bruce, P. G. (1992) *J. Mater. Chem.*, **3**, 379.
Lightfoot, P., Mehta, M. A. and Bruce, P. G. (1993) *Science*, **262**, 883.
Lightfoot, P., Nowinski, J. L. and Bruce, P. G. (1994) *J. Am. Chem. Soc.*, in press.
Lindoy, L. F. (1989) *The Chemistry of Macrocyclic Ligand Complexes*, Cambridge University Press, Cambridge.
Linford, R. G. (ed.) (1987, 1990) *Electrochemical Science and Technology of Polymers*, Vols. 1 and 2.
Liu, H., Okamoto, Y., Skotheim, T., Pak, Y. S. and Greenbaum, S. G. (1985) *Mat. Res. Soc. Symp. Proc.*, **135**, 349.
Liu, K.-J. and Parsons, J. L. (1969) *Macromol.*, **2**, 529.
MacCallum, J. R. and Vincent, C. A. (1987, 1989) *Polymer Electrolyte Reviews*, Vols. 1 and 2, Elsevier, London.
MacCallum, J. R., Smith, M. J. and Vincent, C. A. (1984) *Solid State Ionics.*, **11**, 307.
Matsuura, H. and Miyazawa, T. (1969) *J. Pol. Sci.*, **A-27**, 1735.
Maxfield, J. and Shepherd, I. W. (1975) *Polymer*, **16**, 505.
Nagoka, K., Naruse, H., Shinohara, I. and Watanabe, M. (1984) *J. Polym. Sci. Polym. Let.*, **22**, 659.
Papke, B., Ratner, M. A. and Shriver, D. F. (1981) *J. Phys. Chem. Solids*, **42**, 493.
Papke, B., Ratner, M. A. and Shriver, D. F. (1982) *J. Electrochem. Soc.*, **129**, 1434.
Ratner, M. A. and Shriver, D. F. (1988) *Chem. Rev.*, **88**, 109.
Sanchez, J.-Y., Benrabah, D., Sylla, S., Alloin, F. and Armand, M. (1993) *Ninth International Conference On Solid State Ionics*, Netherlands.
Shriver, D. F., Papke, B. L., Ratner, M. A., Dupon, R., Wong, T. and Brodwin, M. (1981) *Solid State Ionics*, **5**, 83.
Spindler, R. and Shriver, D. F. (1987) in *Conducting Polymers*, Ed. Alcacer, L., D. Reidel, Dordrecht.
Stainer, M., Hardy, L. C., Whitmore, D. H. and Shriver, D. F. (1984) *J. Electrochem. Soc.*, **131**, 784.
Tetters, D. and Frech, R. (1986) *Solid State Ionics*, **18/19**, 271.
Tonge, J. S. and Shriver, D. F. (1987) *J. Electrochem. Soc.*, **134**, 269.
Tonge, J. S. and Shriver, D. F. (1989) in *Polymers For Electronic Applications*, Ed. Lai, J. H., CRC Press, Boca Raton, Florida.
Turowsky, L. and Seppelt, K. (1988) *Inorg. Chem.*, **27**, 2135.
Vallée, A., Besner, S. and Prud'homme, J. (1992) *Electrochimica Acta*, **37–9**, 1579.
Vincent, C. A. (1989) *Prog. Solid State Chem.*, **88**, 109.
Wilson, D. J., Nicholas, C. V., Mobbs, R. H., Booth, C. and Giles, J. R. M. (1990) *Br. Pol. J.*, **22**, 129.
Wright, P. V. (1975) *Br. Polymer J.*, **7**, 319.
Xia, D. W., Soltz, D. and Smid, J. (1984) *Solid State Ionics*, **14**, 221.
Yoshihara, T., Tadokoro, H. and Murahashi, S. (1964) *J. Chem. Phys.*, **41**, 2902.
Zahurak, S. M., Kaplan, M. L., Rietman, E. A., Murray, D. W. and Cava, R. J. (1988) *Macromolecules*, **21**, 654.

6 Polymer electrolytes II: Physical principles

P. G. BRUCE and F. M. GRAY
Department of Chemistry, University of St. Andrews

6.1 Introduction

Within the field of electrochemistry the topic of electrolytes, that is the study of salts dissolved in solvents, is regarded by some as a mature discipline. It is viewed as less exciting than its sister topic of electrodics, which is concerned with the interface between electrolytes and electrodes; however, this is far from true for solid polymer electrolytes. These materials consist of salts dissolved in a solid, coordinating, polymeric solvent. They were first investigated by Fenton, Parker and Wright (1973) and have been intensively studied since Armand (1978) recognised their unique potential as electrolytes with a 'solid-solvent'. In recent years many such electrolytes have been prepared. An introduction to polymer electrolytes including a discussion of the different systems is presented in the preceding chapter by Shriver and Bruce. A sufficient body of knowledge has now accumulated on polymer electrolytes to permit the establishment of the fundamental physical principles on which such materials are based. It is with the physical aspects of solid polymer electrolytes that this chapter deals and in this sense it differs from the previous chapter.

Our fundamental understanding of polymer electrolytes is derived from both the general examination of a wide range of distinct electrolytes and from some very detailed studies of a few model systems. In the latter category electrolytes based on the polyether poly(ethylene oxide), usually abbreviated to PEO [$(\text{—}CH_2\text{—}CH_2\text{—}O\text{—})_n$], still reign supreme. However, since conduction is limited to the amorphous state in polymers, and PEO demonstrates a propensity to crystallise, an all-amorphous polymer host is desirable. A polymer solvent which is proving of great value for the study of polymer electrolytes is methoxy-linked poly(ethylene oxide) [$\text{—}(CH_2\text{—}O\text{—}(CH_2\text{—}CH_2\text{—}O)_m)_n\text{—}$] abbreviated to PMEO (Craven, Mobbs, Booth and Giles, 1986). It may be prepared as an amorphous, high molecular weight (>100000) host with high purity. The residual conductivity of the polymer may be reduced to 10^{-8} S cm^{-1}, some two orders of magnitude lower than PEO (Gray, Shi, Vincent and Bruce, 1991). Since it is a linear chain polymer, it may be easily processed and cast as a film with a variety of salts.

In this chapter we will consider only high molecular weight amorphous solid polymer electrolytes, free from low molecular weight additives or plasticisers. Electrolytes with low molecular weight plasticisers are, in the extreme, best regarded as liquid electrolytes immobilised in a solid polymer matrix. Such systems may differ fundamentally from unplasticised materials. In the succeeding sections we will consider in turn, the energetics of salt dissolution and the structure of the electrolyte formed, the mechanism of ion motion in polymers, and the transport of charge and matter through the electrolyte. A more comprehensive overview of the current understanding of polymer electrolytes than is possible here may be obtained in the reviews cited in Chapter 5.

6.2 *Why do salts dissolve in polymers?*

6.2.1 *Thermodynamics of dissolution*

Dissolution of a salt in a solvent, whether liquid or solid, must be accompanied by a reduction in the Gibbs free energy of the system at constant temperature and pressure. Since $\Delta G = \Delta H - T\Delta S$ we must consider both changes in enthalpy and entropy on dissolution. Considering the second term in the equation first: the overall entropy change ΔS is composed of two parts, a positive entropy change, ΔS_s, due to dissordering of the ions as they pass from the regular crystalline lattice into the polymer, and a negative entropy change, ΔS_p, as the polymer chains stiffen due to coordination around the ions (reflected in an increase in T_g with salt content). The ions may be coordinated by one chain or more leading to the possibility of intra- or interchain cross-linking or both. This reduction in entropy of the solvent is probably more acute than is the case for simple low molecular weight liquid solvents such as tetrahydrofuran where the majority of the solvent is relatively unaffected by solvation. Since the individual entropy changes associated with the salt and polymer are in competition, the overall change ($\Delta S = \Delta S_s + \Delta S_p$) may be positive or negative. It now appears that a negative entropy of dissolution may be common in polymer electrolytes, contrary to previous belief. Evidence in support of this view will be discussed in more detail later in the chapter. In general, however, the individual entropy changes of the salt and polymer are likely to vary less between different systems than the individual enthalpy changes. Therefore we will focus on ΔH as the main factor controlling whether dissolution is likely to occur for any given salt–polymer combination. Several factors must be considered:

(1) the lattice energy of the salt – positive ΔH_s;
(2) the creation of suitable sites in the polymer – positive ΔH_p;
(3) cation solvation, formation of coordinate bonds between the cations and suitable coordinating atoms on the polymer, e.g. ether oxygens – negative ΔH_{s-p};
(4) electrostatic interactions between the dissolved ions – negative ΔH_i.

The lattice energy will vary a great deal from salt to salt. Higher charges on the ions and smaller ionic radii lead to larger lattice energies. The enthalpy required to create sites in the polymer will depend on the strength of association between groups on the polymer chains. ΔH_p is expected to be a less significant factor than ΔH_s.

In water and other hydrogen bonded solvents both the cations and anions are solvated; in contrast the anions are barely solvated in polymer electrolytes and other non-hydrogen bonded solvents such as tetrahydrofuran or acetonitrile. The anions are stabilised in the polymer more by attraction to the cations than by interaction with the polymer chains. Thus the solvation enthalpy of a salt in a polymer depends largely on the strength of the coordinate bond formed between the groups on the polymer chains and the *cations* of the salt. In view of the cation–anion interactions just referred to, ΔH_i is more significant than in aqueous electrolytes. The ion–ion interactions can be sufficiently strong to form ion pairs or larger ion clusters.

The electrostatic interactions between the ions when dissolved in the polymer are a legacy of the electrostatic interactions which existed in the crystalline salt. As a result of the persistence of strong ion–ion interactions when a salt is dissolved in a polymer, dissolution may be regarded as a process in which some of the electrostatic interactions (bonds) between ions in the salt are replaced by interactions between the cation and the polymer.

It is evident that, in practice, the major factors determining whether a salt will dissolve in a polymer are the lattice energy of the salt and the solvation of the cations by the polymer chains.

6.2.2 Which salts dissolve in which polymers

We are now in a position to consider which cations and anions dissolve in which polymers. Beginning with the anions. Because of weak anion solvation, salts containing polyatomic anions with a monovalent charge, which is distributed over the anions, are the best candidates (Armand and Gauthier, 1989). As an example, LiF, which is highly soluble in water, in part because of strong solvation of the F^- ions by water molecules, is insoluble in poly(ethylene oxide), whereas $LiClO_4$ is highly soluble in this polymer. Furthermore, in general, the salts with large monovalent anions will have low lattice energies, which, in turn, aid dissolution. The most common anions used are ClO_4^-, $CF_3SO_3^-$, $(CF_3SO_2)_2N^-$, $(CF_3SO_2)_3C^-$, BPh_4^-, AsF_6^-, PF_6^- and SCN^-. Salts containing monatomic anions are also soluble, provided they are large and polarisable. Hence I^- and Br^- based salts dissolve, but only a few chlorides are soluble and the fluorides are insoluble.

Since the solvation enthalpy of a salt in a polymer depends on the cation–polymer interactions, dissolution only occurs if atoms which are capable of coordinating the cations are available on the polymer chains. As an illustration, $LiClO_4$ dissolves easily in PEO $(—CH_2—CH_2—O—)_n$ whereas it is virtually insoluble in polyethylene $(—CH_2—CH_2—)_n$. Both are linear chain polymers, they differ only in the presence of an ether oxygen between pairs of C_2H_4 groups in PEO. The ether oxygens are excellent coordinating groups for many cations by virtue of the strong coordinating bonds which they form. In low molecular weight liquid solvents such as tetrahydrofuran, each molecule contains only one coordinating atom. In this case the solvation of the cation depends primarily on the number of molecules which may pack around the cation. However, in a high molecular weight polymer, the cations may frequently be coordinated by atoms on the same chain. In other words, it is important that the chain can wrap around the cation without excessive strain and in such a way that coordinating atoms are available to the cation. It has been shown that the repeat unit $(—CH_2—CH_2—O—)_n$ provides just the right spacing between coordinating ether oxygens for maximum solvation of the cations. Polyethers with $(—CH_2—O—)_n$ or $(—CH_2—CH_2—CH_2—O—)_n$ repeat units are much weaker solvents, even although they contain the same coordinating atoms (Cowie and Cree, 1989; Armand, Chabagno and Duclot, 1979). The requirement for the polymeric solvent to provide a suitable cavity lined with coordinating atoms within which the cation can fit as if a hand in a glove highlights the similarity between polyether solvents and macrocyclic ligands such as the crown ethers. As in the case of the macrocycles, if the polymer chain conformation already provides a suitable cavity then the entropy loss on coordination is minimised.

Other coordinating groups which are found widely in polymer electrolytes are

$$—\ddot{N}R— \qquad —\ddot{N}H— \text{ and } —\ddot{\underset{..}{S}}—$$

The strength of interactions between a cation and the different coordinating groups may be classified according to the hard/soft acid base theory (HSAB) (Pearson, R. G., 1963; Bruce, Krok and Vincent, 1988). Non-polarisable ligands of high electronegativity such as the ether oxygens are classified as hard, whereas the more polarisable $—\ddot{\underset{..}{S}}—$ groups are considered to be soft. Similarly, small polarising cations such as Mg^{2+} are hard, in contrast to soft cations such as Hg^{2+}. The strongest interactions occur by matching like with like; therefore for polyether solvents the strongest solvation is expected with the hard cations, e.g. Mg^{2+} or Ca^{2+}. This probably accounts for the solvation of salts such as $MgCl_2$ and $MgBr_2$ where the particularly strong cation solvation is sufficient to overcome the disadvantage of being associated with Cl^- and Br^- ions.

Dissolution is not restricted to salts of mono- or even divalent cations, salts containing trivalent cations are also soluble, e.g. $Eu(ClO_4)_3$ and $La(ClO_4)_3$ (Bruce, Nowinski, Gray and Vincent, 1990; Mehta, 1993). Even although such salts have a higher lattice energy than the corresponding salts with a monovalent cation, the strength of the cation–polymer interaction is also considerably higher, sufficiently so to more than compensate for the high lattice energy.

6.2.3 *Evidence for cation solvation*

The direct evidence on which our view of cation solvation in polymer electrolytes is based comes mainly from spectroscopic techniques. IR and Raman studies have been carried out on a variety of systems (see Chapter 5, Torell and Schantz, 1989; and Frech, Manning, Teeters and Black, 1988). Low frequency vibrational modes, around 860–870 cm^{-1}, associated with the cation–ether oxygen interactions in PEO based systems have been observed; they are absent in PEO itself.

EXAFS (extended X-ray absorption fine structure) and related techniques provide data on the immediate environment around a given ion (Koningsberger and Prins, 1988). Unfortunately not all ions may be probed, but for those which can, it is a valuable tool. EXAFS is particularly suited to the determination of bond lengths, but is less reliable as a method for obtaining precise coordination numbers and geometry (Latham, Linford and Schlindwein, 1989). It is at its most powerful when data are available on model systems which have been well characterised by another technique.

X-ray and neutron diffraction methods are supreme in their ability to reveal precise and detailed information on the structure of materials, including the coordination around ions in crystalline phases. Unfortunately ion transport in polymers is confined largely to the amorphous phases of polymer–salt complexes. However, it is possible to prepare predominantly crystalline and amorphous complexes in many systems indeed, in some cases, both crystalline and amorphous forms may be prepared with the same composition. In view of this crystallographic studies of crystalline polymer– salt complexes are of immense value, since they can reveal the basic principles governing both ion–polymer interactions and polymer electrolyte structure in general. Few such studies have so far been carried out. Full crystal structures have been obtained for complexes in the PEO:NaI and PEO:NaSCN systems by single crystal methods (Chatani and Okamura, 1987; Chatani, Fujii, Takayanagi and Honma, 1990). The full crystal structures of PEO_3:$LiCF_3SO_3$, PEO_3:$NaClO_4$, PEO_4:KSCN, PEO_4:NH_4SCN and PEO_4:RbSCN have been obtained from powder diffraction data (Lightfoot, Mehta and Bruce, 1992, 1993; Lightfoot, Nowinski and Bruce, 1994; Thomson, Lightfoot, Nowinski and Bruce, 1994). The crystal structures and their implications for polymer electrolyte structure in general are discussed in Chapter 5. The fact that the helical conformation of PEO chains in the pure polymer is retained in the complexes, such that cations ranging in size from Li^+ to Rb^+ fit inside the turns of the helices, reinforces a statement made in Section 6.2.2 concerning the importance of coordinating the cations while at the same time adopting a conformation for the polymer chains which is not excessively strained.

6.2.4 *Ion association*

The presence of ion association in polymer electrolytes was alluded to in Section 6.2.1. Here we will consider the nature of the clusters and the structural evidence for their existence. The involvement of associated species in transport processes is discussed later in Section 6.4.

Whereas the dissolution of a salt in a polymer is controlled largely by the local cation–polymer interactions, i.e. by cation solvation, this does not in itself determine the nature of the species found in solution. The dielectric constant of a polymer host such as poly(ethylene oxide) lies between 5 and 8; this is very low compared with water which possesses a dielectric constant of 78. Water is therefore a highly polarisable medium, due to the relatively large permanent dipole moment of the H_2O molecule. The oriented dipoles associated with the H_2O molecules surrounding each ion reduce the effective field of the ions and hence their mutual interaction. In many cases these interactions are sufficiently weak to be treated by Debye–Hückel theory. In the case of a solid polyether this does not happen to the same extent, the ions interact strongly resulting in the formation of ion clusters. Consider a salt MX consisting of M^+ and X^- ions. The simplest species that may form is an ion pair which comprises a single cation and anion separated by a polymer chain $[MX]_s^0$. This is termed a *solvent separated ion pair* (Fig. 6.1). The subscript s denotes separation of the ions by part of the solvent and the superscript reminds us that the species carries no charge. Larger aggregates may also form, most notably ion triples $[M_2X]_s^+$ and $[MX_2]_s^-$ but yet larger, charged and neutral, clusters are possible. Of course such ion association is not restricted to uni–univalent salts, indeed it is likely to be more prominent in the case of salts containing di- or trivalent cations. In such systems ion pairs and triples will no longer be respectively neutral and charged. With a salt MX_2 containing divalent cations and monovalent anions the ion pair will carry a net positive charge of unity.

The formation of solvent separated ion aggregates is largely determined

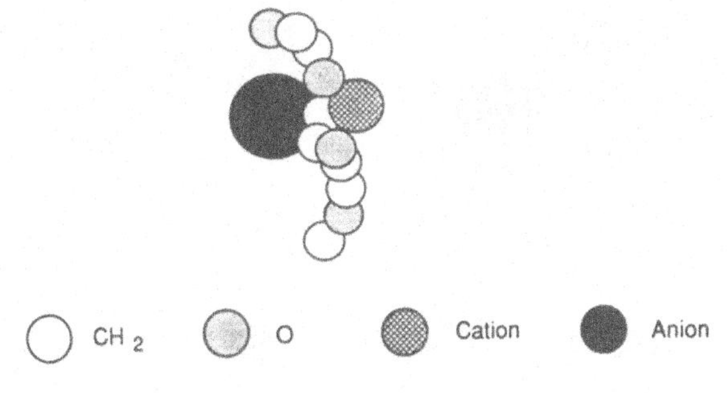

Fig. 6.1 Solvent separated ion pair in a polyether.

by coulombic forces, i.e. the ionic charges and the dielectric constant of the medium. However, a second class of ion aggregate may form in which the solvent no longer separates the ions; such species are termed *contact ion aggregates*, e.g. $[MX]^0$, $[M_2X]^+$, $[MX_2]^-$ (Fig. 6.2). The absence of a subscript indicates direct cation–anion interaction. In low molecular weight solvents such as tetrahydrofuran, formation of a contact ion pair from its solvent separated counterpart depends on the energetics involved in displacing the heteroatom of the separating solvent molecule (in this case an ether oxygen) from its position in the coordination sphere of the cation and replacing it with the anion. Both enthalpy and entropy contributions must be taken into account; however, in general high salt concentrations will favour formation of such contact pairs. In polymer electrolytes the situation is similar, but an additional factor must be considered. The solvating heteroatoms are constrained, being localised on the polymer chain. As a result of steric considerations and to avoid undue strain in the polymer chains the coordination sphere may not be saturated with heteroatoms. Empty sites around the cation could then be occupied by anions without the need to displace other atoms (Bruce, 1989). Since, in addition, the anions are barely solvated by the polymer chains, formation of contact ion pairs may be particularly facile in some polymer electrolytes. The crystal structures of the PEO:salt complexes appear to support this view in that for all systems studied so far the cations fit within the PEO helix leaving two sites for anions to directly coordinate each cation.

Both X-ray diffraction and Raman spectroscopy have been applied in an attempt to probe the nature of ion association. The latter technique relies on monitoring changes in the vibrational modes of the anions, particularly $CF_3SO_3^-$ or ClO_4^-, anions as their immediate environment changes. It is important to note that only contact ion clusters will be

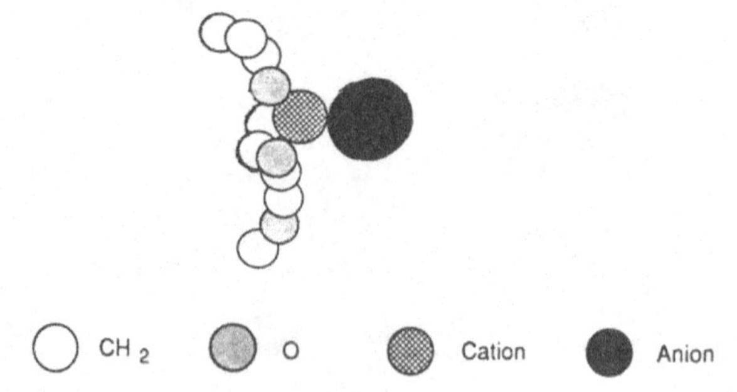

Fig. 6.2 Contact ion pair in a polyether.

detected by this technique. To the spectroscopist solvent separated ion clusters are indistinguishable from free ions; the term 'free ion' in papers reporting spectroscopic studies must be treated with caution. Displacement of the equilibrium between contact ion pairs, triples and possibly larger clusters with changing salt concentration in the range 30:1 to 5:1, has been noted in both poly(ethylene oxide) and poly(propylene oxide) based electrolytes (Torell and Schantz, 1989). It is suggested that ion triples replace ion pairs as the salt concentration increases. The spectroscopic data also indicate that ion aggregation increases with increasing temperature consistent with a negative entropy of salt dissolution. More detailed information may be obtained from the reference cited. EXAFS also reveals the existence of contact ion pairing. For example Latham *et al.* (1989) have noted the direct coordination of Zn^{2+} ions by I^- in PEO:ZnI_2 electrolytes.

6.2.5 *Negative entropy of dissolution*

In Section 6.2.1 we considered the thermodynamics of salt dissolution in a high molecular weight polymer. Two contributions to the overall entropy change were noted, a positive entropy change as the ions enter solution from the crystal lattice and a negative contribution due to the pinning of the polymer chains by the ions. Whereas in some systems the overall entropy change on dissolution may be positive it appears that in many polymer systems the loss of entropy of the chains outweighs the gain due to ion disorder. This situation is more likely to occur in the case of polymer electrolytes rather than for low molecular weight solvents because the loss of entropy of the solvent is greater for polymer chains. Furthermore, the gain in entropy of the ions on dissolution is likely to be less than anticipated for a fully dissociated salt due to ion association. These factors can combine to induce a negative entropy of dissolution (Cameron and Ingram, 1989). Since the free energy of dissolution, ΔG, is given by:

$$\Delta G = \Delta H - T\Delta S$$

If ΔS is negative, dissolution can only occur if ΔH is negative and greater in magnitude than $T\Delta S$. As the temperature of the system is raised the second term grows in magnitude and we can anticipate a temperature at which ΔG becomes positive. The salt will then precipitate out. Such a situation has already been observed. It has been very clearly demonstrated using variable temperature X-ray diffraction that crystalline $Ca(CF_3SO_2)_2$

precipitates from amorphous PEO–$Ca(CF_3SO_3)_2$ on heating (Mehta, Lightfoot and Bruce, 1993). The process of salt precipitation is often incorrectly referred to as salting-out. In fact, the process of salting-out refers to the precipitation of a non-electrolyte solute from a solvent when a salt is added to the same solvent, and is not associated with the solubility of the salt itself (Kortüm, 1965).

6.3 *Mechanisms of ionic conduction*

It is important to draw a distinction immediately between ion transport in low molecular weight, liquid, solvents and in high molecular weight, solid, polymers. In the former medium ions move with their solvent sheath intact and transport is related to the macroscopic viscosity of the electrolyte. This is also true for low molecular weight polymers. For example, in the case of linear chain polyethers this situation pertains up to a molar mass of 3200 (Bruce and Vincent, 1993). Above this limit, ions must become at least partially desolvated to move, i.e. the ion transport is decoupled from the *macroscopic* viscosity of the electrolyte, since the polymer chains become increasingly entangled and cannot therefore move over long distances with the ions. A distinctive solid polymer transport mechanism is identified which is, however, intimately linked to the *microscopic* viscosity of short segments of the polymer chains. This is reflected in the observation that the curved log σ vs $1/T$ plots typical of ion transport in an amorphous polymer (see Chapter 5) frequently parallel the temperature dependence of local relaxations of polymer chain segments, as observed by the dielectric or mechanical relaxations (Fontanella, Wintersgill, Smith, Semancik and Andeen, 1986). Because ions must dissociate from their coordination spheres to move in solid polymers those cations which bind tightly (see discussion of hard soft acid base theory in Section 6.2.2) are immobile. All ions are mobile in low molecular weight solvents. In order to form a solid polymer electrolyte with mobile cations, a balance must be struck between a cation–polymer bond which is strong enough to promote salt dissolution but weak enough to permit cation mobility. PEO:$Hg(ClO_4)_2$ is an excellent example of this compromise. It is the very distinctive transport through an entangled polymer environment which we are considering in this section. Polymer electrolytes can therefore be considered as extremely viscous fluids where the local motion of the polymeric solvent serves to transport the ions. During early investigations of these materials in the late 1970s, ion-hopping through a rigid polymer framework, in which the cations were thought to move along channels within the PEO helices, was believed to be responsible for ion transport. Various experimental techniques applied to polymer electrolytes have subsequently shown the key role of a dynamic polymer environment.

The electrolyte concentration is very important when it comes to discussing mechanisms of ion transport. Molar conductivity–concentration data show conductivity behaviour characteristic of ion association, even at very low salt concentrations (0.01 mol dm^{-3}). Vibrational spectra show that by increasing the salt concentration, there is a change in the environment of the ions due to coulomb interactions. In fact, many polymer electrolyte systems are studied at concentrations greatly in excess of 1.0 mol dm^{-3} (corresponding to ether oxygen to cation ratios of less than ~20:1) and charge transport in such systems may have more in common with that of molten salt hydrates or coulomb fluids. However, it is unlikely that any of the models discussed here will offer a unique description of ion transport in a dynamic polymer electrolyte host. Models which have been used or developed to describe ion transport in polymer electrolytes are outlined below.

6.3.1 *Temperature and pressure dependence of ionic conductivity*

The dc conductivity, σ, of a homogeneous polymer electrolyte, at temperature T, and pressure P, can be expressed in general terms as

$$\sigma(T, P) = \sum_i c_i q_i u_i, \tag{6.1}$$

where c_i is the concentration of charge carriers of type i, q_i is the charge and u_i the mobility. The summation includes all charged species, including single cations and anions, and ion clusters. (The summation also includes electrons and holes, but experimentally it has been shown that these do not contribute to the conductivity in polymer electrolytes.) As the salt concentration is increased, so the concentration and mobility and indeed the nature of the charge carriers will be expected to change. However, by making the assumption that the salt dissolved in the polymer, MX, fully dissociates into cations M^+ and anions X^- which do not interact with each other, then the problem of developing models which are capable of describing the conductivity is simplified to one of describing the ion mobility. The validity of this assumption will be discussed further in the section on ion association.

Empirical relationships

For transport in amorphous systems, the temperature dependence of a number of relaxation and transport processes in the vicinity of the glass transition temperature can be described by the Williams–Landel–Ferry (WLF) equation (Williams, Landel and Ferry, 1955). This relationship was originally derived by fitting observed data for a number of different liquid systems. It expresses a characteristic property, e.g. reciprocal dielectric relaxation time, magnetic resonance relaxation rate, in terms of shift factors, a_T, which are the ratios of any mechanical relaxation process at temperature T, to its value at a reference temperature T_s, and is defined by

$$\log a_T + \text{const} = \log\left[\frac{\eta(T)}{\eta(T_s)}\right]$$

$$= -\frac{C_1(T - T_s)}{(C_2 + T - T_s)} \qquad (6.2)$$

where η is the viscosity (microviscosity for high molecular weight polymers), T_s is a reference temperature and C_1 and C_2 are constants which may be obtained experimentally. These 'universal' constants are independent of the measured property and take the values $C_1 = 8.9$ and $C_2 = 102$ K. Although T_s is an arbitrary temperature, it is often taken to be 50 K above the glass transition temperature and so the shift factor can be expressed as

$$\log a_T = \frac{-17.4(T - T_g)}{51.6 + T - T_g} \qquad (6.3)$$

However, because T_g measurements are kinetically determined, this is a less accurate form of the equation. Very often it is observed that the measured shift factors, defined for different properties, are independent of the measured property. In addition, if for every polymer system, a different reference temperature T_s is chosen, and a_T is expressed as a function of $(T - T_s)$, then a_T turns out to be nearly universal for all polymers. Williams, Landel and Ferry believed that the universality of the shift factor was due to a dependence of relaxation rates on free volume. Although the relationship has *no free volume basis*, the constants C_1 and C_2 may be given significance in terms of free volume theory (Ratner, 1987). Measurements of shift factors have been carried out on crosslinked polymer electrolyte networks by measuring mechanical loss tangents (Cheradame and Le Nest, 1987). Fig. 6.3 shows values of $\log a_T$ for

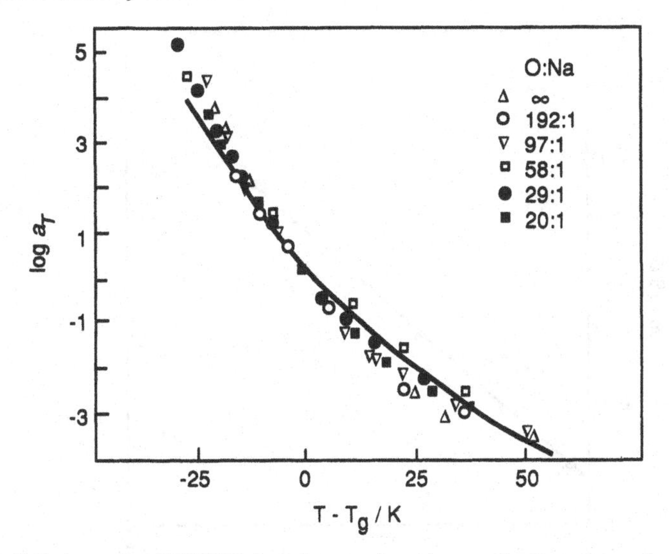

Fig. 6.3 Shift factor for PEO(400) based network and network electrolytes, shown as a function of reduced temperature.

poly(ethylene oxide) based, cross-linked networks as a function of reduced temperature, $T - T_g$. The fitted line corresponds to $C_1 = 10.5$ and $C_2 = 100$. In Fig. 6.4, the shift factor is shown to correlate with the ionic conductivity suggesting that ionic motion is indeed promoted by local polymer segmental motion. By coupling the WLF equation with the Stokes–Einstein equation for the diffusion coefficient, D, and then with the Nernst–Einstein relationship

$$u = qD/kT, \tag{6.4}$$

where q is the charge, assuming that the salt fully dissociates in the polymer, i.e. a strong electrolyte, the temperature dependence of the conductivity can be written in the WLF form

$$\log \frac{\sigma(T)}{\sigma(T_s)} = \frac{C_1(T - T_s)}{C_2 + (T - T_s)}. \tag{6.5}$$

Eqn (6.5) holds reasonably well for a number of polymer electrolyte systems and a decrease in the T_g leads to an increase in conductivity.

The WLF relation was an extension of the Vogel–Tamman–Fulcher (VTF) empirical equation (Vogel, 1921; Tamman and Hesse, 1926; Fulcher, 1925) which was originally formulated to describe the properties of supercooled liquids and, given in its original form, is

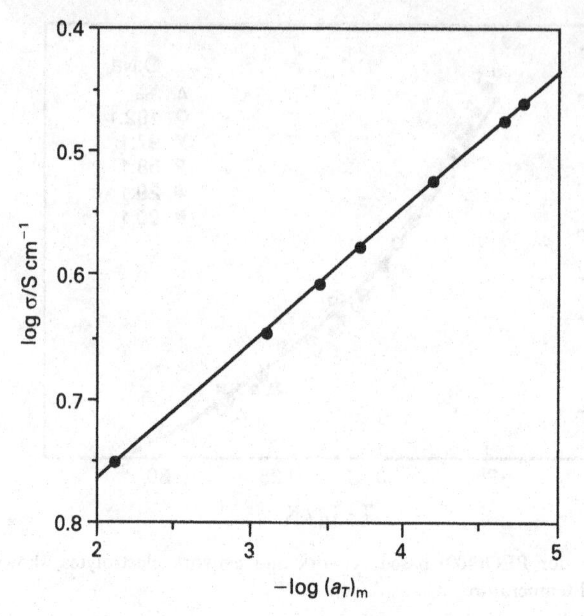

Fig. 6.4 Correlation between a_T, the WLF shift factor and the conductivity for PEO networks.

$$\eta^{-1}(T) = A \exp[-B/R(T - T_0)]. \quad (6.6)$$

Here T_0 is $T_s - C_2$ and A is a prefactor proportional to $T^{1/2}$ which is determined by the transport coefficient (in this case η^{-1}) at the given reference temperature. The constant B has the dimensions of energy but is not related to any simple activation process (Ratner, 1987). Eqn (6.6) holds for many transport properties and, by making the assumption of a fully dissociated electrolyte, it can be related to the diffusion coefficient through the Stokes–Einstein equation giving the form to which the conductivity, σ, in polymer electrolytes is often fitted,

$$\sigma = \sigma_0 \exp[-B/R(T - T_0)]. \quad (6.7)$$

The form of this equation suggests that thermal motion above T_0 contributes to relaxation and transport processes and that for low T_g, faster motion and faster relaxation should be observed.

Eqn (6.7) may be expressed in a number of slightly different forms which depend on the model and assumptions made in the original derivation. If ionic diffusion is considered to be an activated process as, for example, in the case of glasses and ceramics, then included in the preexponential term of Eqn (6.7) is the attempt frequency, ν_0, for ion mobility. Several expressions have been proposed for the attempt frequency but in the simplest case it may be assumed to be temperature independent and ν_0 is a constant. In transition state theory, it is presumed ν_0 represents a fully excited vibrational degree of freedom so that $\nu_0 = kT/h$. Following the approach of Cohen and Turnbull (1959), one might imagine that the ion moved about its cage of neighbouring atoms with the velocity of an ideal gas molecule. This kinetic theory gives $\nu_0 = (8kT/\pi m)^{1/2}/2d$, where m is the mass of the gas particle and d is the cage diameter. Thus the preexponential factor of Eqn (6.7) may be written as σ_0/T^m where $m = 0$, 1 or $\frac{1}{2}$. Somewhat surprisingly, the conductivity variation over a large temperature range ($\sim 300\,°\mathrm{C}$) for a number of glasses has been found to best fit the equation where the preexponential factor was temperature independent. Considering the short temperature range accessible to polymer electrolytes, there is no experimental justification for the inclusion of a temperature term in the preexponential factor. It is almost always possible to obtain a reasonable fit of experimental data for amorphous materials to a VTF/WLF type equation. However, significant discrepancies can arise in parameters if the fittings are made over different temperature ranges.

Free volume based models

The free volume approach was discussed briefly in the context of ion transport in glasses (Chapter 4): it is also one of the simplest ways of understanding polymer segment mobility. It has been used for the analysis of ionic motion in fused salts and fluids, as well as polymer electrolytes, but for none of these is the model wholly satisfactory. Although both the WLF and VTF equations are empirical, the free volume theory can be expressed in the VTF form (Ratner, 1987). Conversely, however, if the temperature dependence of the conductivity of a polymer electrolyte can be described accurately by the VTF equation, it does not follow that the system's behaviour is governed by free volume rearrangements. In the free volume model of Cohen and Turnbull, motion is assumed to be a non-thermally activated process but occurs as a result of redistribution of the free volume. Basically, it states that as the temperature is increased, local voids are produced by the expansion of the material and therefore polymer segments (or ionic species or solvated molecules) can move into this free volume. A distribution function for the size of voids in the given material is derived, and the probability of this distribution is maximised. With viscosity proportional to the inverse of the diffusion coefficient from the Stokes–Einstein equation (assuming a strong electrolyte) and nearly independent of temperature at constant volume, and if the transport or relaxation rate is determined by the rate at which voids of certain minimum critical volume V^* are created, then

$$\eta^{-1} \sim \exp(-\gamma V^*/V_f), \tag{6.8}$$

where V_f is the free volume per mole, V^* is the critical volume per mole and γ is a constant which allows for overlap of the free volume. If V_f is expanded around a temperature T_0, the temperature at which the free volume vanishes, then

$$V_f = V_f(T_0) + (T - T_0)\left(\frac{\delta V_f}{\delta T}\right)_{T_0}. \tag{6.9}$$

It may be shown (Ratner, 1987) that assuming the Nernst–Einstein relationship (Eqn (6.4)), a free volume expression for the conductivity in the form of Eqn (6.7) may be derived, provided the electrolyte is fully dissociated and

$$B = \gamma V^*/(\delta V_f/\delta T)_{T_0}. \tag{6.10}$$

The constant B is not an activation barrier but is related to the critical free volume for transport, V^*, and to expansivity. In polymer electrolytes, V^* is generally taken as fixed by the size of the polymer segment rather than the motion of the ion since the polymer strands must move before either cations or anions can be transported.

Miyamoto and Shibayama (1973) proposed a model which is essentially an extension to free volume theory, allowing explicitly for the energy requirements of ion motion relative to counter ions and polymer host. This has been elaborated (Cheradame and Le Nest, 1987) to describe ionic conductivity in cross-linked polyether based networks. The conductivity was expressed in the form

$$\sigma = \sigma_0 \exp\left(-\frac{\gamma V^*}{V_f} - \frac{\Delta E}{RT}\right), \tag{6.11}$$

which is a combination of Arrhenius and free volume behaviour. The activation energy, ΔE, is given as

$$\Delta E = E_j + W/2\varepsilon, \tag{6.12}$$

where E_j is the energy barrier for cooperative ion transfer from one hole site to another, W is the dissociation energy of the salt and ε is the dielectric constant of the matrix. (Note the similarity to recent ideas on transport in inorganic glasses above T_g, Chapter 4.) If ΔE is relatively small, then the second term in Eqn (6.11) tends to unity and free volume behaviour predominates. By reducing the data so as to eliminate the free volume behaviour (the first term in Eqn (6.11)) an activation energy of $\Delta E \sim 20\ \mathrm{kJ\ mol^{-1}}$ can be deduced for sodium tetraphenyl borate doped PEO networks. It has also been shown that the reduced conductivity is a linear function of the WLF shift factor, i.e.

$$\ln\left[\frac{\sigma(T)}{\sigma(T_g)}\right] = -\left(\frac{V_i^*}{V_p^*}\right)\ln a_T, \tag{6.13}$$

where V_i^* and V_p^* are the critical free volumes for ions and polymer segment motion respectively. V_i^*/V_p^* is not found to vary greatly from unity, suggesting that both properties relate to the same process. Watanabe and coworkers (Watanabe and Ogata, 1987) have also studied the conductivity and elastic modules of salt-containing polyether networks and have related these in a similar manner. However, when the conductivities were plotted as a function of salt concentration at various reduced temperatures the increase in conductivity was larger than that predicted on the assumption of complete dissociation. This was considered to be a consequence of the fact that the activation energy, ΔE, of Eqn (6.11) was not negligible. It was shown that the critical volumes for conductivity and mechanical relaxation were virtually identical and that ΔE could be extracted from

$$\frac{T}{T_g}\left[\frac{\sigma(T)}{\sigma(T_g)}\bigg/\frac{1}{a_T}\right] = A\exp\left(\frac{-\Delta E}{RT}\right). \tag{6.14}$$

Free volume models have been very extensively used in polymer electrolyte studies. While successful in rationalising many of the properties of these materials, in particular the temperature dependence of conductivity, they are inadequate for several reasons. The major weakness is that they ignore the kinetic effects associated with macromolecules. In addition, the models do not relate directly to a microscopic picture and therefore do not predict straightforwardly how such variables as ion size, polarisability, ion pairing, solvation strength, ion concentration, polymer structure or chain length will affect the conduction process. Fig. 6.5 expresses some of the limitation effects. The conductivities of a number of PEO based networks containing different salts have been plotted at a reduced temperature ($T_g + T$) as a function of the lattice energy of the incorporated salt. If free volume theory were applicable, then the conductivity would be expected to be constant for constant ($T_g + T$). This is clearly not so, due largely to the omission of the effects of ionic interactions from the model.

Configurational entropy model

Another group of theories based on the configurational entropy model of Gibbs *et al.* (Gibbs and di Marzio, 1958; Adams and Gibbs, 1965) goes some way to overcoming the deficiencies of the free volume models, described above. As for free volume models, the configurational entropy model discusses only the properties of the polymer. The mass transport mechanism in this model is assumed to be a group cooperative rearrangement of the chain, giving the average probability of a rearrangement as

$$W = \exp\left(\frac{-\Delta\mu S_c^*}{kTS_c}\right), \tag{6.15}$$

where S_c^* is the minimum configurational entropy required for rearrangement and is often taken as $S_c^* = k \ln 2$, S_c is the configurational entropy at temperature T and $\Delta\mu$ is the free energy barrier per mole which opposes the rearrangement. As the glass transition temperature is approached, the molecular relaxation time becomes longer and equilibrium cannot be maintained. Eventually the dynamic configurational entropy will reach zero at a temperature T_0, often found to be approximately 50 K below T_g. By relating W to the inverse relaxation time and assuming ΔC_p, the

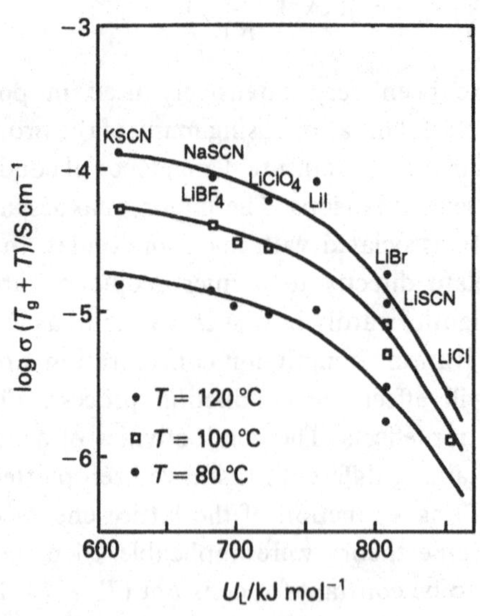

Fig. 6.5 The conductivity of PEO based networks containing different salts plotted at a reduced temperature ($T_g + T$) as a function of the lattice energy of the incorporated salt.

heat capacity difference between the liquid and glass states, to be temperature independent, a WLF type of relationship can be given:

$$-\log a_T = \frac{a_1(T - T_s)}{a_2 + T - T_s}. \tag{6.16}$$

The constants a_1 and a_2 correspond to

$$\left.\begin{aligned} a_1 &= 2.30\Delta\mu S_c^*/\Delta C_p k T_s \ln\left(\frac{T_s}{T_0}\right), \\ a_2 &= T_s \ln\left(\frac{T_s}{T_0}\right)\Big/\left[1 + \ln\left(\frac{T_s}{T_0}\right)\right], \end{aligned}\right\} \tag{6.17}$$

with a_2 slightly temperature dependent. A VTF form can also be written

$$W = A \exp\left(\frac{-K_\sigma}{T - T_0}\right), \tag{6.18}$$

with $K_\sigma = \Delta\mu S_c^*/k\Delta C_p$. Shriver *et al.* (Papke, Ratner and Shriver, 1982; Shriver, Dupon and Stainer, 1983) applied the configurational entropy model to polymer electrolytes. They used the assumption that $\Delta C_p = B/T$, where B is a constant and showed that

$$\sigma(T) = A \exp\left(\frac{-K_\sigma}{T - T_0}\right), \tag{6.19}$$

where $K_\sigma = \Delta\mu S_c^* T_0/kB$. The approach appears to have merit and analyses of several polymer electrolyte systems give reasonable values for the activation energy and values close to 50 K for $(T_g - T_0)$. Certain aspects of conductivity are also implied by the configurational entropy approach. For example, the fall in conductivity with increasing pressure, the molecular weight independence of conductivity in amorphous polymers above a certain molecular weight and the fact that if T_0 is lowered, the conductivity at a fixed reduced temperature $(T - T_0)$ should increase. Angell and Bressel (1972) have rationalised the conductivity maxima in conductivity/concentration plots on the basis of an isothermal version of the VTF equation

$$\sigma = AX \exp\left[\frac{-K}{Q(X_0 - X)}\right], \tag{6.20}$$

where $Q = (\mathrm{d}T_g/\mathrm{d}X)$, X is the mole fraction of salt and X_0 is the mole fraction of salt at $T = T_g$. Cowie (Cowie, Martin and Firth, 1988) has reported reasonably good agreement between experimental data for an amorphous comb-type polymer-$LiClO_4$ electrolyte and the conductivity and maxima predicted by Eqn (6.20) (Fig. 6.6). The pre-exponential factor, A, was found to decrease with increasing temperature to a particular temperature and then show a slight fall. This may imply some temperature dependence of the cation-to-polymer interaction equilibrium.

The configurational entropy model describes transport properties which are in agreement with VTF and WLF equations. It can, however, predict correctly the pressure dependences, for example, where the free volume models cannot. The advantages of this model over free volume interpretations of the VTF equation are numerous but it lacks the simplicity of the latter, and, bearing in mind that neither takes account of microscopic motion mechanisms, there are many arguments for using the simpler approach.

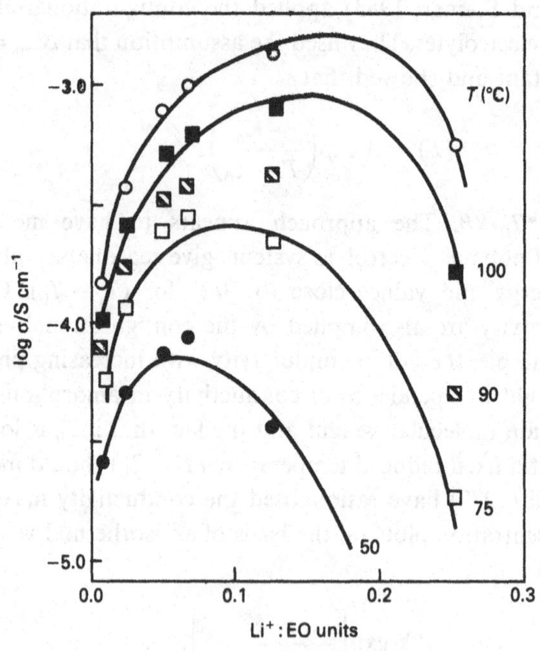

Fig. 6.6 Variation of log σ (at reduced temperature $T - T_g$) with salt concentration for a series of amorphous polymer-$LiClO_4$ systems. The solid lines are obtained from calculations using Eqn (6.20).

6.3.2 The dynamic response of polymer electrolytes

Although the VTF and WLF equations, along with free volume or configurational entropy approaches may describe some of the transport properties in polymer electrolytes, they are not based on a microscopic treatment and therefore local mechanistic information is lost. Experimentally, a number of techniques can be employed to probe characteristic time scales for segmental motion. Techniques such as mechanical relaxation, dielectric relaxation and loss, high frequency ac conductivity, Brillouin scattering, nuclear magnetic resonance (NMR) relaxation and quasielastic neutron scattering (QENS) have all been applied to polymer electrolytes to study the frequency dependent properties of both polymeric and ionic species (MacCallum and Vincent, 1987, 1989; Gray, 1991).

A method of characterising transport mechanisms in solid ionic conductors has been proposed which involves a comparison of a structural relaxation time, τ_s, and a conductivity relaxation time, τ_σ. This differentiates between the amorphous glass electrolyte and the amorphous polymer electrolyte, the latter being a very poor conductor below the T_g. A decoupling index has been defined where

$$R_\tau = \tau_s/\tau_\sigma. \tag{6.21}$$

The behaviour of R_τ as a function of temperature was investigated in different ionic materials. In glass materials such as $LiAlSiO_4$, R_τ is very near to unity at high temperatures well above the glass transition temperature. As the molten glass is cooled below T_g, it falls out of equilibrium at a particular temperature and the structural relaxation time becomes progressively longer. At temperatures well below T_g, the ratio R_τ in these glasses can be very large, typically 10^{12}. In polymer electrolytes, the transport and relaxation are very closely related to one another with R_τ close to unity, as a result of the very similar mechanisms for conductivity and structural relaxation. For concentrated polymer–salt complexes, R_τ is generally found to be less than unity, often of the order of 0.1. This suggests that relaxation processes that do permit rearrangement of the polymeric structure do not necessarily permit ions to move. This could arise from interionic interactions, resulting in ion immobilisation or coulomb drag. Figure 6.7 shows R_τ data for a number of compounds and demonstrates the disparity between solvent-assisted motion in the polymer electrolytes and activated hopping in the vitreous material.

Motion of ions in polymer electrolytes is strongly dependent on segmental motion of the polymer host. Based on this and assuming a weak dependence of the conductivity on interionic interactions, Druger, Nitzan and Ratner proposed a microscopic model (Ratner, 1987; Ratner and Nitzan, 1989) to describe the transport mechanism. This is known as the dynamic bond percolation (DBP) theory. For conductivity in polymer electrolytes, cation and anion motions are considered to be fundamentally different. The former is visualised as the making and breaking of coordinate bonds with motion between coordinating sites, while anion motion is regarded as a hopping between an occupied site and a void which is large enough to contain the ion. Conductivity is visualised as being due to a combination of cooperative motion with the occasional independent ion movement; the time scale for the latter is much shorter than for polymer relaxation.

The simplest model involves ion hopping between sites on a lattice (not fixed as in a solid electrolyte such as AgI) with the ions obeying a hopping type equation

$$\dot{P}_i = \sum_j P_j W_{ji} - P_i W_{ij}. \tag{6.22}$$

P_i is the probability of finding a mobile ion at site i, W_{ij} is the probability per unit time that an ion will hop from site j to site i and is equal to zero

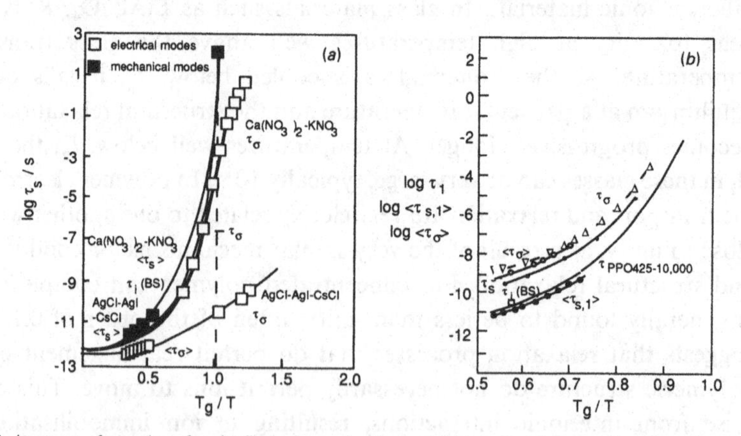

Fig. 6.7 Average shear (mechanical) and conductivity (electric modules) relaxation times, τ_s and τ_σ, for ionically conductive materials. (*a*) For glassy materials $\tau_\sigma \sim \tau_s$ above T_g and $\tau_\sigma \ll \tau_s$ below T_g. (*b*) For polymer electrolytes ($(PPO)_{13}$:$NaCF_3SO_3$) shown), $\tau_\sigma \geq \tau_s$ for $T > T_g$.

except between neighbouring sites. It is then assumed that W_{ij} may take two values,

$$\begin{rcases} W_{ij} = 0, \text{ probability } 1 - f \\ W_{ij} = w, \text{ probability } f \end{rcases} \quad (6.23)$$

with W_{ij} equal to zero, if all sites are already filled. Jumps are available with a relative probability f ($0 \leq f \leq 1$). Because of polymer motion, the configuration is continually changing and sites move with respect to each other. Therefore hopping probabilities readjust or renew their values on a time scale τ_{ren}, which is determined by the polymer motion. The W_{ij} values are thus fixed by the parameters w, f, and τ_{ren}. These can be related to system parameters such as ion size, free volume, temperature and pressure. Unlike static percolation models, no threshold for diffusion exists in renewal models (as found experimentally) but a marked change in the diffusion coefficient is found. In a comparison with free volume theory, f was identified with $\exp(-\gamma\eta^*/\eta_f)$ and w with the ion velocity divided by a lattice constant. τ_{ren} is the characteristic relaxation time corresponding to configurational or orientational changes. A number of results follow from this model:

(*a*) For an observation time $\gg \tau_{ren}$, with $f > 0$, ionic motion is always diffusive.

(*b*) For very fast renewal times, motion corresponds to hopping in a homogeneous system with an effective hopping rate, wf. If the renewal rate is very slow, then observed motion is that of a static bond percolation model. As the renewal time increases, diffusion coefficients and therefore conductivity increase until the rate-determining step changes to that of ion hopping and ion transport is independent of segmental motion.

(*c*) Frequency-dependent properties such as spectroscopic behaviour, dielectric relaxation and frequency dependent conductivity may be described by a DBP model. In Fig. 6.8, substantially different behaviour for the mean polymer host and the polymer ion conductor is observed. At frequencies above ~10 GHz, these two responses are essentially identical and may reflect displacive motion of ions or dipoles within the polymer host. For lower frequencies, only the ions continue to respond diffusively over lengthening time scales. Thus the response of the pure polymer falls to zero while that of the electrolyte levels out at low frequency, corresponding to the dc diffusion arising from the renewal process. These results are compared with experimental findings.

In addition to specific applications, the dynamic bond percolation model has been extended to focus on the importance of lattice considerations. The role of correlations among different renewal processes and the

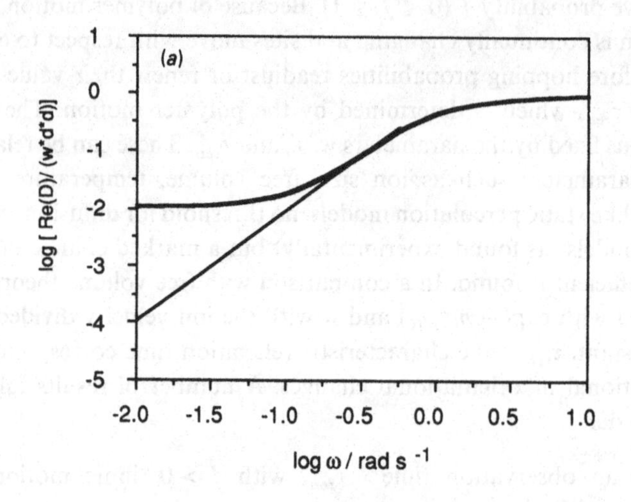

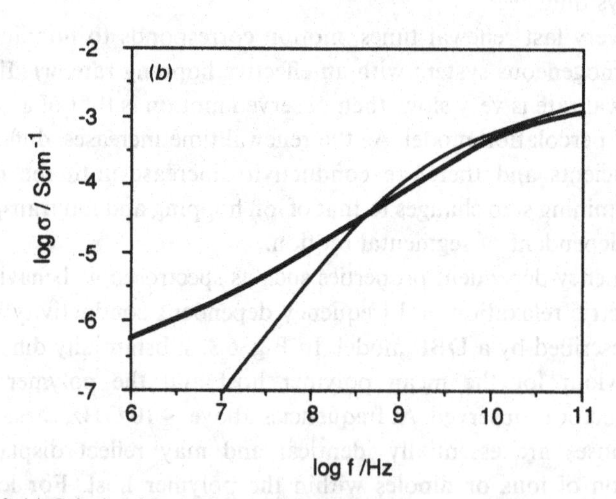

Fig. 6.8 (*a*) Calculated frequency-dependent conductivity for a simple dynamic percolation model. Lower line represents the diffusion coefficient without renewal, upper that with renewal. (*b*) Frequency-dependent conductivity for pure PEO (bold) and PEO–NaSCN at 22 °C. Only ions are able to diffuse long distances, corresponding to renewal diffusion.

effects of renewal on diffusion at the percolation threshold transition have been highlighted. Discussions on these aspects have been given by Ratner and Nitzan (1989).

A number of important issues involving the structure and dynamics of ionically conducting polymers have yet to receive thorough theoretical consideration. For example, in the case of multivalent cations, some systems exhibit cation transport whereas others do not, due to strong cation solvation. Therefore a term associated with ion–polymer dissociation must be important in systems which are on the borderline between these two extremes. This term is likely to be of the form $\exp(-\Delta H_{diss}/2RT)$. The most important deficiency in the models developed so far concerns the failure to take account of interactions between the mobile ions. As the ionic concentration in polymer electrolytes is frequently greater than $1.0\,\mathrm{mol\,dm^{-3}}$ and the mean distance between ions of the order of 0.5–0.7 nm, then relatively stong coulombic interactions exist which must affect ion motion. Ratner and Nitzan have begun to address this problem from a theoretical viewpoint (Ratner and Nitzan, 1989) although it has not been fully developed yet to give a complete description of conduction in ion associated polymer electrolytes. The interactions between ions which lead to ion association are discussed further in the following section.

6.4 *Ion association and ion transport*

We have already discussed ion association in Section 6.2. In that section we referred to evidence for the existence of ion clusters from static techniques such as IR, Raman, EXAFS and X-ray diffraction. In this section we examine ion association from the point of view of dynamics, concentrating in particular on electrochemical measurements which reveal the presence of ion clusters. Because ion association is so intimately connected to the transport of matter and charge through polymer electrolytes, it seems appropriate to consider these two topics in the same section.

6.4.1 *Ion association*

Molar conductivity

The nature of ion association in non-aqueous liquid electrolytes has traditionally been investigated by measuring the molar conductivity, Λ, as a function of salt concentration, c (Kortüm, 1965). The molar conductivity is defined as σ/c where σ is the conductivity of the electrolyte. Therefore, Λ is the conductivity per unit of salt concentration. For a simple uni–univalent salt, MX, which dissociates fully into cations M^+ and anions X^-, σ is given by

$$\sigma = ce(u_+ + u_-), \tag{6.24}$$

where c is the salt concentration, e is the charge on an electron, and u_+, u_- represent the ion mobilities. It follows that

$$\Lambda = e(u_+ + u_-) \tag{6.25}$$

and thus provided the ion mobilities are invariant with concentration, Λ should not vary. Λ has been determined as a function of salt concentration for $LiClO_4$ in the *solid* amorphous polyether, methoxylinked poly(ethylene oxide), average molar mass 100 000 (Gray *et al.*, 1991). Careful purification of this polymer has permitted conductivity measurements to be extended down to less than 10^{-3} mole dm^{-3}. The results are presented in Fig. 6.9 spanning the concentration range from 10^{-3} to 0.25 mol dm^{-3}. Beginning at the low concentrations, the sharp decrease in molar conductivity with increasing salt concentration could not be accounted for by changes in the mobility of the ions. Indeed measurement of the glass transition temperature, T_g, (217 K) indicated that it, and hence the ion mobility, are constant up to 0.1 mole dm^{-3} (Gray *et al.*, 1991). The sharp decrease in Λ has its origin in ion association. At extremely low salt concentrations, several orders of magnitude lower than those presented in Fig. 6.9, the salt exists in the form of isolated Li^+ and $ClO_4{}^-$ ions. As the concentration increases, mutual interactions between the ions are sufficiently strong to promote the formation of ion pairs, which are in equilibrium with the free ions:

$$Li^+ + ClO_4{}^- \rightleftharpoons [LiClO_4]^0$$

By the law of mass action the concentration of these ion pairs will grow at the expense of the free ions, as the overall salt concentration increases. Since the ion pairs carry no charge the conductivity per unit salt concentration, i.e. the molar conductivity, must fall as observed. As the salt concentration is further increased a minimum in Λ is eventually reached at which the concentration of ion pairs is a maximum. Two mechanisms have been postulated for the increase in ion concentration, and hence Λ, beyond the minimum. The first mechanism assumes that at higher salt concentrations triple ions will form which will be in rapid equilibrium with the ion pairs and, being charged, will result in an increase in Λ. Alternatively the increasing salt concentration may, due to long range (Debye–Hückel like) ion interactions, lower the activity of the ions, which, in turn, will stabilise free ions compared with the neutral ion pairs resulting in a shift of the dynamic equilibrium towards the free ions, i.e. redissociation. Triple ion formation is probably more important than activity effects, in view of the fact that the increase in Λ begins at such a low salt concentration, 10^{-2} mol dm^{-3}. The formation of ion aggregates is also consistent with the classical studies in non-aqueous liquid electrolytes with a low dielectric constant (Fuoss and Kraus, 1933). Two possible mechanisms of charge transport involving triple ions may be envisaged, both may in fact be operative. The first involves long range transport of the triple ions as discrete entities through the polymer, facilitated as usual by the segmental motion of the chains as described for free ions in

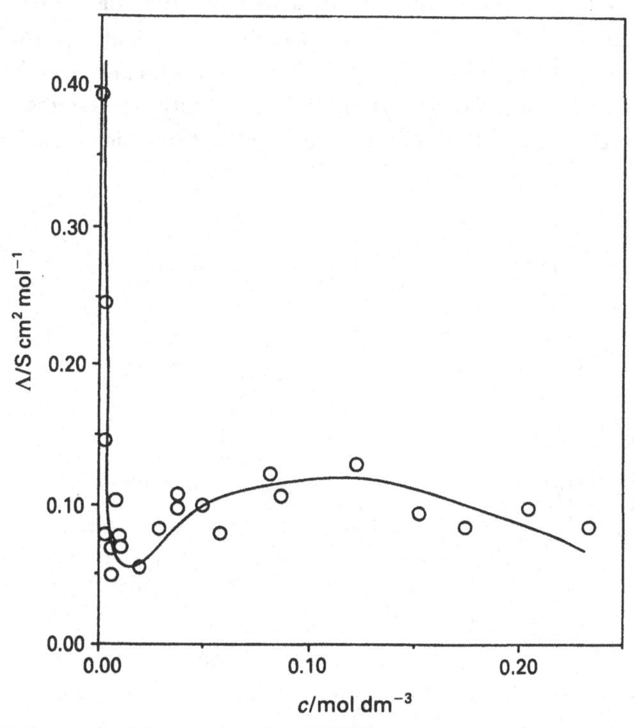

Fig. 6.9 Molar conductivity as a function of $LiClO_4$ concentration in oxymethylene-linked poly(ethylene oxide) at 298 K (Gray *et al.*, 1991).

Section 6.3. This is more likely for the anion triples $[MX_2]^-$ than the cation triples $[M_2X]^+$ since motion of the latter requires the simultaneous dissociation of two M^+ cations from a chain or chains. The second mechanism is also facilitated by chain motion but in this case the triple ion which moves with the chain as it describes its segmental motion will, when it encounters an ion pair, transfer an ion to the pair, forming a new triple and leaving an ion pair behind. The distinction between the two mechanisms is not in the segmental motion but in the fact that in one case the entire triple is 'passed on' whereas in the other mechanism only a single cation or anion is transferred. In the second mechanism the triple ions and ion pairs may be *immobile*. As the concentration of the salt increases and ion aggregates come into closer proximity, the second mechanism is likely to become more important. The second mechanism is equally applicable to cation or anion triples and may, at relatively high concentrations, also involve transfer of individual ions between larger aggregates which are themselves immobile.

The results presented in Fig. 6.9 for a solid polymer electrolyte follow the same trend as that obtained for electrolytes based on low molecular weight liquid polymers, Fig. 6.10 (MacCallum, Tomlin and Vincent, 1986; Cameron, Ingram and Sorrie, 1987). The liquid polymer systems are very similar to conventional non-aqueous electrolytes which also show a

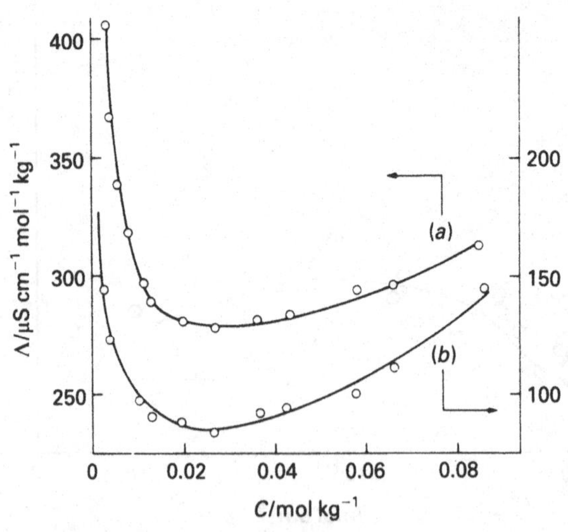

Fig. 6.10 Molal conductivity as a function of concentration for (*a*) $LiClO_4$ and (*b*) $LiCF_3SO_3$ in PEO (400) at 25 °C (MacCallum, Tomlin and Vincent, 1986, with permission).

minimum in Λ at low concentrations. Thus we have established a close link between ion association in low dielectric constant electrolytes with solid and liquid solvents, at least at concentrations below approximately 0.1 mole dm^{-3}.

The measurement of Λ vs concentration provides no evidence as to the nature of the ion pairs which form, i.e. whether they are contact or solvent separated species. Also, the mobility of the ion pairs does not influence the results. Contact ion pairs are likely to be more mobile than those separated by solvent since the latter include a section of at least one polymer chain. However, it is possible to envisage mechanisms, involving concerted motion of the cation and anion of a solvent separated pair, which would allow the effective movement of the neutral pair. This is also true for contact vs solvent separated triples. Measurements to be discussed below, involving the dc polarisation of cells, are capable of distinguishing between mobile and immobile pairs.

dc polarisation

Molar conductivity measurements are equally applicable to both solid and liquid electrolytes. In contrast, the measurement of current flowing through an electrochemical cell on a time scale of minutes or hours while the cell is perturbed by a constant dc potential is only of value for solid solvents (Bruce and Vincent, 1987) where convection is absent. Because of the unique aspects of dc polarisation in a solid solvent this topic is treated in some detail in this chapter. Let us begin by considering a cell of the form:

$$\overset{+}{M}/M^{+}X^{-}/\overset{-}{M}$$

consisting of metal electrodes M between which is placed an electrolyte, MX, that fully dissociates into M^+ cations, and X^- anions. The cations are electroactive towards the electrodes, the X^- anions are not. On application of a dc potential M^+ cations are transported under the influence of the electric field towards the negative cathode while at the same time X^- anions are transported by the field towards the positive anode; this is termed migration (Fig. 6.11(*a*)). The initial current due to the ion migration when divided by the applied potential yields the bulk conductivity of the electrolyte. At longer times (in the region of a few seconds) cations arriving at the cathode, and those left behind by anions migrating away from it, are consumed, while an equivalent number of cations are produced at the anode. Electrolysis is occurring and concentration gradients are formed in the vicinity of each electrode. Transport near the electrodes is now influenced by diffusion down the concentration gradients as well as by migration (Fig. 6.11(*b*)). The current flowing through the cell will change with time due to the continuous extension of the diffusion layers out into the bulk electrolyte. Such semi-infinite diffusion results in the current decaying with a square root dependence on time. In the case of liquid electrolytes the extent of the diffusion layers is limited by the onset of convection, so that diffusion is confined to a region near the electrodes. The onset of such convection limits the normal time scale of the experiment in liquids, usually to some tens of seconds unless the convection is controlled by, for example, the use of a rotating disc electrode. In contrast, with a solid solvent the diffusion layers continue to grow towards each other until they merge and a steady state is reached (Fig. 6.11(*c*)). In this condition the net anion flux is zero since anion migration due to the field from right to left is exactly balanced by diffusion from left to right down the concentration gradient. The current in the steady state is due only to the migration and diffusion of cations. Such steady state measurements can reveal the nature of ion association; we will therefore consider them in more detail. Expressions have been derived for the steady state current, I_s^+, flowing through such a solid polymer electrolyte cell and for the potential, ΔV, across the cell, in terms of the salt concentrations at the anode and cathode, c_a and c_c (Bruce, Evans and Vincent, 1987).

$$\Delta V = \frac{2RT}{F} \ln\left(\frac{c_a}{c_c}\right) \tag{6.26}$$

$$I_s^+ = -2FD_+(c_a - c_c) \tag{6.27}$$

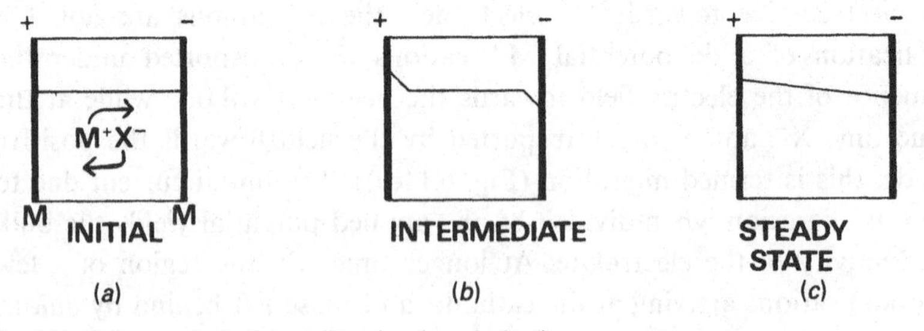

Fig. 6.11 Polymer electrolyte cell under dc polarisation.

where D_+ is the diffusion coefficient of the cations and the other terms have their usual meaning. The expression for potential follows the form of the Nernst equation, and that for the current is related to Fick's first law of diffusion. The derivation of these equations is presented in the appendix to this chapter; it is based on the assumption of a fully dissociated ideal electrolyte, i.e. ion interactions are ignored, and reversible electrode reactions are assumed. In a dc polarisation experiment the steady state current is measured at a given potential, therefore we wish to relate I_s^+ and ΔV directly. In order to do this it is necessary to eliminate the concentration terms, and this is possible only for small ion concentration differences between the anode and cathode. Under these circumstances c_a/c_c approaches unity and thus the $\ln(c_a/c_c)$ term may then be linearised, permitting elimination of c_a and c_c between the two expressions. Further, the ratio of $I_s^+/\Delta V$ is then independent of the applied voltage and may be regarded as the effective conductivity of the cell in the steady state. On the other hand, when ΔV exceeds about 20 mV, the ion concentration difference at anode and cathode is no longer sufficiently small to permit this approximation. A consequence of this is that at higher applied voltages, the steady state current increases more slowly than that of an ohmic system and hence the effective conductivity falls below the limiting value measured at low ΔV.

Although examining an ideal electrolyte is helpful in developing our understanding of dc polarisation, polymer electrolytes are not ideal systems since interactions between the ions of the salt are always likely to be significant in a medium of such low permittivity. It is therefore necessary to take into account two effects:

(1) long range interactions between the ions;
(2) long lived associated species, i.e. ion pairs, triples or higher aggregates.

For a fully dissociated but non-ideal polymer electrolyte (i.e. long range ion interactions are present but not ion association) the following expressions for the steady state potential ΔV, and current I_s^+, may be derived, again assuming reversible electrode behaviour:

$$\Delta V = (1 + G)(d \ln a_\pm / d \ln c)(RT/F) \ln(c_a/c_c) \quad (6.28)$$

$$I_s^+ = -(1 + G)(d \ln a_\pm / d \ln c) F D_+ (c_a - c_c) \quad (6.29)$$

The equations are similar to those for an ideal electrolyte but with the addition of two terms. G depends on the coupling of the cation and anion fluxes and is related to the cross coefficient in the transport equations of irreversible thermodynamics from which these equations are derived, $a_{\pm}$ is the ionic activity and $d \ln a_{\pm}/d \ln c$ is the thermodynamic enhancement factor (Bruce, 1991). The details of the derivation of these equations from irreversible thermodynamics are given in the reference. For dilute electrolytes, G and $d \ln a_{\pm}/d \ln c$ both tend to unity and thus I_s^+ remains linearly related to ΔV only up to a maximum of about 20 mV. At higher ion concentrations, however, both G and $d \ln a_{\pm}/d \ln c$ exceed unity with the result that the predicted linearity limit increases above the 20 mV value.

The situation in an electrolyte which contains mobile associated species is complex. Hardgrave (1990) has identified 16 separate situations involving free ions, ion pairs and triple ions. Bruce, Hardgrave and Vincent (1989) have developed equations relating the steady state current and the applied voltage for the simplest case of an associated electrolyte, namely one in which ion pairs are the only associated species, and where the ions are assumed to behave ideally. For such a system, the voltage and steady state current are given by the following expressions:

$$\Delta V = (2RT/F)[\ln(c_a/c_c) + KD_0(c_a - c_c)/D_-] \qquad (6.30)$$

$$-I_s^+ = 2FD_+(c_a - c_c) - (FKD_0/D_-)(c_a^2 - c_c^2)(D_+ + D_-) \qquad (6.31)$$

where K is the association constant for the formation of the ion pair MX, D_0 is the ion pair diffusion coefficient, and D_- is the diffusion coefficient of the anion. Again it is possible to predict the voltage range for which a cell containing such an electrolyte would show an ohmic response. If D_0 is relatively large, the migration of the anion in one direction at steady state is balanced by the diffusion of both the anion and the ion pairs (each of which carries the X^- constituent) in the opposite direction. This permits large potential differences to be applied without necessarily inducing a large ion concentration gradient. Under these circumstances the logarithmic term may be linearised and the ohmic range may be extended to several volts!

The results of dc polarisation measurements on the cell

$$Li(s)/LiClO_4 \text{ in solid poly(ethylene oxide)}/Li(s)$$

at 120 °C and for a range of potentials are presented in Fig. 6.12 (Bruce, Hardgrave and Vincent, 1992a). For each salt concentration, the cell was polarised at a constant dc voltage until a steady state current was reached

The experiment was then repeated at a series of increasing applied voltages, with the steady state current being recorded in each case. The applied voltage was always corrected for the resistance of the passivating layers on the Li electrodes by the use of ac impedance measurements, as described previously (Bruce *et al.*, 1987) so that the experimental results could be compared with the equations for reversible electrodes. A plot was then constructed of steady state current against corrected applied voltage and the limit of linearity obtained (Fig. 6.12). In Fig. 6.13 data are presented in which the limit of linearity of the I_s^+ vs ΔV response is plotted for a range of lithium perchlorate concentrations in poly(ethylene oxide) at 120 °C. Dividing Fig. 6.13 into low and high concentration regions and considering first the former, it is evident that the limit of linearity far exceeds 20 mV despite the fact that at these low concentrations the electrolyte is likely to approach ideal behaviour (i.e. long range ion interactions will be small). These results can be rationalised by postulating the existence of electrically neutral $LiClO_4$ ion pairs. As noted above, provided that the diffusion coefficient and the concentration

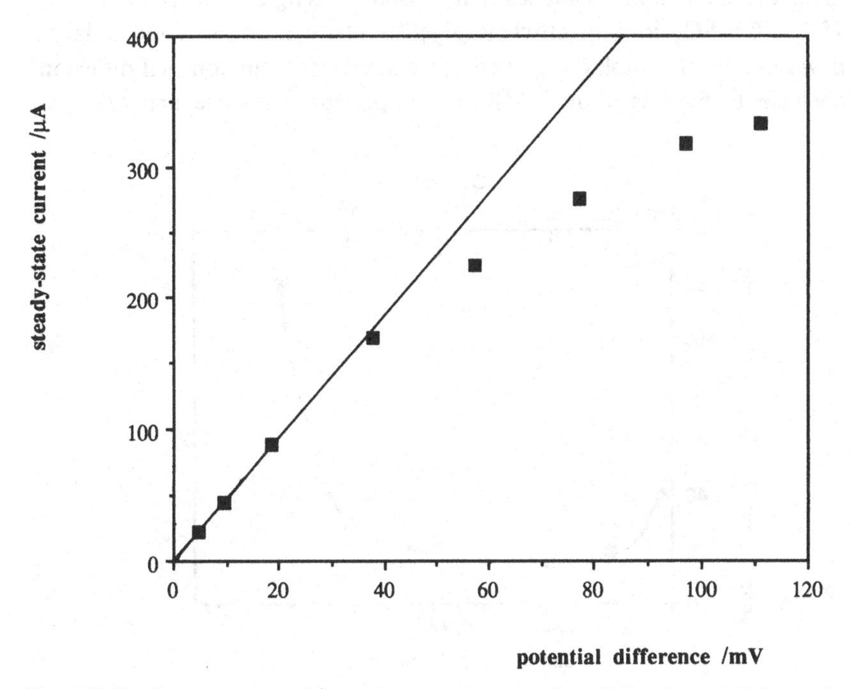

Fig. 6.12 Steady state current, I_s^+ as a function of corrected applied voltage, ΔV, for a cell based on 50:1 PEO–$LiClO_4$ at 120 °C (Bruce *et al.*, 1992a).

of the ion pairs are comparable with or greater than the corresponding values for the ions, an extended linearity range is expected. This hypothesis is in agreement with the explanation offered earlier in this subsection for the measurements of equivalent conductance at very low concentrations in an amorphous high molecular weight methoxy-linked poly(ethylene oxide). However, unlike the measurement of Λ, the dc polarisation results require the ion pairs to be *mobile* so that important additional information is gained from the dc measurements.

In the O:M range of 100:1 to 18:1 (0.2–1.3 mole dm^{-3}) the linearity limit corresponds to a system where mass transport is dominated by charged ionic species (simple or complex). It is possible to obtain $d \ln a_{\pm}/d \ln c$ values from published data (Bouridah, Dalard, Deroo and Armand, 1986) but G has only been estimated from measurements on aqueous systems (Katchalsky and Curran, 1965). Using reasonable approximations, however, it is not possible to explain the linear region of 120 mV found at the concentration of 15:1 simply by deviations from ideality. Again, the simplest explanation is that some form of ion pair is involved. This finding matches that of Boden, Leng and Ward (1991) for 15:1 $LiCF_3SO_3$ in poly(ethylene glycol). These authors noted a large deviation in the molar conductivity calculated from ion self-diffusion coefficients measured by NMR in comparison with the experimental

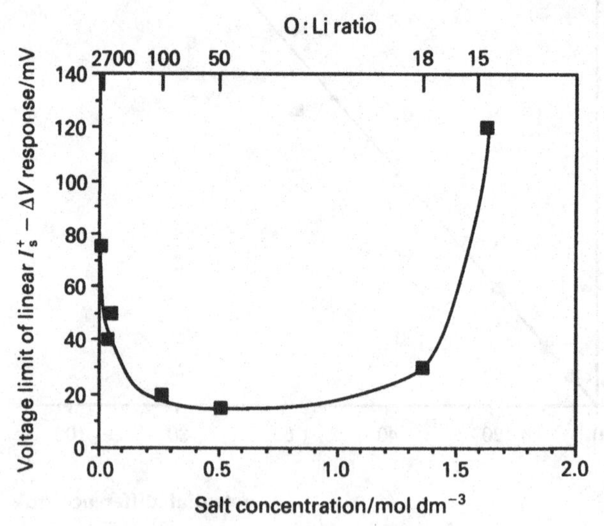

Fig. 6.13 I_s^+ vs ΔV linearity limits from the current/voltage behaviour of Li/$LiClO_4$–PEO/Li cells at 120 °C (Bruce *et al.*, 1992a).

values. They explained the deviation as being due to 'correlated diffusive motion of pairs of positively and negatively charged ions' – i.e. of ion pairs. At these high concentrations ion pairs, triples or larger clusters even if incapable of long range transport can contribute to an enhancement of the relative permittivity (a rise in permittivity from 3.1 in pure dimethyl carbonate, to 6.3 in a 0.29 mol dm^{-3} $LiClO_4$ solution, has been noted previously (Delsignore, Faber and Petrucci 1985)). Under these circumstances redissociation of the aggregates on increasing the concentration in these highly concentrated electrolytes may occur. However, it might also be borne in mind that the concept of discrete aggregates, including ion pairs, loses physical significance when the ions are separated by only a few Å as is the case at salt concentrations equivalent to 18 ether oxygens per Li^+ ion. While the evidence supporting ion association at low salt concentrations is strong, it must be stressed that the results of the dc polarisation measurements of high salt concentrations are based on certain assumptions, in particular the value of G. Other interpretations of the results at high concentrations cannot be unequivocally rejected.

6.4.2 *Transport*

Transport, in the context of electrolytes, refers to the movement of the species which form when a salt is dissolved in a solvent. As an illustration, if a uni–univalent salt MX dissolves in a polymer forming the species M^+, X^-, $[MX]^0$, $[M_2X]^+$, $[MX_2]^-$, then transport measurements are directed towards an understanding of the movement of each of the five species, which may vary in both their mobility and relative concentration. A distinction is drawn between charged and neutral species: the former can be transported by both an electric field and a concentration gradient whereas the latter can be transported only by a gradient in concentration. The terms transport number and transference number, both of which are used in this section, are often confused elsewhere. First they both refer exclusively to the transport of charged species. If a potential is applied to an electrolyte and the current measured, then the transport number, t, of any charged species is the proportion of the current carried by that species or equivalently the proportion of the overall electrolyte conductivity due to the species. The sum of the transport numbers for all the species present is equal to unity. A transference number, T, on the other hand refers to the proportion of the current carried by a *constituent* of the salt (Spiro, 1971). Taking our uni–univalent MX salt which is composed of the basic

Table 6.1. *Methods for the measurement of transport in polymer electrolytes*

I	II	III
Hittorf/Tubandt	radiotracer	dc polarisation
concentration cell	PFG NMR	ac methods
centrifugal	Cottrell	

constituents, $\mathbf{M}^+$ and $\mathbf{X}^-$, then when dissolved in the polymer the $\mathbf{X}^-$ constituent exists as an X^- species but also as part of the $[M_2X]^+$ cation and $[MX_2]^-$ anion triples. The constituent is also present in the $[MX]^0$ ion pair; however this does not influence the transference number. The $\mathbf{X}^-$ constituent is transported by all three charged species they in turn carry the charge on the constituent around in the cell. For the passage of one Faraday of charge across the cell, the net number of Faradays carried by the $\mathbf{X}^-$ constituent is its transference number. The sum of the cation and anion transference numbers is equal to unity. For a salt dissociating fully into species M^+ and X^- then the transport and transference numbers of the cations or of the anions are identical.

Conductivity measurements are often the first to be carried out on an electrolyte; however, they provide information only on the total transport of charge. Even in a fully dissociated electrolyte, such measurements do not differentiate between the current carried by the cations and the anions. Transport or transference measurements attempt to probe more deeply into the movement of species in electrolytes.

The different techniques which have been applied to determine transport in polymer electrolytes are listed in Table 6.1. For a fully dissociated salt all the techniques yield the same values of t (small differences may arise due to second order effects such as long range ion interactions or solvent movement which may influence the different techniques in different ways). In the case of associated electrolytes, any of the techniques within one of the three groups will respond similarly, but the values obtained from different groups will, in general, be different. Space does not permit a detailed discussion of each technique, this is available elsewhere (see Bruce and Vincent (1989) and the references cited therein). However, we will consider one technique from each group to illustrate the differences. A solid polymer electrolyte containing an associated uni–univalent salt is assumed.

Hittorf/Tubandt measurements

This method involves the passage of a measured quantity of charge through a cell and subsequent determination of changes in the composition of the electrolyte in the vicinity of the anode and cathode. The technique uses cells of the type

$$\mathrm{M(s)/polymer{-}MX(s)/polymer{-}MX(s)/polymer{-}MX(s)/M(s)}$$

in which all three electrolyte compartments are identical. On passage of 1 F of charge, 1 mol of M is stripped from the anode and is deposited on the cathode. Assuming that the electrolyte contains M^+, X^- and $[MX]^0$ then current is carried by the motion of M^+ towards the cathode and of X^- towards the anode. Provided that the central compartment remains invariant throughout the experiment, neutral ion pairs are not involved in the flux between the electrode compartments and the central compartment. Measurement of the change in salt concentration in the cathode compartment gives directly the change in the concentration of X. In the case of a completely dissociated electrolyte, this would lead to a direct determination of the anion transport number. This is also the situation for an electrolyte with M^+, X^- and $[MX]^0$. However, for an associated electrolyte with M^+, X^-, $[MX]^0$, $[M_2X]^+$ and $[MX_2]^-$ species, analysis of the cathode compartment leads only to the determination of the net transfer of the X^- constituent of the salt due to the transport of $[M_2X]^+$ into the compartment and the transport of X^- and $[MX_2]^-$ out. For the system considered, the transference number for the X^- constituent, T_{X^-} (given by the change in the number of moles of X in the cathode compartment) may be related to the individual transport number for X-containing species by

$$T_{X^-} = t_{X^-} + 2t_{MX_2^-} - t_{M_2X^+}). \tag{6.32}$$

Similarly

$$T_{M^+} = (t_{M^+} + 2t_{M_2X^+} - t_{MX_2^-}) \tag{6.33}$$

and,

$$T_{M^+} + T_{X^-} = 1. \tag{6.34}$$

To date there have been few reliable measurements of Hittorf transference numbers in solid polymer electrolytes because of experimental difficulties in applying the technique. Leveque, Le Nest, Gandini and Cheradame (1983) have, however, applied it to highly cross-linked networks where cells could be formed using a series of non-adherent thin sections. When the anion was immobilised by attachment to the polymer host, the transport number of the cation was shown to be equal to unity. Recently, Bruce, Hardgrave and Vincent (1992b) have successfully carried out transference number measurements on poly(ethylene oxide) containing $LiClO_4$, demonstrating that Hittorf measurements may be made on non-cross-linked polymers. For this polymer electrolyte at 120 °C, and with an ethylene oxide to lithium ratio of 8:1, $T_+ = 0.06 \pm 0.05$. This is consistent with the view that the anions are much more mobile than the cations, since the former species are more loosely bound to the polymer chains than the latter.

Radiotracer measurements

These measurements are based on following the progress of radioactively labelled nuclei of the salt constituents as they diffuse through the polymer electrolyte, the labelled salt having first been deposited as a thin layer on the polymer surface. Experiments using serial-sectioning techniques have been performed by Chadwick, Strange and Worboys (1983) to study, for example, PPO–NaSCN polymer electrolytes by measuring the distribution of ^{22}Na and ^{14}C. If we limit our consideration to the situation of a polymer electrolyte containing free ions, M^+ and X^-, in equilibrium with ion pairs, $[MX]^0$, then on introducing radioactively labelled nuclei of constituent $\mathbf{M}^+$, they are partitioned between the M^+ and $[MX]^0$ species, and similarly labelled nuclei of constituent $\mathbf{X}^-$ are distributed between the ion pairs and the X^- anions. All the labelled species will diffuse throughout the polymer. When the concentration or mobility of the ion pairs is negligibly small compared with that of the ions then the measured diffusion coefficients of $\mathbf{M}^+$ and $\mathbf{X}^-$ correspond respectively to that of the M^+ and X^- ions. In this case the cation transport number may be defined as

$$t_+ = D_+/(D_+ + D_-) \tag{6.35}$$

However, when the concentration or mobility of ion pairs is significant compared with the individual ions then the measured diffusion coefficients for both constituents approach that of the ion pairs and not the free ions and as a consequence the apparent t_+, and hence t_-, approach 0.5. In fact it is no longer valid to apply the above equation in order to determine transport numbers. Generally, in the presence of mobile ion pairs or more complex mobile ion clusters, diffusion coefficients and t_+ measurements obtained from the techniques in this group do not correspond to individual species within the electrolyte.

The pulsed field gradient nuclear magnetic resonance (PFG NMR) technique is experimentally distinct from the radiotracer technique but the principle is very similar. A fraction of the nuclei of each constituent is labelled by flipping their spins and monitoring the diffusion of these species. Both techniques are influenced by the presence of neutral associates unlike the Hittorf/Tubandt method.

The current fraction

The dc polarisation technique has been described in Section 6.4.1. For an electrolyte with no mobile associated species (such as $[MX]^0$ or $[M_2X]^+$) the ratio of the steady state current I_s^+ to the initial current, I_0, for the above cell at low values of the polarisation voltage, ΔV, is equal to the transport number of the cation, t_+. If the electrolyte contains mobile ion pairs or other uncharged associated species, the interpretation of the current ratio is more complicated, since in the steady state, transport of the cation constituent across the polymer film and hence the steady state current includes a contribution from the diffusion of the uncharged species in the concentration gradient. Nonetheless whatever the state of the electrolyte, the ratio is of considerable value in characterising ion transport in thin film cells which are similar in nature to those in practical electrochemical devices, since in such devices it is the net passage of the electroactive constituent of the salt, e.g. lithium across the cell during discharge, which is important, not the bulk electrolyte conductivity.

Since the electrolyte may contain associated species, we choose to define the general term *current fraction* as I_s^+/I_0, assuming that interfacial resistances, which may change during the course of an experiment, have been allowed for. Because the steady state current is not a linear function of the applied potential difference above some undefined potential, the above parameter is generally potential-dependent. However, because electrolytes display a linear, steady state, current–applied potential difference response up to at least 20 mV we may define a limiting current fraction, F_+, as

$$F_+ = \lim_{\Delta V \to 0} I_s^+/I_0 \qquad (6.36)$$

This parameter is then a potential-independent property of the electrolyte at a given temperature over some unspecified potential difference range.

The limiting current fraction is the maximum fraction of the initial current which may be maintained at steady state in the absence of interfacial resistances. In specific circumstances this parameter may be equal to the transport or transference number of particular species, but without *a priori* knowledge of the species present in an electrolyte it is preferable that F_+ values are referred to, rather than t_+ or T_+ values. For polyether electrolytes containing $LiClO_4$ values of 0.2–0.3 are often observed.

6.4.3 Summary

A combination of electrochemical and spectroscopic measurements indicate that ion association is prevalent in solid polymer electrolytes as it is in liquid electrolytes with a low dielectric constant. For simple uni–univalent salts in polyethers measurement of molar conductivity vs salt concentration indicates that at very low concentrations ($\ll 10^{-3}$ mole dm^{-3}) free ions dominate, these are increasingly replaced by ion pairs on raising the salt concentration, up to 10^{-2} mole dm^{-3}. These ion pairs are mobile as indicated by the wide linearity range of the dc polarisation experiments. At higher concentrations (up to $\sim 10^{-1}$ mole dm^{-3}) triple ions appear to replace ion pairs. The triple ions may move as a discrete entities or transfer a single cation or anion to a neighbouring ion pair. The second mechanism is likely to become more important with increasing concentration. Dc polarisation measurements above ~ 1 mole dm^{-3} also suggest that neutral associates, i.e. pairs or larger clusters, may again be present; however, at such high concentrations the system is best regarded as a solvated molten salt. As a result of ion association, techniques aimed at determining transport or transference numbers do not all yield similar results but application of several different techniques for determining transport can yield useful information on the nature of association in polymer electrolytes.

Appendix

Derivation of expressions for the potential, ΔV, and steady state currents, I_s^+

Consider a cell consisting of a polymer electrolyte, in which the salt, MX, dissociates fully into M^+ and X^- ions, the two electrodes, M, are reversible to M^+. The cell constant is unity, and the electrolyte is ideal.

In the steady state the flux of X^- ions is zero and their electrochemical potential is everywhere equal.

$$\mathrm{d}\tilde{\mu}_- = RT\,\mathrm{d}\ln c - F\,\mathrm{d}\phi = 0, \tag{A6.1}$$

where $\tilde{\mu}_-$ is the electrochemical potential of the anions, c the salt concentration (electroneutrality imposes, $c = c_+ = c_-$) and ϕ the potential at any point in the electrolyte.

The gradients of electrical and chemical potentials are related as

$$\frac{\mathrm{d}\phi}{\mathrm{d}x} = \frac{RT}{F}\frac{\mathrm{d}\ln c}{\mathrm{d}x} = \frac{RT}{F}\left(\frac{1}{c}\right)\frac{\mathrm{d}c}{\mathrm{d}x}, \tag{A6.2}$$

where x is any position between the electrodes from cathode to anode. Also

$$I_s^+ = (I_s^+)^{\mathrm{d}} + (I_s^+)^{\mathrm{m}}, \tag{A6.3}$$

where d and m refer to the currents due to diffusion and migration. Now $(I_s^+)^{\mathrm{d}}$ may be obtained from Fick's first law and $(I_s^+)^{\mathrm{m}}$ from the usual equation for conductivity in a potential gradient, so that

$$I_s^+ = -FD_+\frac{\mathrm{d}c}{\mathrm{d}x} - \left(\frac{F^2D_+}{RT}\right)c\frac{\mathrm{d}\phi}{\mathrm{d}x}, \tag{A6.4}$$

where D_+ is the cation diffusivity. Hence, on substituting for $\mathrm{d}\phi/\mathrm{d}x$ from Eqn (A6.2)

$$I_s^+ = -2FD_+\,\mathrm{d}c/\mathrm{d}x$$

This equation has a number of important implications for the steady state current, I_s^+: since $I_s^+ = 2(I_s^+)^{\mathrm{d}} = 2(I_s^+)^{\mathrm{m}}$, we have $(I_s^+)^{\mathrm{d}} = (I_s^+)^{\mathrm{m}}$. At steady state the current flow is constant throughout the electrolyte: furthermore, since the electrolyte is ideal, D_+ is also constant. Hence from Eqn (A6.4), the concentration varies linearly with x

$$\mathrm{d}c/\mathrm{d}x = \Delta c/\Delta x = c_{\mathrm{a}} - c_{\mathrm{c}}, \tag{A6.5}$$

since

$$\Delta x = 1.$$

Now the concentration, c, at any point in the electrolyte is given by

$$c = (c_{\mathrm{a}} - c_{\mathrm{c}})x + c_{\mathrm{c}}. \tag{A6.6}$$

It may also be noted that

$$c_{\mathrm{a}} + c_{\mathrm{c}} = 2c_0. \tag{A6.7}$$

We may now calculate the potential drop across the electrolyte in the steady state.

$$\frac{d\phi}{dx} = -(I_s^+)^m\left(\frac{RT}{F^2D_+}\right)\frac{1}{c} = -(I_s^+)^m\left(\frac{RT}{F^2D_+}\right)\frac{1}{(c_a - c_c)x + c_c}, \quad (A6.8)$$

$$\therefore \quad \phi_a - \phi_c = \Delta\phi = (I_s^+)^m\left(\frac{RT}{F^2D_+}\right)\int_0^1 \frac{dx}{(c_a - c_c)x + c_c}$$

$$= -(I_s^+)^m\left(\frac{RT}{F^2D_+}\right)\frac{1}{(c_a - c_c)}\ln\left(\frac{c_a}{c_c}\right). \quad (A6.9)$$

But

$$(I_s^+)^m = (I_s^+)^d = -FD_+\,dc/dx = FD_+(c_a - c_c), \quad (A6.10)$$

$$\therefore \quad \Delta\phi = (RT/F)\ln(c_a/c_c).$$

Also the concentration difference between the anode and cathode generates a Nernst potential, $\Delta E = RT/F\ \ln(c_a/c_c)$, therefore

$$\Delta V = 2\Delta\phi = (2RT/F)\ln(c_a/c_c) \quad (A6.11)$$

and from Eqn (A6.10)

$$I_s^+ = -2FD_+(c_a - c_c).$$

References

Adams, G. and Gibbs, J. H. (1965) *J. Phys.*, **43**, 139.

Angell, C. A. and Bressel, R. D. (1972) *J. Phys. Chem.*, **76**, 3244.

Armand, M. B. and Gauthier, M. (1989) in *High Conductivity Solid Ionic Conductors: Recent Trends and Applications*, Ed. Takahashi, T. World Scientific, Singapore, p. 114.

Armand, M. B., Chabagno, J. M. and Duclot, M. (1978) *2nd Int. Conference on Solid Electrolytes*, St. Andrews.

Armand, M., Chabagno, J. M. and Duclot, M. (1979) *Fast Ion Transport in Solids*, Eds. Vashishta, P., Mundy, J. N. and Shenoy, G. K., North-Holland, Amsterdam, p. 131.

Boden, N., Leng, S. A. and Ward, I. M. (1991) *Solid State Ionics*, **45**, 261.

Bouridah, A., Dalard, F., Deroo, D. and Armand, M. B. (1986) *Solid State Ionics*, **18/19**, 287.

Bruce, P. G. (1989) *Faraday Discuss. Chem. Soc.*, **88**, p. 91.

Bruce, P. G. (1991) *Synthetic Metals*, **45**, 267.

Bruce, P. G., Evans, J. and Vincent, C. A. (1987) *Polymer*, **28**, 2324.

Bruce, P. G., Hardgrave, M. T. and Vincent, C. A. (1989) *J. Electroanal. Chem.*, **271**, 27.

References

Bruce, P. G., Hardgrave, M. T. and Vincent, C. A. (1992a) *Electrochimica Acta*, **37**, 1517.
Bruce, P. G., Hardgrave, M. T. and Vincent, C. A. (1992b) *Solid State Ionics*, **53/56**, 1087.
Bruce, P. G., Krok, F. and Vincent, C. A. (1988) *Solid State Ionics*, **27**, 81.
Bruce, P. G., Nowinski, J. L., Gray, F. M. and Vincent, C. A. (1990) *Solid State Ionics*, **38**, 231.
Bruce, P. G. and Vincent, C. A. (1987) *J. Electroanal. Chem.*, **225**, 1.
Bruce, P. G. and Vincent, C. A. (1989) *Faraday Disc. Chem. Soc.*, **88**, 43.
Bruce, P. G. and Vincent, C. A. (1993) *J. Chem., Soc., Faraday Trans*, **89**, 3187.
Cameron, G. G. and Ingram, M. D. (1989) *Polymer Electrolyte Reviews*, Vol. 2, Eds. MacCallum, J. R. and Vincent, C. A., Elsevier Applied Science, London, p. 157.
Cameron, G. G., Ingram, M. D. and Sorrie, G. A. (1987) *J. Chem. Soc., Faraday Trans.*, **83**, 3345.
Chadwick, A. V., Strange, J. H. and Worboys, M. K. (1983) *Solid State Ionics*, **9/10**, 1155.
Chatani, Y. and Okamura, S. (1985) *Polymer*, **28**, 1815.
Chatani, Y., Fujii, Y., Takayanagi, T. and Honma, A. (1990) *Polymer*, **31**, 2238.
Cheradame, H. and Le Nest, J. F. (1987) *Polymer Electrolyte Reviews – 1*, Eds. MacCallum, J. R. and Vincent, C. A., Elsevier Applied Science Publishers, London, p. 103.
Cohen, M. H. and Turnbull, D. (1959) *J. Chem. Phys.*, **31**, 1164.
Cowie, J. M. G. and Cree, S. H. (1989) *Ann. Rev. Phys. Chem.*, **40**, 85.
Cowie, J. M. G., Martin, A. C. S. and Firth, A. M. (1988) *British Polymer J.*, **20**, 247.
Craven, J. R., Mobbs, R. H., Booth, C. and Giles, J. R. M. (1986) *Macromol. Chem. Rapid Commun.*, **7**, 81.
Delsignore, M., Faber, H. and Petrucci, S. (1985) *J. Phys. Chem.*, **89**, 4968.
Fenton, D. E., Parker, J. M. and Wright, P. V. (1973) *Polymer*, **14**, 489.
Fontanella, J. J., Wintersgill, M. C., Smith, M. K., Semancik, J. and Andeen, C. G. (1986) *J. Appl. Phys.*, **60**, 2665.
Frech, R., Manning, J., Teeters, D. and Black, B. E. (1988) *Solid State Ionics*, **28/30**, 954.
Fulcher, G. S. (1925) *J. Amer. Ceram. Soc.*, **8**, 339.
Fuoss, R. M. and Kraus, C. A. (1933) *J. Am. Chem. Soc.*, **55**, 2387.
Gibbs, J. H. and di Marzio, E. A. (1958) *J. Chem. Phys.*, **28**, 373.
Gray, F. M. (1991) *Polymer Electrolytes: Fundamentals and Technological Applications*, VCH Publishers, New York.
Gray, F. M., Shi, J., Vincent, C. A. and Bruce, P. G. (1991) *Phil. Mag. A.*, **64**, 1091.
Hardgrave, M. T. (1990) PhD Thesis, University of St. Andrews.
Katchalsky, A. and Curran, P. F. (1965) *Non-Equilibrium Thermodynamics in Biophysics*, Harvard University Press, Harvard.
Koningsberger, D. C. and Prins, R. (Eds.) (1988) *X-Ray Absorption*, Wiley, Chichester.
Kortüm, G. (1965) *Treatise on Electrochemistry*, Elsevier, Amsterdam.

Latham, R. J., Linford, R. G. and Schlindwein, W. S. (1989) *Faraday Discuss. Chem. Soc.*, **88**, 103.
Leveque, M., Le Nest, J.-F., Gandini, A. and Cheradame, H. (1983) *Macromol. Chem. Rapid Commun.*, **4**, 497.
Lightfoot, P., Mehta, M. A. and Bruce, P. G. (1992) *J. Mater. Chem.*, **2**, 379.
Lightfoot, P., Mehta, M. A. and Bruce, P. G. (1993) *Science*, **262**, 883.
Lightfoot, P., Nowinski, J. N. and Bruce, P. G. (1994) *J. Am. Chem. Soc.* in press.
MacCallum, J. R. and Vincent, C. A. (1987) *Polymer Electrolyte Reviews – 1*, Elsevier Applied Science Publishers, London.
MacCallum, J. R. and Vincent, C. A. (1989) *Polymer Electrolyte Reviews – 2*, Elsevier Applied Science Publishers, London.
MacCallum, J. R., Tomlin, A. S. and Vincent, C. A. (1986) *Eur. Polymer J.*, **22**, 787.
Mehta, M. A. (1993) PhD Thesis, University of St. Andrews.
Mehta, M. A., Lightfoot, P. and Bruce, P. G. (1993) *Chem. of Materials*, **5**, 1338.
Miyamoto, T. and Shibayama, K. (1973) *J. Appl. Phys.*, **44**, 5372.
Papke, B. L., Ratner, M. A. and Shriver, D. F. (1982) *J. Electrochem. Soc.*, **129**, 1694.
Pearson, R. G. (1963) *J. Am. Chem. Soc.*, **85**, 97.
Ratner, M. A. (1987) *Polymer Electrolyte Reviews – 1*, Eds. MacCallum, J. R. and Vincent, C. A., Elsevier Applied Science Publishers, London, p. 173.
Ratner, M. A. and Nitzan, A. (1989) *Farad. Discuss. Chem. Soc.*, **88**, 19.
Shriver, D. F., Dupon, R. and Stainer, M. (1983) *J. Power Sources*, **9**, 383.
Spiro, M. (1971) in *Physical Methods of Chemistry*, Eds. Weissberger, A. and Rossiter, B. W., Vol. 1, Part II A, Ch. IV, Wiley-Interscience, New York.
Tamman, G. and Hesse, W. (1926) *Z. Anorg. Allg. Chem.*, **156**, 245.
Thomson, J. B., Lightfoot, P., Nowinski, J. L. and Bruce, P. G. (1994) personal communication.
Torell, L. M. and Schantz, S. (1989) *Polymer Electrolyte Reviews – 2*, Eds MacCallum, J. R. and Vincent, C. A., Elsevier Applied Science, London, p. 1.
Vogel, H. (1921) *Phys. Z.*, **22**, 645.
Watanabe, M. and Ogata, N. (1987) *Polymer Electrolyte Reviews – 1*, Ed. MacCallum, J. R. and Vincent, C. A., Elsevier Applied Science Publishers, London, p. 39.
Williams, M. L., Landel, R. F. and Ferry, J. D. (1955) *J. Amer. Chem. Soc.*, **77**, 3701.

7 Insertion electrodes I: Atomic and electronic structure of the hosts and their insertion compounds

W. R. McKINNON
National Research Council of Canada, Ottawa

7.1 Important aspects of ionic and electronic structure

Intercalation or insertion compounds are solids made of host atoms and guest atoms (or molecules). The host atoms provide a lattice or framework; the guest atoms occupy sites within this framework. Two properties distinguish intercalation compounds from other solids: the guests are mobile, moving between sites in the host lattice; and the guests can be added to the host or removed from it, so the concentration of guests can change. These two properties are exploited when intercalation compounds are used as electrodes in electrochemical cells.

There are many kinds of guests, ranging from protons through to divalent cations (Bruce, Krok, Nowinski, Gibson and Tavakkolik, 1991) and on to large organic molecules. I shall limit this chapter to alkali metal guests, and almost exclusively to lithium. There are also many kinds of hosts, and I shall focus on transition metal oxides or chalcogenides (sulphides, selenides, and tellurides). I shall also mention graphite intercalation compounds, but only briefly, for there are detailed review articles (Solin, 1982; Dresselhaus and Dresselhaus, 1981). Even within these limitations, I can only discuss a small fraction of the known compounds. The examples chosen illustrate the fundamental chemistry and physics of intercalation, which is the primary aim of this chapter.

In alkali metal intercalation compounds, the guest is ionised in the host, donating its outer s electron to the host's electronic energy levels. Thus there are two aspects to consider: the sites where the ion resides, and the energy levels or bands that the electron occupies. Guests such as water that remain neutral will only be discussed in the section on cointercalation. In some hosts, notably graphite, some guests accept electrons from the host to become negative ions. Much of the theory in this chapter could be applied to these acceptors with a few obvious changes.

7.1.1 *Sites for ions*

In a transition metal chalcogenide or oxide, positive guests like Li^+ occupy sites surrounded by negative chalcogen or oxygen ions, and distance themselves as far as possible from the positive transition metal ions. Since Li^+ has a filled outer core of electrons (unlike the transition metal ions in many of the hosts), the geometry of the site is not important as long as the anions are distributed evenly around the site. Thus Li^+ surrounded by four anions would prefer the anions to form a tetrahedron rather than a square.

In the compounds to be discussed here, the guests spend most of their time localised on a given site, only occasionally hopping from one site to another. Such compounds can be described by lattice-gas models. This description breaks down for guest ions so large that they fill several sites at once, as for Cs in graphite. There, it is more appropriate to consider the Cs as a liquid, weakly interacting with the periodic potential of the graphite lattice (Clarke, Caswell and Solin, 1979).

The sites available for ions depend on the structure, and a given structure can contain different types of sites. As an example, two types of site lie between close-packed chalcogen (or oxygen) planes, as shown in Fig. 7.1(*a*): octahedral sites, surrounded by six chalcogens, and tetrahedral sites, surrounded by four. Li generally prefers these two types of site. Larger ions like Na in Na_xTiS_2 (Moline *et al.*, 1984) may occupy trigonal prismatic sites, which are found when one close-packed layer sits directly over another (Fig. 7.1(*b*)). The size of an ion in a solid is an empirical quantity, and published ionic radii (Shannon, 1976) give a good indication whether or not an ion can fit into a given site.

Taken together, the sites for the guests form a lattice, which can be one-, two-, or three-dimensional, depending on how the sites are connected. The host, too, can be classified by the dimensionality of its strong bonds. Fig. 7.2 shows four possible geometries. If the host consists of chains weakly bonded together (Fig. 7.2(*a*)), the guests can move in all directions between the chains. This is a one-dimensional host with a three-dimensional lattice of guest sites. Conversely, the host can be strongly bonded in all directions and contain tunnels of sites (Fig. 7.2(*b*)). When the host is layered, so are the sites (Fig. 7.2(*c*)). Finally, a three-dimensional host can contain a three-dimensional lattice of sites (Fig. 7.2(*d*)). Section 7.2 discusses examples of all four cases.

The dimensionality of the host influences various properties. Layered hosts can accommodate large guests as well as small, and Li ions may carry large solvent molecules with them when they enter the layers from the solution in an electrochemical cell. Such cointercalation, discussed in Section 7.5.2, is not a problem in three-dimensional hosts, where the sites are usually too small to accommodate organic molecules.

7.1.2 *'Sites' for electrons (electronic structure)*

To understand what happens to the electron donated to the host, we have to consider the host's electronic structure. In transition metal chalcogenides or oxides, electron p orbitals from the chalcogen or oxygen overlap with s, p, and d orbitals from the transition metal, forming bonding and antibonding levels. The periodicity of the solid spreads

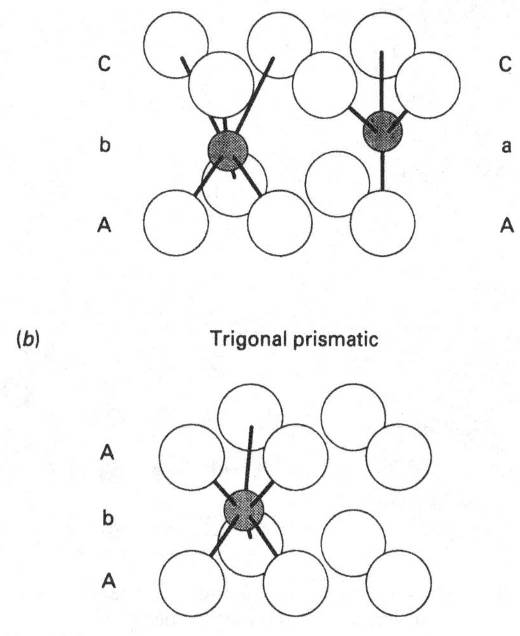

Fig. 7.1 Sites for intercalated ions between close-packed layers of anions: (*a*) octahedral and tetrahedral sites; (*b*) trigonal prismatic sites.

these levels into bands. Although each band contains contributions from both types of atom, it is convenient to label a band by the atom and orbital that contribute the most weight to it, and to neglect the mixed nature of the bands where possible.

Fig. 7.3 shows schematically the relevant energy bands. The oxygen or chalcogen p bands are filled, and the transition metal d bands are empty or partially occupied. The outer s orbitals of the guest produce bands (not shown) above the transition metal states. The electrons from the guests are added to the d bands as intercalation proceeds, so the Fermi energy (the energy separating full and empty states) moves upward relative to the bottom of the d bands.

In Fig. 7.3 the d bands are shown split into sub-bands. The lower d bands are derived from the d orbitals that overlap least with the chalcogen p orbitals. (This is because the d bands are formed from antibonding combinations of d and p orbitals, and those with more overlap are pushed higher in energy.) The details of the splitting depend on how the

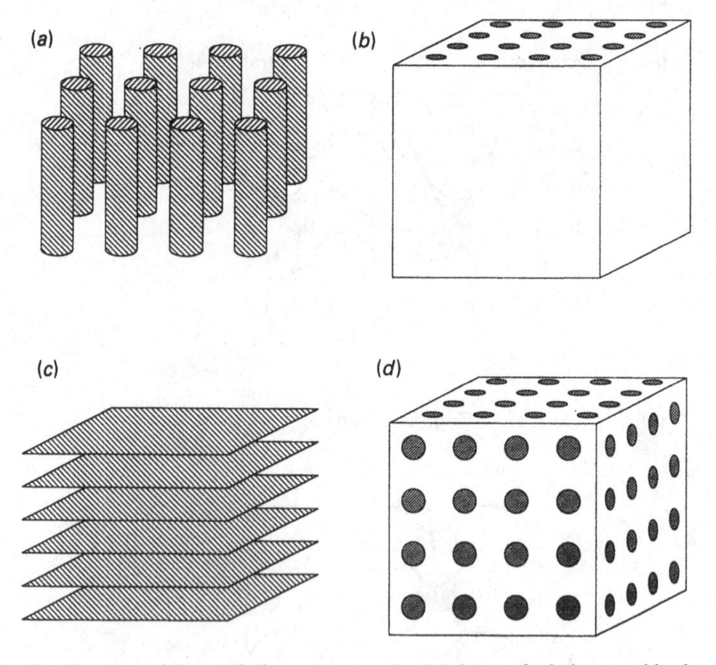

Fig. 7.2 Classification of intercalation compounds: (*a*) host of chains weakly bonded together; (*b*) three-dimensional host with one-dimensional lattice of sites for guest ions; (*c*) layered host (two-dimensional host and two-dimensional lattice of sites); (*d*) three-dimensional host with three-dimensional lattice of sites.

chalcogens coordinate the transition metal. For octahedral coordination, the lower band contains six states and the upper one four per transition metal, counting spin up and down states separately. For trigonal prismatic coordination, the lower band splits further, and the lowest band has two states. Compounds like TiS_2 or MoO_3, where the d bands are empty, are semiconductors or insulators. In MoS_2, Mo contributes four of its six outer electrons to the sulphur bands, because each sulphur is two electrons short of having a full outer shell. That leaves two d states occupied, so the 2H form of MoS_2, where Mo lies in a trigonal prism, is a semiconductor because the lowest d band is filled. The 1T form, where Mo lies in an octahedron, is a metal because this splitting has disappeared. The 1T form of MoS_2 was first made in an electrochemical cell, where Li was added to the 2H form. The driving force for the transformation is probably the energy gained by eliminating the gap between the d states when electrons occupy the second d band (Py and Haering, 1983).

Although this chapter discusses compounds where the band picture of Fig. 7.3 holds, this picture breaks down for smaller transition metal atoms, especially those toward the end of the first row of the periodic table (Zaanen, Sawatsky and Allen, 1985). There, a localised description is more appropriate; the d states remain atomic-like d orbitals, and the materials can be non-metallic even though the band picture indicates a partially

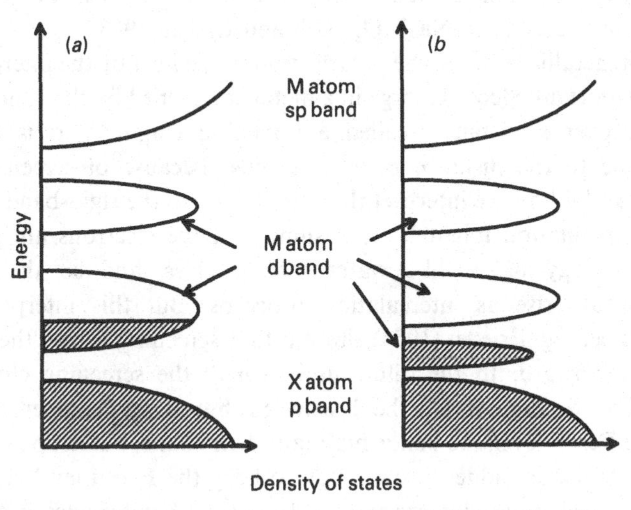

Fig. 7.3 Electron bands for transition metal oxides and chalcogenides. M = metal, X = chalcogen or oxygen. The filling of the bands is appropriate for MoS_2.

occupied, metallic band. In this case, we adopt an atomic description, and say that electrons from the guest change the valence of the transition metal (Rouxel, 1989).

The discussion so far has ignored the influence that the charge of the intercalated ion has on the energy bands. Because the ion is closer to the chalcogens of the host than to the transition metals, its positive charge lowers the energy of the p electrons relative to the energy of the d electrons. This lowering can be seen in calculations of the band structures of $LiTiS_2$ and TiS_2 (Umrigar, Ellis, Wang, Krakauer and Posternak, 1982; see also Fig. 6 in McKinnon, 1987). Thus the bands are modified by the guests, even though the electron states introduced by the guests lie far above the Fermi energy. To a first approximation, though, the shape of the d bands near the Fermi energy is not changed. It is reasonable to adopt a rigid-band model, and assume that the shape (but not necessarily the position on the energy scale) of the d bands is unchanged (rigid) as intercalation proceeds.

The influence of the guest ion's charge is strongest near the guest ion itself. The guest ion polarises the chalcogen atoms around it, and those, in turn, affect the neighbouring transition metal ions, lowering the energy of their d states. When electrons are first added to an empty d band, this lowering can pull states out of the band, localising electrons until the material becomes metallic at a critical concentration of electrons (Mott and Davis, 1979). This so-called Mott transition has been observed by electron spectroscopy in Na_xWO_3 (Hill and Egdell, 1983).

Even in metallic systems, the polarisation of the host by the intercalated ion has important effects. Charge in a metal is screened by the conduction electrons, over a distance (called a screening length) that is usually comparable to the distance between atoms. Because of screening, we must be careful how we interpret the word 'rigid' in the rigid-band model. Our first inclination might be to assume that the electrons are poured into the energy bands like water into a glass, and so the Fermi energy should rise as intercalation proceeds. But this interpretation ignores screening. Friedel (1954) showed that screening causes the bands to shift downward. In the dilute limit, where the screening clouds of different ions do not overlap, the downward shift exactly compensates the rise in the Fermi level; the bands maintain their shape, but drop in energy as the electrons are added just enough to keep the Fermi level constant. This is the appropriate interpretation of the rigid-band model in metallic compounds.

7.2 Examples of host compounds

This section discusses several examples of intercalation compounds. The compounds are classified according to Fig. 7.2.

7.2.1 One-dimensional host, three-dimensional network of sites

These hosts consist of chains weakly bonded to one another. One example is a series of compounds of Mo and Se, in which cubes of Mo and Se are arranged in chains, and can accommodate guest ions between the chains (Chevrel and Sergent, 1982). In these compounds, it is possible to remove the intercalated ions and separate the chains in solution (Tarascon *et al.*, 1985).

Organic compounds like polyacetylene are another example of this class of intercalation compound, as discussed in Chapter 5.

7.2.2 Three-dimensional structure, with one-dimensional tunnels

In many oxides, transition metals are octahedrally coordinated by oxygen, and the octahedra can join together to form chains along one direction. Fig. 7.4 shows the arrangement in the rutile structure, viewed from the side (*a*) and above (*b*). Octahedra in different chains are joined by sharing one oxygen; they are said to share corners. The tunnels between the octahedra are denoted as 1×1, because they have one octahedron along each side. In other compounds, such as hollandite and ramsdellite, chains are wider than one octahedron, creating larger tunnels, such as 2×2 (*c*), or 2×1 (*d*). Structures with larger tunnels are usually unstable without ions in the tunnels.

As a specific example, Li can be intercalated into several oxides with the rutile structure (Murphy, DiSalvo, Caricles and Waszczak, 1978). In Li_xMoO_2, the structure changes from monoclinic to octahedral and back to monoclinic as x goes from 0 to 1 (Dahn and McKinnon, 1985). In both monoclinic structures, Mo atoms are shifted to form Mo–Mo pairs along the chains (Cox, Cava, McWhan and Murphy, 1982), and these pairs disappear in the octahedral structure at intermediate x (Dahn and McKinnon, 1985). The rutile Li_xWO_2 has been considered as a replacement for metallic Li in electrochemical cells (Murphy *et al.*, 1978).

7.2.3 Two-dimensional systems: layered host, layers of sites for guests

Many layered compounds are made of close-packed anions, with transition metals in octahedral or trigonal prismatic sites, as shown in Fig. 7.5. Adjacent layers of anions are only weakly coupled together, and so various sizes of guest ions can be inserted between them. The kinds of site between these layers have already been illustrated in Fig. 7.1.

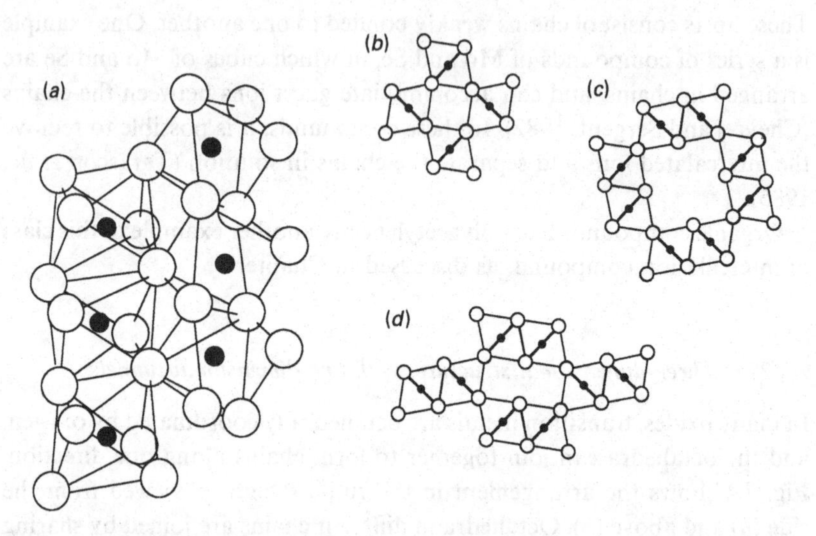

Fig. 7.4 Tunnel structures based on rutiles. (*a*) Rutile-like chains, showing alternately short and long metal–metal distances, as in MoO_2. (*b*)–(*d*) show the chains viewed from above; (*b*) 1 × 1 tunnels as in rutile (TiO_2) or MoO_2; (*c*) 2 × 2 tunnels as hollandite ($BaMn_8O_{16}$); (*d*) 2 × 1 tunnels as ramsdellite (MnO_2).

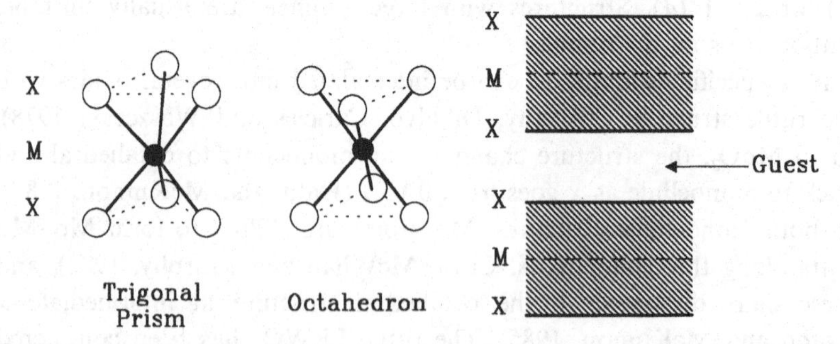

Fig. 7.5 Layered structures. Guest ions can intercalate between the X–M–X sandwiches shown at the right. Within the sandwiches, the M atoms are coordinated in trigonal prisms or octahedra by the X atoms.

The structure of layered compounds often can be described by the ABC notation, shown in conjunction with the sites between close-packed layers in Fig. 7.1. Given a close-packed layer of atoms, with atoms at positions we call A, there are three possible places to put the atoms of the next layer: directly above the first layer (also in the A positions), or fitted into the interstices of the first layer, either in B or C positions. In this notation, a trigonal prism sandwich is denoted AbA, and an octahedral sandwich is AbC, where the uppercase letters denote the layers of chalcogen or oxygen, and the lowercase letters the layers of transition metal atoms.

Perhaps the most well known of the lithium intercalation compounds is Li_xTiS_2. Both Li (for $x < 1$) and Ti are octahedrally coordinated by S (Fig. 7.1); in the ABC notation, the structure is AbC(b)AbC, where the letter in parentheses denote lithium atoms. This structure is also called the 1T form, because of its trigonal (T) symmetry and the single layer per unit cell. The electrochemical behaviour of Li in TiS_2 is described below in connection with staging.

A given compound is sometimes found with different stacking sequences, or poly types, which may differ in the coordination of the transition metal atoms, or just in the stacking of the different sandwiches (Hulliger, 1976). Intercalation can induce transitions between polytypes. The transition from a trigonal prismatic compound to an octahedral one can be explained by the band structure in Fig. 7.3, as mentioned above in connection with Li_xMoS_2. (In the ABC notation, the structure goes from the hexagonal 2H form, AbA BaB to the 1T form, AbC AbC.) Intercalation can also change the stacking sequence without changing the coordination of the transition metal ion. An example is Li_xZrS_2, where the Li causes layers to shift from 1T, or AbC(b)AbC(b)AbC, to the rhombohedral 3R form, AbC(a)BcA(b)CaB (Whittingham and Gamble, 1975). The Li and Zr are farther from one another in the 3R form than in the 1T form, so the change in structure can be explained as a way of reducing the Coulomb repulsion between Li and Zr ions.

Li generally occupies octahedral sites in these compounds at low concentrations. But there is only one octahedral site per transition metal in a compound like VSe_2, and so beyond $x = 1$ at least some of the Li must occupy tetrahedral sites. Since the tetrahedral sites are above and below the planes of the octahedral sites, the Li–Li repulsion is reduced if the Li already in octahedral sites shift to tetrahedral sites beyond $x = 1$. Thus the structure goes from AbC(b)AbC at $x = 1$ to AbC(a,c)AbC at $x = 2$. That is one way to rationalise why all the Li occupy tetrahedral sites in the compound Li_2VSe_2 (Tigchelaar, Wiegers and van Bruggen, 1982).

The layered oxides $LiNiO_2$ and $LiCoO_2$ also have the 3R form AbC(a)BcA(b)CaB. In this structure, the oxygen lattice considered alone has cubic close packing ACBACB (or equivalently ABCABC). As a result, these compounds are closely related to cubic compounds. To visualise the structure of $LiNiO_2$ or $LiCoO_2$, for example, start from cubic NiO or CoO (AbCaBcAbCaB), and replace every second layer of Ni or Co by Li. In the case of Ni this replacement may be incomplete, and the Li layers may contain residual Ni (Dahn, von Sacken and Michal, 1990b).

More complicated layered structures are also possible. Examples are MoO_3 (Hulliger, 1976), vanadium oxides like V_2O_5 (Murphy, Christian, DeSalvo, Carides and Waszczak, 1981), and the phospho-sulphides like $NiPS_3$ (Brec, 1986). Some of these more complicated structures cannot be described by the ABC notation of close-packed layers, and so are represented by coordination polyhedra. As an example, Fig. 7.6(*a*) compares part of the structure of V_2O_5 and (*b*) δ-LiV_2O_5, where each V is surrounded by a pyramid of five oxygen. (Note the oxygens are at the corners of the pyramids, but are not shown.) Introducing Li causes the layers to shift, producing the smaller sites for Li in Fig. 7.6(*b*) (Cava *et al.*, 1986).

7.2.4 *Three-dimensional systems*

The structure of a Chevrel compound like Mo_6Se_8 is based on a cube with Se on the corners and Mo in the centre of the faces (Fig. 7.7). The cubes are arranged on a lattice that is almost cubic, but are rotated so that the Mo in one cube bonds to the Se in an adjacent one. Large guests occupy the sites shown in Fig. 7.7. For smaller guests, or for x beyond 1 in $Li_xMo_6Se_8$, these sites split into a ring of six sites (too close together to all be occupied at once), and there is a second surrounding ring of another six (Yvon, 1982).

The electronic structure of Chevrel compounds is similar to that in Fig. 7.3(*a*), but with different numbers of states in the upper and lower bands. Mo–Mo bonds split the d states into two bands (Nohl, Klose and Andersen, 1982). The lower band contains 24 states per Mo_6Se_8 cluster, because there are 12 Mo–Mo bonds in each cluster (the 12 edges of the Mo_6 octahedron). Each Mo contributes six electrons to the bands, for a

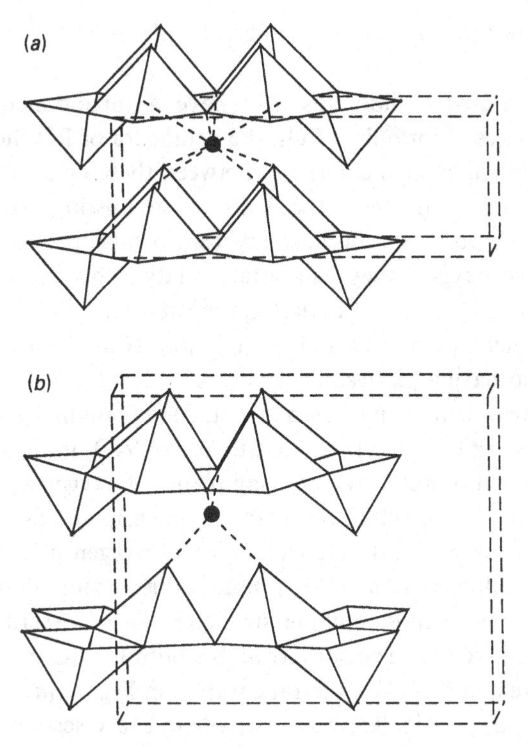

Fig. 7.6 Partial structures of (*a*) V_2O_5 and (*b*) LiV_2O_5. The oxygens sit at the corners of the five-sided prisms, and each prism holds one vanadium. The solid circle in (*a*) is a possible site for Li, and in (*b*) is the actual site for Li. The dashed box is the unit cell, which is twice as large in (*b*) as in (*a*) because of the displacement of the layers. (Not all the atoms in the unit cell are shown.) Based on the structure of Cava *et al.* (1986).

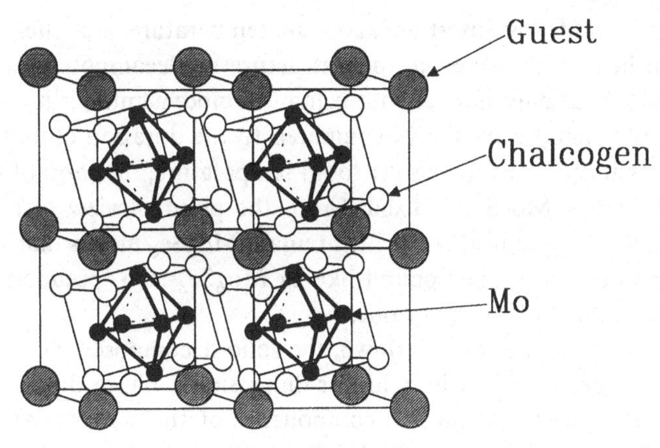

Fig. 7.7 Structure of Chevrel compounds like Mo_6Se_8.

total of 36, but 16 of those go to fill the Se p bands, leaving 20 in the lower d band.

Perovskites and related compounds also have a three-dimensional structure. In perovskites of formula ABO_3, the octahedra of BO_3 lie on a cubic lattice, and are joined at the corners. Between these octahedra are large sites for the A atoms. In ReO_3, the A atoms are missing, so guests can be added to the A positions. Because adjacent octahedra are joined together by only one oxygen; they can rotate relative to one another, changing the shape of the A site. In Li_2ReO_3, the rotation splits the large A sites into two smaller sites more suitable for Li ions (Cava *et al.*, 1982). Bronzes of WO_3 also have a perovskite structure.

The structure of ReO_3 can often be used as a starting point in describing other structures. For example, the layered structure of V_2O_5 in Fig. 7.6(*a*) can be derived from that of ReO_3 by removing planes of oxygen and then shearing the structure. The structure of the three-dimensional host V_6O_{13} is derived from V_2O_5 by removal of a second set of oxygen planes, and then a second shear. Murphy *et al.* (1981) discuss this in more detail.

Spinels are cubic compounds, but the structure can be described as layers of close-packed oxygen. The sites available both for the transition metals and guests are octahedral and tetrahedral, as in Fig. 7.1(*a*). Unlike the layered compounds in Fig. 7.5, however, where every second metal layer is empty, the metal atoms of the host occupy octahedral or tetrahedral sites in every layer.

7.3 Thermodynamics of insertion, ΔG, ΔS, ΔH

Compounds made by insertion at room temperature are often metastable – if heated, they change their structure or decompose into other compounds. That does not rule out using thermodynamics; it just means that processes happening slowly compared to the duration of an experiment are assumed to be frozen. At room temperature, the ratio of Mo to Se in a host like Mo_6Se_8 is fixed. From the point of view of thermodynamics, the constraint that the host remain Mo_6Se_8 means that we can regard an intercalation compound like $Li_xMo_6Se_8$ as a 'pseudo-binary' compound instead of a ternary one.

When this constraint breaks down, the solid decomposes. The decomposition can be described in a higher-dimensional phase diagram that considers the guest and all the components of the host on an equal footing (Godshall, Raistrick and Huggins, 1980). It can be difficult to distinguish this decomposition from intercalation in an electrochemical cell, especially because the solids formed in this way are usually highly disordered.

7.3.1 *Relation of voltage to chemical potential*

The most important aspect of the thermodynamics of insertion is that the concentration of the guest can change. Consequently, we are interested in changes in the Gibbs free energy G with the number n of intercalated guest atoms. The thermodynamic quantity describing these changes is the chemical potential μ, defined as

$$\mu = \partial G/\partial n. \tag{7.1}$$

The advantage of studying intercalation with electrochemical cells is that μ can be measured directly from the voltage E between the electrodes of such cells.

Consider a cell with some host as one electrode and Li metal as the other. Denote the chemical potential of Li in the host and in Li metal as μ and μ_0, respectively. If the guest has charge ze in the solution of the cell ($z = 1$ for Li), one ion is intercalated for every z electrons passed through the external circuit. Since the electrons move through the potential difference E, the work done on the cell per ion intercalated is $-zeE$. This work must equal the change in free energy of the two electrodes, which is $(\mu - \mu_0)$, so

$$-zeE = \mu - \mu_0. \tag{7.2}$$

Thus measuring the cell voltage at equilibrium vs charge passed between the electrodes is equivalent to measuring the chemical potential as a function of x, the Li content of a compound like $Li_xMo_6Se_8$. Thermodynamics requires that μ increase with concentration of guest ions, and so E decreases as ions are added to the positive electrode.

If the other electrode of the cell were another host, then both μ and μ_0 would vary as the cell is charged or discharged. It is most convenient, however, to measure E against Li metal (or the bulk form of whatever guest is intercalating), because μ_0 is constant, and because the zero on this voltage scale is the limit of intercalation. When E reaches zero, the bulk state of the guest has the same chemical potential as the guest inside the host, so the guest deposits on the host's surface.

Features in $E(x)$ typically occur on the scale of millivolts, and voltage is easily measured to microvolts. Thus $E(x)$ can be measured accurately enough for the derivative $-\partial x/\partial E$ to be calculated. (The negative sign is needed because E decreases with x.) Subtle variations in $E(x)$ are easier to see in the derivative, as Fig. 7.8 shows. (The features in Fig. 7.8 will be discussed in Section 7.4.5.) The measurements in Fig. 7.8 were recorded as a cell discharged at constant current, a procedure that gives results sufficiently close to equilibrium if the cells are discharged slowly. Ten or twenty hours is often slowly enough, although some measurements may need several weeks.

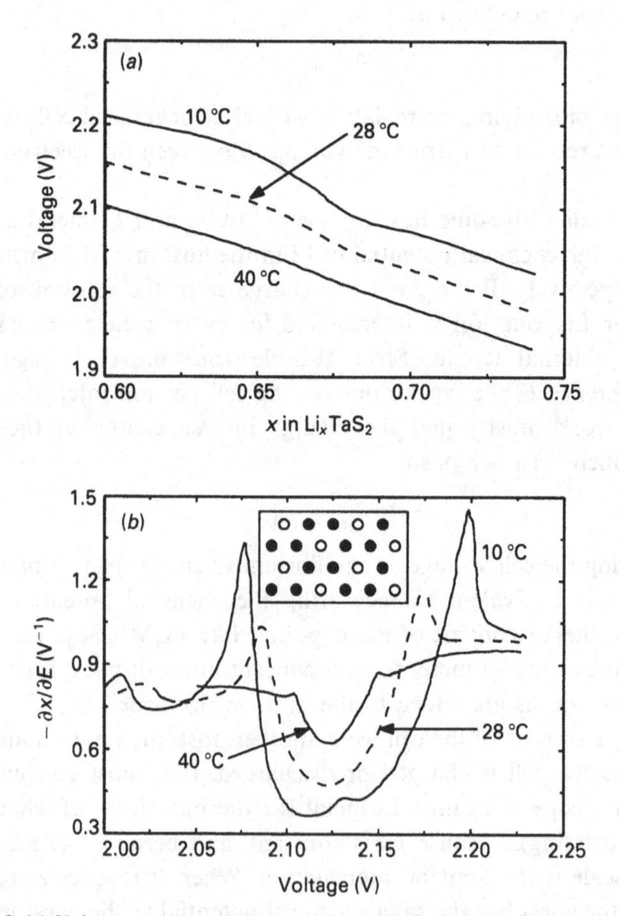

Fig. 7.8 (*a*) Voltage and (*b*) $-\partial x/\partial E$ vs x in Li_xTaS_2 near $x = \frac{2}{3}$. Features in $-\partial x/\partial E$ are due to ordering of Li ions. Data from Dahn and McKinnon (1984). The voltages in (*a*) have been offset for clarity. The insert shows the ordered lattice at $x = \frac{2}{3}$.

Features in $E(x)$ and $-\partial x/\partial E$ reveal a great deal about the thermodynamics of a system. Consider a compound where the free energy G as a function of x has two branches, as shown in Fig. 7.9. If a straight line drawn tangentially to the branches is lower than either branch, the system breaks into two phases between x_1 and x_2, the points where the tangent line touches the two branches. For any x between x_1 and x_2, the system consists of small regions, called domains, of the two phases. Increasing x causes the domains of the phase with larger composition (x_2) to grow at the expense of the phase of lower composition (x_1). Such a transition between two phases is called a first-order transition. Since the compositions of the coexisting phases do not change, the chemical potential is constant in this two-phase region.

In an electrochemical cell, the voltage should be constant in a two-phase region, and $-\partial x/\partial E$ should diverge. In practice, kinetic effects generally cause E to decrease slightly through the two-phase region, and so $-\partial x/\partial E$ has a peak rather than a divergence. Not all plateaux, however, are phase transitions. The plateaux due to first-order phase transitions are usually significantly higher in voltage when a cell is charged than when it is discharged, even at low currents, and this hysteresis can be a good indication of a first-order transition.

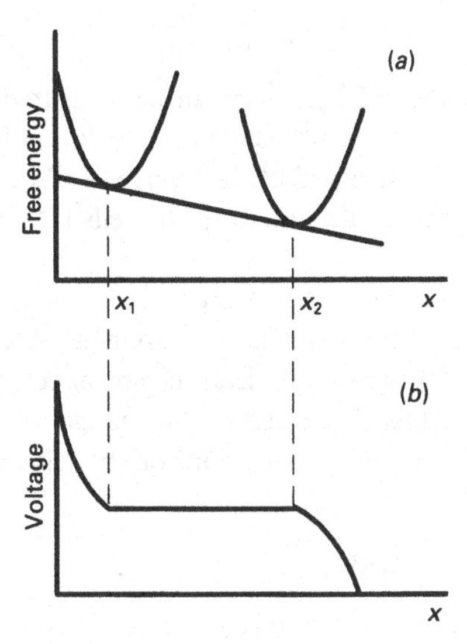

Fig. 7.9 (*a*) Free energy and (*b*) voltage at a first-order phase transition in which phases of composition x_1 and x_2 coexist.

In higher-order or continuous phase transitions, there is no phase coexistence; instead, $-\partial x/\partial E$ diverges at a single composition. This divergence in $-\partial x/\partial E$ is mathematically analogous to the divergences in specific heat or magnetic susceptibility (McKinnon and Haering, 1983). As discussed in Section 7.4.5, the peaks in $-\partial x/\partial E$ in Fig. 7.8 are probably due to continuous transitions.

7.3.2 *Measuring the partial entropy*

Like the free energy, the chemical potential can be separated into a contribution from enthalpy H and another contribution from entropy S; the relationship is

$$\mu = \partial G/\partial n = \partial H/\partial n - T\,\partial S/\partial n, \tag{7.3}$$

where T is the absolute temperature. It is possible to measure one of the two contributions and infer the other, using either temperature derivatives or calorimetry.

The Maxwell relations of thermodynamics relate quantities formed by differentiating G once with respect to one variable and once with respect to another (Huang, 1987). Choosing the two variables to be T and n leads to the following relationship:

$$(\partial\mu/\partial T)_{n=\mathrm{constant}} = (\partial S/\partial n)_{T=\mathrm{constant}}. \tag{7.4}$$

Since E is proportional to μ by Eqn (7.2), $\partial S/\partial n$ can be measured from changes in E with temperature. Such measurements as a function of x are slow because the temperature must be cycled at each value of x, and they are plagued by thermoelectric voltages generated in the cells (Dahn and Haering, 1983).

An alternate method that works at constant temperature, but requires more sophisticated equipment, is a calorimetric measurement. A calorimeter measures the heat Q that flows into an electrochemical cell as the cell is discharged or charged. The heat is related to the change in energy $\mathscr{E}$ of the intercalation compound and the work W_k done on the cell, leading to the following relationship (Dahn *et al.*, 1985):

$$\begin{aligned} \mathrm{d}Q/\mathrm{d}t &= \mathrm{d}\mathscr{E}/\mathrm{d}t - \mathrm{d}W_k/\mathrm{d}t \\ &= T[(\partial S/\partial n) - (\partial S/\partial n)_0]I/e - I\eta, \end{aligned} \tag{7.5}$$

where t is time and I is current. The subscript 0 refers to the Li anode, and η is the overvoltage, the difference between the measured voltage and the equilibrium voltage. A measurement of dQ/dt thus gives the difference between $\partial S/\partial n$ for the two electrodes, if η is small.

7.4 Lattice-gas models

In various forms, lattice-gas models permeate statistical mechanics. Consider a lattice in which each site has two states. If we interpret the states as 'full' or 'empty', we have a lattice–gas model, and an obvious model for an intercalation compound. If the states are 'spin up' and 'spin down', we have an Ising model for a magnetic system; if the states are 'Atom A' and 'Atom B', we have a model for a binary alloy. Many different approximation techniques have been derived for such models, and many lattices and interactions have been considered.

One complication in applying these models to intercalation compounds is in treating the dissociation of the intercalated atom into ions and electrons. The chemical potential can be written as a sum of contributions from ions and electrons, according to

$$\mu = \mu_i + \mu_e \tag{7.6}$$

for a singly charged ion. (The chemical potential of electrons, μ_e, is usually called the Fermi energy.) Although this separation is a standard procedure, it is not unique, because the strong interaction between electrons and ions must be partitioned in some arbitrary way between the two terms in Eqn (7.6). In metals, it is often possible to arrange the interaction terms so that μ_e is constant.

7.4.1 Entropy of ions

Let us begin by assuming μ_e is indeed constant. Solving a lattice–gas model for the ions means finding the energy and entropy of distributing the ions over the available sites. When the ions do not interact with one another, and when all the sites are equivalent, the sites are occupied at random. The entropy of distributing ions randomly over a fraction x of the N_s available sites is (Huang, 1987)

$$S = -N_s k[x \log x + (1 - x) \log(1 - x)], \tag{7.7}$$

where k is Boltzmann's constant. Then the partial entropy is

$$\partial S/\partial n = -k \log[x/(1 - x)]. \tag{7.8}$$

If the energy to put an isolated ion and its electrons into the lattice is denoted ε, then the chemical potential is

$$\mu = \varepsilon + kT \log[x/(1 - x)]. \tag{7.9}$$

(The so-called site energy ε also contains small contributions from the entropy associated with vibrations of the ions in their sites.)

This equation ignores interactions between the ions. The simplest way to treat the interactions is just to add them to this equation, and assume that the ions remain randomly arranged. Suppose U is the total interaction energy that a given ion would feel if all the other sites were full. When only a fraction x of the sites is occupied, it costs an extra energy Ux to add another ion to the lattice, so μ becomes

$$\mu = \varepsilon + kT \log[x/(1 - x)] + Ux. \tag{7.10}$$

This is often called a mean-field expression, because each ion feels the mean interaction or field from its neighbours. It is exact if the interaction between ions extends over a long range, because then the interaction will not cause the ions to rearrange from a random distribution. If the ions do rearrange, the entropy will change and better approximations are needed.

As a function of x, these expressions lead to s-shaped curves like those shown in Fig. 7.10. The scale of the variation is set by kT/e, which is 25 mV at room temperature. The partial entropy diverges near $x = 1$ and $x = 0$, reflecting the impossibility of filling every last site or removing every last ion.

7.4.2 *Entropy of electrons*

The electrons, if they are separated from the ions, will also contribute to the entropy, and one might naively expect an expression similar to Eqn (7.8). Then the chemical potential for an atom would be the sum of two terms like Eqn (7.10), one from ions and one from electrons, and so the entropy term would be doubled. This is not so, however, in metallic intercalation compounds. In metals, the entropy of electrons is small. Electrons added by intercalation do not have a choice of all the empty states in a band, but only those within kT of the Fermi energy. If the Fermi energy is expressed as a temperature T_F and is measured from the bottom of the band, the change in entropy with the number n_e of electrons, $\partial S/\partial n_e$, is of order kT/T_F (Kittel, 1971), not of order k like Eqn (7.8) for the ions. Typically T/T_F is much less than unity, so the partial entropy of electrons is much less than the partial entropy of the ions.

When the added electrons, instead of filling energy bands, are localised on transition metal ions in the host, the compound does not become metallic as intercalation proceeds. Such a localised system is usually described as having a mixture of transition metal valences. It might then be supposed that the entropy of distributing these two valence states over the lattice of transition metal ions is of the same form as Eqn (7.8). This would be true, however, only if the valence states could be distributed independently of the guest ions. But the Coulomb interaction between ions and electrons is large, so the two distributions are probably highly correlated, producing an entropy more complicated than just a doubling of Eqn (7.8). Unfortunately, there are no experiments on these localised systems as detailed as those for the metallic system discussed in the next section.

7.4.3 *$Li_xMo_6Se_8$ as example of a lattice gas*

Although the simple mean-field expression (Eqn (7.10)) for a lattice-gas model has been used to understand intercalation systems qualitatively

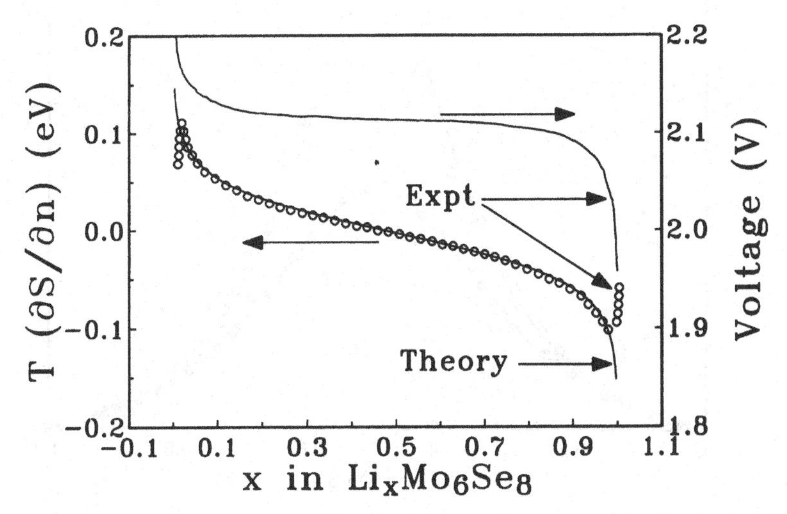

Fig. 7.10 Voltage E and partial entropy $\partial S/\partial n$ in $Li_xMo_6Se_8$. The theory for $\partial S/\partial n$ is for a random distribution of guest ions, with no contribution to the entropy from the electrons. Data from Dahn *et al.* (1985).

(Berlinsky, Unruh, McKinnon and Haering, 1979), it is surprising to find a system where the expression holds almost exactly. Yet a simple mean-field theory describes various aspects of intercalation in $Li_xMo_6Se_8$ for $0 < x < 1$ within experimental error. The voltage is accurately described by Eqn (7.10). The theory is indistinguishable from the voltage curve in Fig. 7.10. Eqn (7.10) also describes the derivative $-\partial x/\partial E$ within experimental error, as Fig. 7.11 shows. There is one parameter in this fit, the interaction energy U, which in Fig. 7.11 is -0.0904 eV.

Fig. 7.10 also compares the partial entropy (measured in a calorimeter) with Eqn (7.8). The shape of $\partial S/\partial n$ is set by k, and cannot be adjusted in comparing theory and experiment. The calculated entropy is for the ions alone, proving that the entropy of the electrons can be neglected in this metallic compound.

The negative value of U in the fit in Fig. 7.11 signifies that the interaction between Li ions is attractive. Under an attractive interaction the ions can cluster, and the compound should separate into two phases at low temperatures. Fig. 7.12 shows one of the Bragg peaks in the X-ray diffraction pattern for $Li_{0.5}Mo_6Se_8$ as it cools (Dahn and McKinnon,

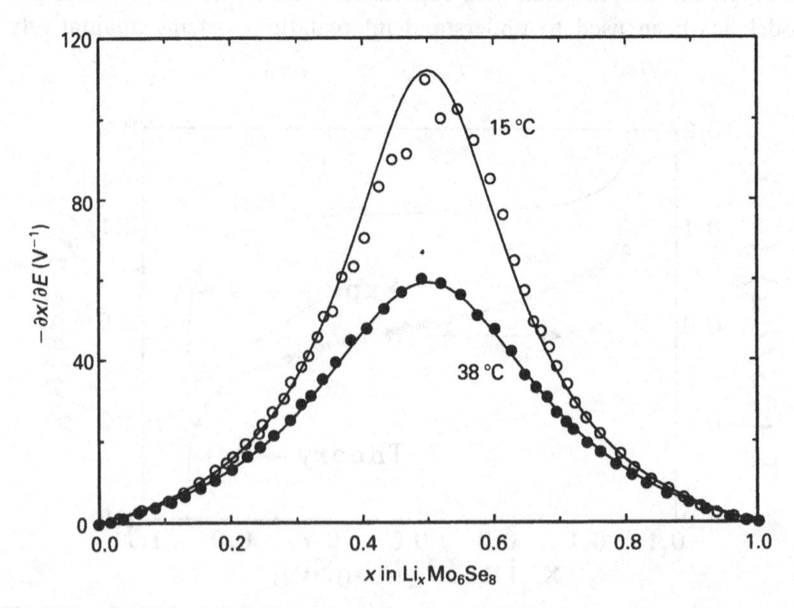

Fig. 7.11 $-\partial x/\partial E$ for $Li_xMo_6Se_8$ at two temperatures. The points are data; the solid lines are fits to the mean-field theory with $U = -0.0904$ eV. From Coleman, McKinnon and Dahn (1984).

1985). The peak splits, indicating separation into two phases, and the peaks move apart, showing that the difference in composition of the two phases increases as temperature decreases. The phase diagram derived from these Bragg peaks is shown in Fig. 7.12(*b*). Phase separation first appears when T is low enough that the entropy lost in forming separate phases is compensated by the energy gained, or when $U \approx kT$. (More precisely, for mean-field theory, when $U = 4kT$.) The solid line from the figure is calculated from the mean-field theory of Eqn (7.10), using the same value of U as calculated from a fit to $-\partial x/\partial E$.

7.4.4 Site energies in lattice-gas models

The magnitude of the voltage in an electrochemical cell is mainly determined by ε in Eqn (7.10); the entropy term gives a variation on the scale of $kT/e = 25$ mV (Fig. 7.10), and U is typically a small fraction of an electron volt. Calculations of ε are complicated by the strong interactions between ions and electrons. We are not yet able to predict site energies reliably from first principles, and so we have to be content to identify different contributions to ε.

The series of compounds $Li_xMo_6Se_zS_{8-z}$ shows the effects of the host's

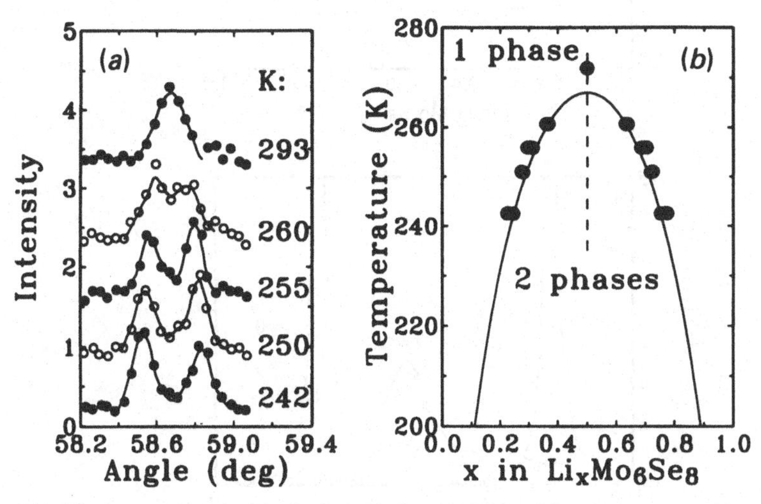

Fig. 7.12 Phase separation in $Li_xMo_6Se_8$. (*a*) Powder X-ray diffraction as a sample with $x = 0.5$ is cooled. (*b*) Phase diagram inferred from results of (*a*) and other Bragg peaks. The dashed line in (*b*) shows the path followed in the measurement of (*a*). Data from Dahn and McKinnon (1985).

anions on ε. Fig. 7.13 shows $E(x)$ and $-\partial x/\partial E$ for several compounds in this series (Selwyn, McKinnon, Dahn and Le Page, 1986). In the pure sulphide ($z = 0$), the peak in $-\partial x/\partial E$ is at 2.45 V; in the pure selenide ($z = 8$) it is at 2.1 V. Thus the effect of replacing S by Se is to lower ε by 0.35 V. In the mixed compounds, however, the curve is not just a shifted version of the pure compounds; it has structure. The curve for $z = 2$ clearly shows three peaks in $-\partial x/\partial E$. Each peak corresponds to a different site energy ε. (Note that in this case, the peaks do not represent phase transitions to an ordered structure; cases of phase transitions are discussed below.)

The appearance of three peaks is the result of an important principle. The interaction between ions or electrons is mainly a local interaction,

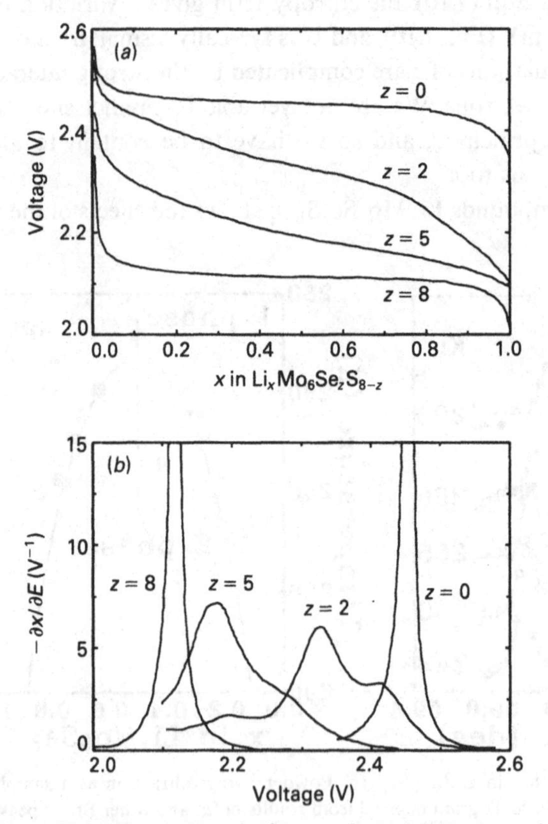

Fig. 7.13 (*a*) Voltage and (*b*) $-\partial x/\partial E$ vs x in $Li_xMo_6Se_zSe_{8-z}$ for several values of z. Selected from the data in Selwyn *et al.* (1986).

occurring over small distances in solids. Thus ε is determined by the local environment of the intercalated ion. In a Chevrel compound, each Li ion has eight S or Se around it, but two are closer than the other six, and these two dominate the anions' contribution to ε. The three peaks for $z = 2$ are from Li surrounded by either two S, two Se, or one S and one Se.

Chevrel compounds have also shed light on the role of electrons in determining ε, in studies of $Li_xRu_zMo_{6-z}Se_8$ (Selwyn and McKinnon, 1987). Since Ru lies two columns to the right of Mo in the transition metal series, each Ru adds two more electrons to the host bands than the Mo it replaces. There are four empty states for electrons in Mo_6Se_8 per formula unit, so the compound with $x = 0$ and $z = 2$ is an insulator (Perrin, Chevrel, Sergent and Fischer, 1980). Similarly, an intercalated compound with $x + 2z = 4$ should also be an insulator. Fig. 7.14 shows $E(x)$ for four compounds of the form $Li_xRu_zMo_{6-z}Se_8$ (Selwyn and McKinnon, 1987). The voltage drops abruptly at $x = 4 - 2z$. The resistivity of the compounds as a function of x has a sharp peak just at that drop (Selwyn and McKinnon, 1987). The resistivity falls again when x exceeds $(4 - 2z)$, indicating that the electrons are filling the upper d band. A similar drop in $E(x)$ also occurs in $Li_xNb_yMo_{1-y}S_2$ at $x = y$, where the

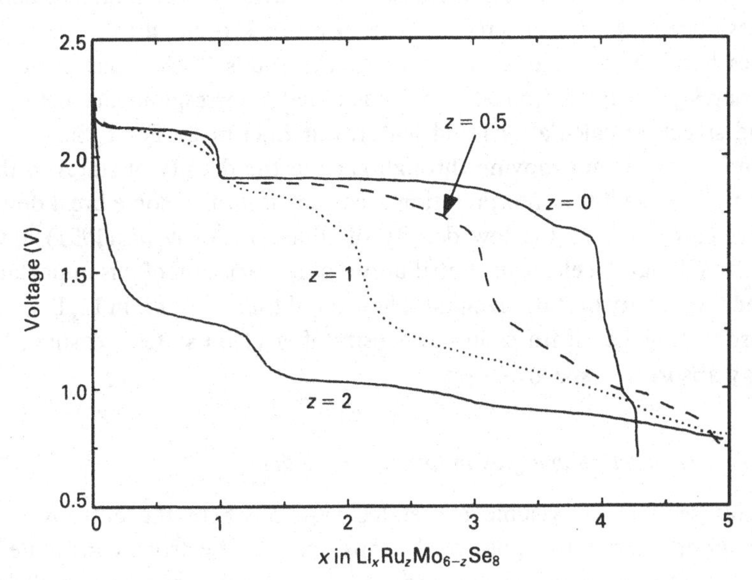

Fig. 7.14 Voltage vs x in $Li_xRu_zMo_{6-z}Se_8$ for various z. From Selwyn and McKinnon (1987).

lower d band in Fig. 7.3(*b*) fills (Py and Haering, 1984). Because it is possible to add electrons to the upper band, the limit $x = 4$ to Li intercalation in Mo_6Se_8 is not electronic, as was originally thought, but rather due to Li–Li interactions (McKinnon and Dahn, 1986). With other guests, however, the jump in chemical potential of almost 1 eV when the band fills could well stop further intercalation.

Another point to note in Fig. 7.14 is that the upper voltage plateau does not change with Ru content until $z = 2$. The voltages of this and other features in $E(x)$ (Selwyn and McKinnon, 1987) do not change with z except at the band edge, where $x + 2z = 4$. Note, though, that when the Fermi level moves through a gap in the density of states, the voltage does drop; the first plateau is lower at $z = 2$ where electrons are added to the upper d band than for $z < 2$ where they are added to the lower band. Fig. 7.15 suggests schematically how this occurs. The bands move down when the Fermi energy is in a band, because of screening; at a gap in the density of states, however, screening breaks down, and the Fermi energy moves up. The chemical potential of electrons thus depends on which band is being occupied, but does not vary for a given band.

Fig. 7.15 is based on an extreme interpretation of Friedel's argument, in which screening only breaks down at a gap in the electron energy bands. The screening length increases as the density of conduction band states decreases, and so screening can also break down if the density of states becomes so small that the screening clouds of electrons begin to overlap significantly. Rouxel (1989) has noted a correspondence between band structure calculations and features in $E(x)$ in $Li_xFeNb_3Se_{10}$. The Fermi energy is not moving through gaps in the density of states in the calculations, so if this interpretation is correct, it implies some breakdown of screening due to the low density of states. Dahn *et al.* (1991) have confirmed that the electron contribution to the variation of μ is important when the density of states is small. They found that $-\partial x/\partial E$ in $Li_xB_zC_{1-z}$ is proportional to features in the density of electron states measured by X-ray absorption spectroscopy.

7.4.5 *Interaction energies in lattice-gas models*

In this section we assume a metallic system where the entropy and interaction energy are those of the ions, and the electrons contribute a constant to ε. The mean-field expression, Eqn (7.10), is exact when all the ions interact equally, regardless of the distance between them. Although such interactions might seem unphysical, elastic interactions between ions do extend through a solid. These interactions were first applied to intercalation compounds in studies of metal hydrides (Wagner, 1978). An intercalated ion expands (or contracts) the lattice, and sets up a strain field that falls off as the inverse square of the distance from the ion. In a lattice with free surfaces, this field produces a second field needed to satisfy the boundary conditions of no stress at the surface. The second field is almost constant through the lattice. It gives rise to an elastic interaction between ions that is attractive, roughly independent of the distance between the ions, and large enough to explain the measured U in $Li_xMo_6Se_8$ (Coleman *et al.*, 1984).

The mean-field theory can also be applied to short range interactions,

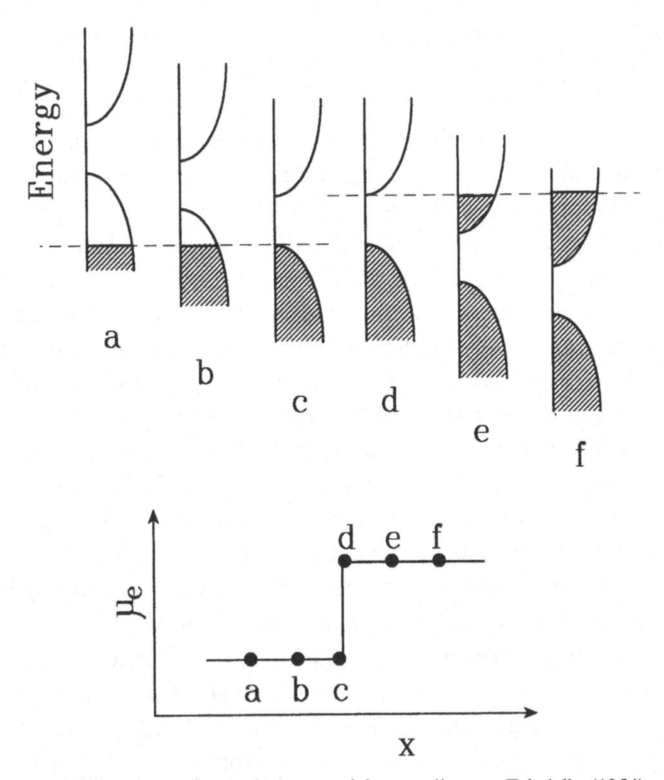

Fig. 7.15 Band filling in an intercalation model according to Friedel's (1954) notion of screening. The upper panel shows the position of the bands at various degrees of filling; the lower panel shows the corresponding values of the electron chemical potential (Fermi energy).

but only approximately. Suppose that U_{nn} is the interaction between two neighbouring ions, and that each ion has γ neighbours. We can use Eqn (7.10) as an approximation to μ if we set $U = \gamma U_{nn}$. Short range attractive interactions are qualitatively the same as long range ones; both cause μ to vary less rapidly with x than when $U = 0$, and both can lead to phase separation. Quantitative details, however, depend on the range of the interactions. Such short range attractive interactions can come both from oscillations in electron density near an intercalated ion – so-called Friedel oscillations – or from the distortion of the lattice (McKinnon and Haering, 1983).

With short range interactions the ions can take advantage of the attractive interactions by forming small clusters, a possibility not available with long range interactions. This short range order pushes the phase separation to lower temperatures than mean-field theory predicts. Short range interactions also lead to different forms of the divergence of various quantities than predicted by mean-field theory (Huang, 1987). Qualitatively, however, mean-field theory works well for attractive interactions.

The only effect of repulsive interactions (positive U) in mean-field theory is to make $E(x)$ vary more rapidly with x. In cubic Li_xTiS_2, mean-field theory accurately describes $E(x)$ (Sinha and Murphy, 1986). Fig. 7.16 shows the experimental results, comparing them to layered Li_xTiS_2, where the mean-field theory is not obeyed. The success of mean-field theory here suggests the repulsive interactions are long ranged. Reducing the range of repulsive interactions, however, can lead to qualitatively new behaviour not predicted by Eqn (7.10), because intercalated ions can avoid a short range repulsive interaction by occupying the sites in an ordered way. The insert to Fig. 7.8 shows a lattice of octahedral sites between two close-packed S layers in Li_xTaS_2. When $\frac{1}{3}$ of the sites are occupied, the intercalated ions can avoid all nearest-neighbour interactions by forming an ordered arrangement. Similarly, they can minimise the repulsion when $\frac{2}{3}$ of the sites are occupied by having the empty sites form the same arrangement. If the interaction is kT or larger, the order will become long range near $x = \frac{1}{3}$ or $x = \frac{2}{3}$, extending through the entire lattice. The transition from short range to long range order is a phase transition, which may be first-order or continuous depending on the lattice of sites and the range of the interactions. In Fig. 7.8(*b*), the peaks and minima in $-\partial x/\partial E$, and their variation with temperature, are just as expected for a lattice–gas model with nearest-neighbour repulsions (Dahn and McKinnon, 1984). The peaks mark the position of the phase transitions, which are continuous transitions in the model. The minima correspond to the composition of the ordered arrangement, $x = \frac{2}{3}$. This correspondence of minima in $-\partial x/\partial E$ to ordered structures is a general feature of lattice–gas models (Berlinsky *et al.*, 1979).

7.4.6 Role of disorder

The features in $-\partial x/\partial E$, particularly peaks due to first-order transitions, are broadened by disorder. Disorder can be incorporated in lattice–gas

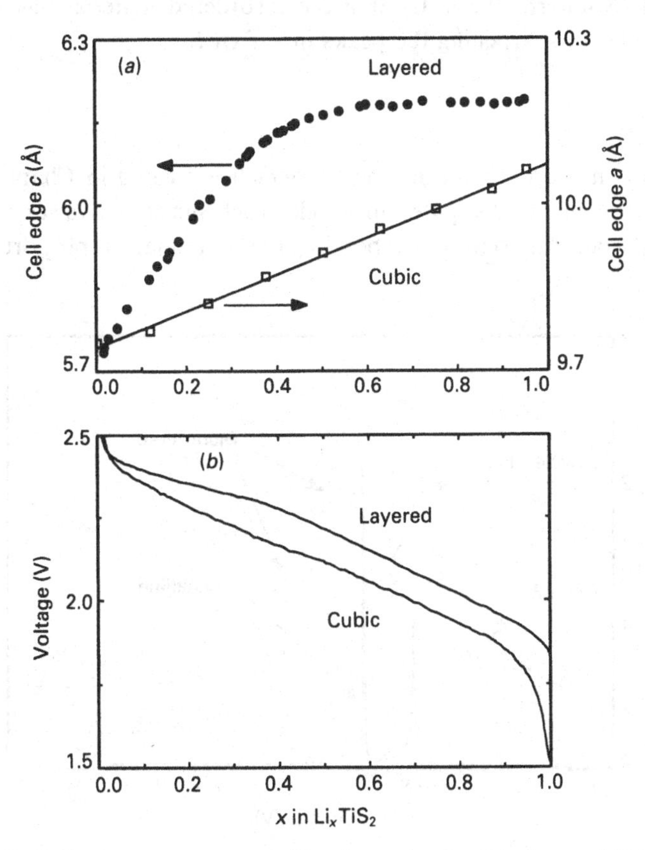

Fig. 7.16 (*a*) Lattice parameters and (*b*) voltage vs x for two forms of Li_xTiS_2. The c parameter for the layered (1T) form is the distance between the layers; the a parameter in the cubic (spinel) form is the edge of the cubic unit cell. Data from Sinha and Murphy (1986), Dahn and McKinnon (1984), and Dahn and Haering (1981b).

models as distributions in site energy. Like a repulsive U in mean-field theory, a distribution in ε causes the voltage to vary more rapidly than for a single ε. It can likewise compensate attractive interactions and suppress first-order phase transitions (Richards, 1984). This suppression has been seen in Li_xC (Dahn, Fong and Spoon, 1990a).

As an example of disorder broadening features in $-\partial x/\partial E$, Fig. 7.17 compares $-\partial x/\partial E$ for crystalline and strongly disordered forms of 1T MoS_2. The crystalline form was made by heating $LiMoS_2$ at high temperatures; the disordered form is produced when Li intercalates into crystalline MoS_2 at room temperature. In the crystalline form, the sharp peaks in $-\partial x/\partial E$ are first-order phase transitions, as confirmed by X-ray diffraction (Mulhern, 1989). Cycling the disordered material has some annealing effects, sharpening the peaks in $-\partial x/\partial E$.

7.4.7 *Hysteresis*

Hysteresis is not a problem of kinetics (which is treated in Chapter 8). Normally, the energy dissipated in an electrochemical cell varies as the square of the current, so the corresponding overpotential η is proportional

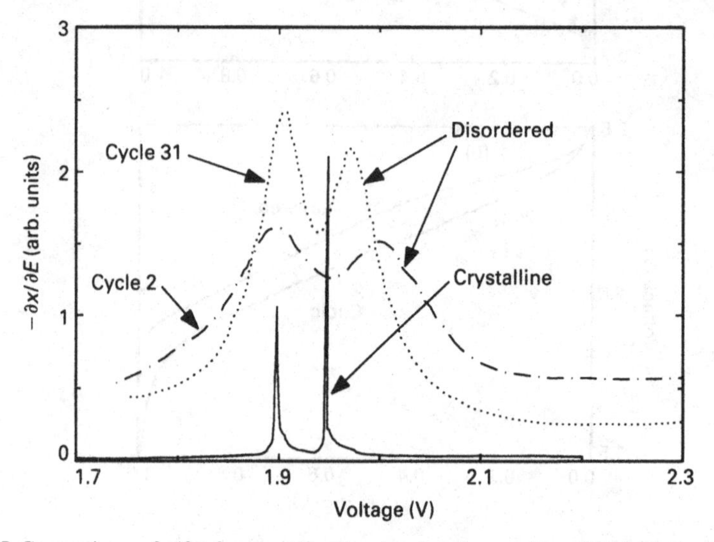

Fig. 7.17 Comparing $-\partial x/\partial E$ for strongly disordered and crystalline 1T-Li_xMoS_2. The disordered form was made by discharging the natural form of MoS_2 to below 1.0 V; the crystalline form was made by a high-temperature reaction of Mo, S, and Li_2S. Data from Mulhern (1989) and Murray and Alderson (1989).

to current. But in systems with two phases separated into domains, dissipation associated with moving domain boundaries can produce hysteresis, in which η is independent of the magnitude of the current (although η does depend on the current's sign). If a cell is charged and discharged over a range of x, the voltage $E(x)$ from charge and discharge, plotted together, forms a loop (called a hysteresis loop) even at low currents. Hysteresis also leads to an extra term in the relation between Q and $\partial S/\partial n$ (Murray, Sleigh and McKinnon, 1991).

7.5 Microstructure – staging and cointercalation

This section looks at two other aspects of intercalation: staging, a particular type of ordering in layered compounds; and cointercalation, where more than one type of guest is intercalated in the same compound.

7.5.1 Staging

When guest ions attract one another, they cluster. If the interaction is strong enough, the compound separates into regions of high and low density of guests. Usually these regions extend over many atomic distances in all directions. In layered compounds, however, the guest ions intercalated in a given layer may attract one another while the ions in different layers repel. This leads to an ordered arrangement known as staging, where the regions of high and low density are one layer thick, and alternate in an ordered way.

In the usual notation (Safran, 1987), a stage-n structure is one where every nth layer contains guests, and the other $(n-1)$ layers are almost empty. The value n can be large; $n = 8$ has been reported in Rb_xC (Underhill, Krapchev and Dresselhaus, 1980). Stage n/m compounds, where m full layers alternate with $(n-m)$ empty layers, have also been observed (Fuerst, Fischer, Axe, Hastings and McWhan, 1983). In principle, it is possible to have an infinite number of even more complicated patterns (for example, three full, two empty, four full, three empty, . . .).

Such complicated structures or large values of n imply a long range interaction U_1 between ions in different layers. We saw in Section 7.4.5 that ordering occurs in lattice–gas models when interactions are of order kT. Interactions between ions in the same layer are usually about kT at room temperature, and it is likely that the interaction between ions in different layers is considerably smaller. Staging can still occur, however, with small values of U_1. Consider the problem in two steps. First, some attractive interaction between ions in a given layer produces clusters of some number of ions, say n_c. Second, these clusters repel one another, not with energy U_1, but with an energy somewhere between $n_c U_1$ and $n_c^2 U_1$, depending on how many ions in one cluster interact with a given ion in the other. Since n_c can be large, this repulsion can be of order kT even if U_1 alone is not.

One obvious choice for the attractive interaction in a given layer is the elastic attraction discussed in Section 7.4.5. The layers may be pushed apart by several angstroms during intercalation, corresponding to large elastic energies. The expansion is often non-linear – the lattice expands faster at low x than at larger x. Because of this non-linear expansion, the elastic energies are more important at small x than at larger x. Fig. 7.16 compares the expansion of the lattice in layered and cubic Li_xTiS_2. A simple model can explain the non-linearity in the layered structure (Dahn *et al.*, 1982). The model considers two elastic forces (modelled as 'springs') – one kind of spring that holds the layers together, a second representing the energy stored in the local compression near an intercalated ion, tending to push the layers apart. When the layers have been pushed apart enough that the second springs are fully extended, the expansion stops. The energies in this model can be incorporated in lattice–gas models, and lead to phase diagrams like those in Fig. 7.18. Note that the staged structures are pushed to lower compositions at higher temperatures. In the shaded regions in the diagram, different staged structures coexist, so that the transitions between the phases are first-order.

The simple models of staging consider layers to be homogeneous, having the same guest concentration everywhere. This would imply that switching from one stage to another should involve emptying and filling entire layers. What is more likely is the Dumas–Herold model (Safran, 1987) shown in the insert to Fig. 7.18, where each layer has the same average composition, but where the clusters (or islands) locally form staged structures. Even so, changing from one stage to another still requires significant rearrangement of ions, and at room temperature will likely occur only slowly. The various stages may not develop fully in an experiment where x changes over its allowed range in hours or even days. It has been argued that in some cases a 'devil's staircase' of infinitely many staged structures would be seen as a function of x if the kinetics did not interfere (Bak and Forgacs, 1985).

Since the transitions between staged structures are first-order, they should produce peaks in $-\partial x/\partial E$ in electrochemical cells. Many layered compounds have peaks in $-\partial x/\partial E$ at low x, over the same range of x as where the distance between the layers changes rapidly with x. Although it was first thought that these peaks were caused by ordering of ions in a given layer, X-ray diffraction has confirmed that they are indeed caused by staging. Fig. 7.19 shows these peaks in $-\partial x/\partial E$ in three layered compounds. Since the peaks are phase transitions, the compounds are mixtures of phases when x lies in the peaks, and are single phases at the minima between the peaks. In TaS_2 (Dahn and McKinnon, 1984) and $NbSe_2$ (Dahn and Haering, 1981a), X-ray diffraction at these minima shows Bragg reflections that correspond to a stage-2 structure. In TiS_2 these extra reflections are not observed, but the observed reflections broaden as x passes through the peaks on either side of the arrow in Fig. 7.19. This broadening indicates that the compound is tending to separate into two phases but cannot fully do so in the time scale of the experiment (Dahn and Haering, 1981b).

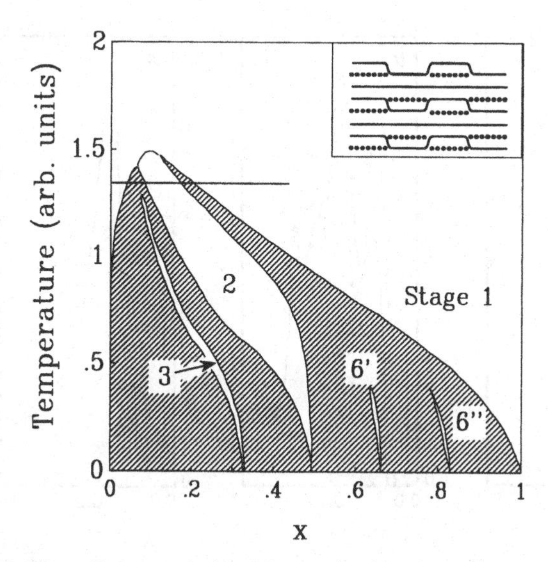

Fig. 7.18 Phase diagram calculated for staging in a layered compound where the layers expand non-linearly with x (from Dahn *et al.* (1982)). The shaded regions are mixtures of phases. The solid line near a temperature of 1.3 represents a series of compounds made at constant temperature. The insert shows the Dumas–Herold picture of islands in a stage-2 structure.

7.5.2 Cointercalation

Cointercalation occurs when two (or more) guests enter the same host. We can distinguish two types of cointercalation: when both guests donate electrons to the host; or when one guest is neutral and surrounds the other, forming a 'solvent cloud' like that around an ion in solution.

For the first case, two types of guest ion, we can generalise the lattice–gas models considered above. The models so far have been binary models: the host (or the sites for the guest ions) is one component, the guest ions the other. With two guests, the obvious generalisation is to a ternary model: the host, and the two types of guest. Ternary phase diagrams are usually drawn as a triangle, as in Fig. 7.20 (McKinnon, Dahn and Jui, 1985) for $Li_xCu_yMo_6S_8$. This phase diagram shows regions of single phase, two phases, and three phases. In two-phase regions, the two phases that coexist are the compositions at the end of lines (shown dotted in the figure) called tie lines. Changing Li content moves the system horizontally in the diagram from one tie line to another, as shown for the dashed line. As Li is added, Cu moves between the two phases, so the

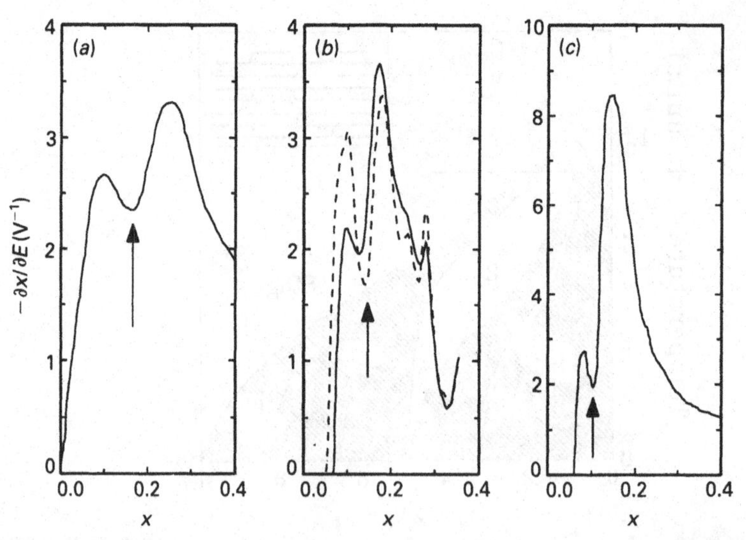

Fig. 7.19 $-\partial x/\partial E$ near $x = 0$ for three layered compounds (*a*) Li_xTiS_2 and (*b*) Li_xTaS_2 (both from Dahn and McKinnon, 1984); (*c*) Li_xNbSe_2 (from Dahn and Haering (1981a)). The arrows indicate the minima corresponding to stage 2 compounds. The dashed line in (*b*) is for a current $\frac{1}{4}$ as large as for the solid line (so the experiment took four times longer).

compositions of the two phases change. As a result, μ_{Li} is not constant in a two-phase region in a ternary phase diagram. It is, however, constant in a three-phase region, such as the triangle surrounded by the two-phase regions in the diagram. Fig. 7.20(*b*) shows $-\partial x/\partial E$ for the compound along the dashed line in the phase diagram. The arrows indicate the points where the compound enters and leaves the two-phase regions. The sudden changes in $-\partial x/\partial E$ at these points are predicted by simple lattice–gas models (McKinnon *et al.*, 1985).

In the second type of cointercalation, guest ions in the host are surrounded by a solvation cloud of another guest. This solvation can occur in electrochemical cells, where the Li ions come from a solution where they are surrounded with molecules of the solvent. In principle, this case is no different from the first one, except that the solvent does not donate electrons to the host. In practice, the solvent molecules are generally large, so this case only occurs in materials that can expand considerably – generally, layered compounds. Measuring the partial pressure of the solvent as a function of density gives the chemical potential of the solvent (Johnston, 1982), and the resulting curves could, in principle, be used to construct ternary phase diagrams.

The first type of cointercalation is a possible means of improving a given host as an electrode; adding a second larger guest might prop the lattice apart and improve the diffusion of the first. In electrochemical cells,

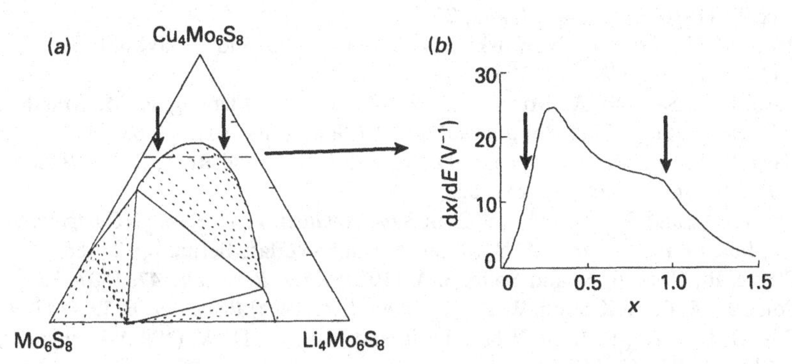

Fig. 7.20 (*a*) Pseudo-ternary phase diagram of $Li_xCu_yMo_6S_8$ (McKinnon *et al.*, 1985). Dotted lines are tie lines, marking two-phase regions. The central triangle enclosed by the two-phase region is a three-phase region. The dashed line gives the path followed by a compound with $y = 2.5$ as x varies. $-\partial x/\partial E$ for this compound is shown in (*b*). The arrows relate the features in $-\partial x/\partial E$ to where the sample crosses boundaries between two-phase and single-phase regions.

however, solvent cointercalation is usually a problem. It can destroy electrodes because it expands the host lattice so much, and can dry out the cell if too much solvent leaves the solution.

7.6 Future prospects

Many of the intercalation batteries being studied now are so-called 'rocking chair' or 'lithium ion' cells, in which both electrodes are lithium intercalation compounds. Thus there is a need for electrodes with a low voltage vs lithium (for the anode) as well as those with a high voltage (for the cathode). Early studies of intercalation compounds for electrodes were part of a search for cathodes, not anodes, and may have passed over materials suitable for anodes.

On a more fundamental side, much of this chapter has focused on lattice–gas models applied to intercalation systems. The application of such models to metallic intercalation compounds is understood, and indeed the models describe some intercalation compounds quantitatively. But more study is needed of systems where the density of states is low, or where the band picture breaks down.

References

Bak, P. and Forgacs, G. (1985) *Phys. Rev. B*, **32**, 7535–7.
Berlinsky, A. J., Unruh, W. G., McKinnon, W. R. and Haering, R. R. (1979) *Solid State Comm.*, **31**, 135–8.
Brec, R. (1986) *Solid State Ionics*, **22**, 3–30.
Bruce, P. G., Krok, F., Nowinski, J. L., Gibson, V. C. and Tavakkolik, K. (1991) *J. Mat. Chem.*, **1**, 705.
Cava, R. J., Santoro, A., Murphy, D. W., Zahurak, S., Fleming, R. M., Marsh, P. and Roth, R. S. (1986) *Journal of Solid State Chemistry*, **65**, 63–71.
Cava, R. J., Santoro, A., Murphy, D. W., Zahurak, S. and Roth, R. S. (1982) *J. Solid State Chem.*, **42**, 251–62.
Chevrel, R. and Sergent, M. (1982) in *Superconductivity in Ternary Compounds I*, Eds. O. Fischer and M. B. Maple, Springer-Verlag, Berlin, pp. 25–86.
Clarke, R., Caswell, N. and Solin, S. A. (1979) *Phys. Rev. Lett.*, **47**, 1407–10.
Coleman, S. T., McKinnon, W. R. and Dahn, J. R. (1984) *Phys. Rev. B*, **29**, 4147–9.
Cox, D. E., Cava, R. J., McWhan, D. B. and Murphy, D. W. (1982) *J. Phys. Chem. Solids*, **43**, 657–66.
Dahn, D. C. and Haering, R. R. (1981a) *Solid State Comm.*, **44**, 29–32.
Dahn, J. R. and Haering, R. R. (1981b) *Solid State Comm.*, **40**, 245–8.
Dahn, J. R. and Haering, R. R. (1983) *Can. J. Phys.*, **61**, 1093–8.
Dahn, J. R. and McKinnon, W. R. (1984) *J. Phys. C*, **17**, 4231–43.
Dahn, J. R. and McKinnon, W. R. (1985) *Phys. Rev. B*, **32**, 3003–5.

References

Dahn, J. R. and McKinnon, W. R. (1987) *Solid State Ionics*, **23**, 1–7.

Dahn, J. R., Dahn, D. C. and Haering, R. R. (1982) *Solid State Comm.*, **42**, 179–83.

Dahn, J. R., Fong, R. and Spoon, M. J. (1990a) *Phys. Rev. B*, **42**, 6424–32.

Dahn, J. R., McKinnon, W. R., Murray, J. J., Haering, R. R., McMillan, R. S. and Rivers-Bowerman, A. H. (1985) *Phys. Rev. B*, **32**, 3316–18.

Dahn, J. R., von Sacken, U. and Michal, C. A. (1990b) *Solid State Ionics*, **44**, 87–97.

Dahn, J. R., Reimers, J. N., Tiedje, T., Gao, Y., Sleigh, A. K., McKinnon, W. R. and Cramm, S. (1991) *Phys. Rev. Lett.*, **68**, 835–8.

Dresselhaus, M. S. and Dresselhaus, G. (1981) *Adv. Phys.*, **30**, 139–326.

Friedel, J. (1954) *Adv. Phys.*, **3**, 446–507.

Fuerst, C. D., Fischer, J. E., Axe, J. D., Hastings, J. B. and McWhan, D. B. (1983) *Phys. Rev. Lett.*, **50**, 357–60.

Godshall, N. A., Raistrick, I. D. and Huggins, R. A. (1980) *Mat. Res. Bull.*, **15**, 561–70.

Hill, M. D. and Egdell, R. G. (1983) *J. Phys. C*, **16**, 6205–20.

Huang, K. (1987) *Statistical Mechanics*, 2nd edition, Wiley, New York.

Hulliger, F. (1976) *Structural Chemistry of Layer-Type Phases*, Reidel, Dordrecht.

Johnston, D. C. (1982) *Mat. Res. Bull.*, **17**, 13–23.

Kittel, C. (1971) *Introduction to Solid State Physics*, 4th edition, Wiley, New York.

McKinnon, W. R. and Dahn, J. R. (1986) *J. Phys. C*, **19**, 5121–33.

McKinnon, W. R., Dahn, J. R. and Jui, C. C. H. (1985) *J. Phys. C*, **18**, 4443–58.

McKinnon, W. R., Dahn, J. R., Murray, J. J., Haering, R. R., McMillan, R. S. and Rivers-Bowerman, A. H. (1986) *J. Phys. C*, **19**, 5135–48.

McKinnon, W. R. (1987) in *Chemical Physics of Intercalation; NATO ASI Series, Series B: Physics*, **172**, 181–94.

McKinnon, W. R. and Haering, R. R. (1983) in *Modern Aspects of Electrochemistry*, Vol. 15, Eds. R. E. White, J. O'M. Bockris and B. E. Conway, Plenum, New York, pp. 235–304.

Moline, P., Trichet, L., Rouxel, J., Berthier, C., Chabre, Y. and Segransan, P. (1984) *J. Phys. Chem. Solids*, **45**, 105–12.

Mott, N. F. and Davis, E. A. (1979) *Electronic Processes in Non-Crystalline Materials*, Clarendon Press, Oxford.

Mulhern, P. J. (1989) *Can. J. Phys.*, **67**, 1049–52.

Murphy, D. W., DiSalvo, F. J., Carides, J. N. and Waszczak, J. V. (1978) *Mat. Res. Bull.*, **13**, 1395–402.

Murphy, D. W., Christian, P. A., DiSalvo, F. J., Carides, J. N. and Waszczak, J. V. (1981) *J. Electrochem. Soc.*, **128**, 2053–60.

Murray, J. J. and Alderson, J. E. A. (1989) *J. Power Sources*, **26**, 293–9.

Murray, J. J., Sleigh, A. K. and McKinnon, W. R. (1991) *Electrochimica Acta*, **36**, 489–98.

Nohl, H., Klose, W. and Andersen, O. K. (1982) *Superconductivity in Ternary Compounds I*, Eds. O. Fischer and M. B. Maple, Springer-Verlag, Berlin, pp. 25–86.

Perrin, A., Chevrel, R., Sergent, M. and Fischer, O. (1980) *J. Solid State Chem.*, **33**, 43–7.
Py, M. A. and Haering, R. R. (1983) *Can. J. Phys.*, **61**, 76–84.
Py, M. A. and Haering, R. R. (1984) *Can. J. Phys.*, **62**, 10–14.
Richards, P. M. (1984) *Phys. Rev. B*, **30**, 5183–9.
Rouxel, J. (1989) in *Solid State Ionics*, Eds. G. Nazri, R. F. Huggins and D. F. Shriver, Materials Research Society, Pittsburgh, pp. 431–42.
Safran, S. E. (1987) *Solid State Phys.*, **40**, 183–246.
Selwyn, L. S. and McKinnon, W. R. (1987) *J. Phys. C*, **20**, 5105–23.
Selwyn, L. S., McKinnon, W. R., Dahn, J. R. and Le Page, Y. (1986) *Phys. Rev. B*, **33**, 6405–14.
Shannon, R. D. (1976) *Acta Crystallographica A*, **32**, 751–67.
Sinha, S. and Murphy, D. W. (1986) *Solid State Ionics*, **20**, 81–4.
Solin, S. A. (1982) *Adv. Chem. Phys.*, **49**, 455–532.
Tarascon, J. M. and Colson, S. (1989) *Mat. Res. Soc. Symp. Proc.*, **135**, 421–9.
Tarascon, J. M., DiSalvo, F. J., Chen, C. H., Carroll, P. J., Walsh, M. and Rupp, L. (1985) *J. Solid State Chem.*, **58**, 290–300.
Tigchelaar, D., Wiegers, G. A. and van Bruggen, C. F. (1982) *Revue de Chimie Minerale*, **19**, 352–9.
Umrigar, C., Ellis, D. E., Wang, D. E., Krakauer, H. and Posternak, M. (1982) *Phys. Rev. B*, **26**, 4395–50.
Underhill, C., Krapchev, T. and Dresselhaus, M. S. (1980) *Synthetic Metals*, **2**, 47–55.
Wagner, H. (1978) in *Hydrogen in Metals I*, Eds G. Alefeld and J. Volkl, Springer-Verlag, Berlin, pp. 5–51.
Whittingham, M. S. and Gamble, F. R. (1975) *Mat. Res. Bull.*, **10**, 363–72.
Yvon, K. (1982) in *Superconductivity in Ternary Compounds I*, Eds. O. Fischer and M. B. Maple, Springer-Verlag, Berlin, pp. 25–86.
Zaanen, J., Sawatzky, G. A. and Allen, J. W. (1985) *Phys. Rev. Lett.*, **55**, 418–21.

8 Electrode performance

W. WEPPNER
Christian-Albrechts-University, Kaiserstr. 2, D-24143 Kiel, Germany

8.1 Electrodes: Ionic sources and sinks

The thermodynamics of insertion electrodes is discussed in detail in Chapter 7. In the present chapter attention is focused mainly on the general kinetic aspects of electrode reactions and on the techniques by which the transport of species within electrodes may be determined. The electrodes are treated in a general fashion as exhibiting mixed ionic and electronic transport, and attention is concentrated on the description of the coupled transport of these species. In this context it is useful to consider that an electronically conducting lead provides the electrons at the electrodes and compensates the charges of the ions transferred by the electrolyte.

Because of the high mobility of the mobile (electroactive) species in the electrolyte, there is a tendency for these ions to move from the electrode of higher activity to the electrode of lower activity. The displacement of the ions comes to a stop when an electrical field is built up which drives the ion i in the opposite direction

$$\text{grad}\,\phi = -\frac{1}{z_i q}\,\text{grad}\,\mu_i, \tag{8.1}$$

where ϕ, μ, z and q are the electrostatic potential, the chemical potential (per ion), the charge number and the elementary charge, respectively. The integral of Eqn (8.1) between both electrodes is the galvanic cell voltage. In most good solid electrolytes, the chemical potential of the mobile ions i is essentially constant throughout the bulk electrolyte because of the high ionic disorder. There is no major electrostatic potential drop within the electrolyte in this case. Also, an electrical field may not be built up within the electrodes because of the high electronic (metallic) conductivity. The electrostatic potential drop occurs nearly exclusively at the interfaces between the electrodes and the electrolyte. This is an ionic analogue of semiconductor junctions. The flux due to the electrical field at the interface

$$j_E = -\frac{\sigma_i}{z_i q} \operatorname{grad} \phi \tag{8.2}$$

compensates the flux due to diffusion

$$j_D = -\frac{c_i D_i}{kT} \operatorname{grad} \mu_i = -\frac{\sigma_i}{z_i^2 q^2} \operatorname{grad} \mu_i \tag{8.3}$$

(Section 8.2) caused by the different concentrations of the mobile ions in the electrode and in the electrolyte.

The chemical potential μ is related to the activity a by

$$\mu = \mu^\circ + kT \ln a \tag{8.4}$$

where μ°, k and T are the chemical potential in the standard state, Boltzmann's constant and the absolute temperature, respectively. The activity is related to the concentration c by $a = \gamma c$, where γ is the activity coefficient. The electrical migration is proportional to the concentration of the species, whereas the diffusion is only proportional to the activity *gradient*, or concentration *gradient* if γ is constant. The galvanic cell voltage therefore shows a logarithmic dependence on the activity (or concentration) of the electroactive species in the electrodes.

The cell voltage measured using two metallic conducting leads connected to the electrodes also includes electrostatic potential gradients which compensate gradients of the chemical potential of the electrons in the electrodes analogous to the Eqn (8.1). The measurable cell voltage is then given by the difference of the chemical potentials of the neutral electroactive species μ_{ix}

$$E = \frac{1}{z_i q}(\mu_{ix}^{\mathrm{l}} - \mu_{ix}^{\mathrm{r}}) = \frac{kT}{z_i q} \ln \frac{a_{ix}^{\mathrm{l}}}{a_{ix}^{\mathrm{r}}}, \tag{8.5}$$

where l and r stand for the left and right hand electrode, respectively (Nernst's law). Accordingly, the electrodes provide a driving force for the mobile ions in the electrolyte which generates the galvanic cell voltage E.

One may also look at the effect of the electrodes from the point of view of energy balance. Measurement of voltages always requires at least a small electrical current. This corresponds in the case of galvanic cells to the transfer of electroactive species from one electrode to the other. The corresponding chemical work is the change of the Gibbs energy, ΔG, which is $\mu_{ix}^{l} - \mu_{ix}^{r}$ in the case of the transfer of a single ion, if the temperature and total pressure are kept constant. This corresponds to the electrical work $z_i qE$ of z_i electrons through the external electrical circuit. The result of this energy balance is the same as given in Eqn (8.5). More details concerning the thermodynamics of insertion electrodes may be obtained from Chapter 7.

In order to exhibit a high cell voltage E (e.g. for a battery), the change of the Gibbs energy when an electroactive species moves from one electrode to the other has to be high, or the chemical potential of the neutral electroactive component should differ greatly for the two electrodes. Generally, the ΔG value and the chemical potential depend on the composition of the electrodes which changes during the discharge of a galvanic cell and may result in a change of the cell voltage. This may be disadvantageous in the case of a battery. However, when the number of components equals the number of phases of the electrode, the chemical potential and therefore the cell voltage are independent of the composition according to Gibbs' phase rule

$$f + p = c + 2, \tag{8.6}$$

where f, p and c are the degree of freedom, the number of phases and the number of components, respectively, if the temperature and total pressure are kept constant. The relative amounts of the p phases may change, but the same phases are in equilibrium with each other. Often in practice the effective number of thermodynamically active components is much smaller than the real number of components, because an equilibrium may not be reached within reasonable periods of time because of the low mobility of the species. For example, yttria stabilised zirconia is a ternary compound which may be treated as a binary one because of the low diffusion coefficients of Zr and Y which have been estimated to be below 10^{-50} cm^2 s^{-1}. The ratio of these two components will not change during the reequilibration of phases.

We have seen that the cell potential is generated at the interfaces between the electrodes and the electrolyte. Therefore, the composition of the electrode at this interface is important and this does not have to be identical with the bulk composition. In fact, large deviations have been observed due to segregation of some of the components of the electrode and especially due to impurities at the surface. If the surface of the electrode is equilibrated with the bulk, both have the same chemical potential of the electroactive component if that is sufficiently mobile in the electrode. A problem may occur, however, if the chemical potential of the electrons changes drastically between the surface and the bulk of the electrode. Another reason for changes at the potential generating interfaces is the chemical reactivity between the two phases in contact with each other. This is a most critical issue in many cases. It is very common to use phases which may not be considered to be really thermodynamically stable when in contact with each other. The cell potential is always generated at the transition from predominantly ionic to electronic conductivity. As long as this interface is in equilibrium with the bulk of the electrode for the electroactive component, there is no variation of the cell voltage. But, in many cases it is expected that the chemical potential will drop over the reaction product layer and an erroneous result will be obtained. The chemical potential at the surface of the (electronically conducting) reaction product is measured.

In addition to the thermodynamic aspects of electrodes as described so far there are important kinetic requirements of the ionic sources and sinks. In order to pass a sufficiently large current through a galvanic cell, the electroactive species have to be supplied to the electrolyte at one electrode and incorporated into the other electrode sufficiently quickly. Polarisations (decreases of the galvanic cell voltage by a current flux) occur within the electrodes because of local inhomogeneities. Chemical diffusion is the driving force for the motion of the ions within the electrodes which limits the current. The cell voltage measures the chemical potential of the electroactive component at the electrode surfaces. The difference in electrode potentials of the electroactive species between the bulk electrode and its surface is established in order to provide a sufficiently large concentration gradient to generate a diffusional flux as required by the electrical current. In addition to this bulk transport, polarisation effects may occur by the transfer of the ions through segregation barriers and reaction products. Also, transfer across the interfaces may be a limiting factor.

Electrolytes as well as electrodes should have high diffusivities for the electroactive species in order to allow high currents. The two phases should have very different electronic properties, however. For electrolytes the conductivity of the electrons must be low. This has to be mainly caused by electronic species of low mobility rather than low concentration. Otherwise, small variations in the stoichiometry of the electrolyte, e.g. by the application of two different electrodes, would result in a relatively large change in the concentration which could cause high electronic conductivity. Also, impurities would readily alter a low electron concentration which again would result in a high electronic conductivity. From a practical point of view when fabricating cells it is important to be able to tolerate a comparatively large amount of impurities.

In electrodes, however, the electronic conductivity should be large. It will be shown in Section 8.3 that a favourable situation is a low concentration of electronic species with high mobility. This allows high effective diffusion rates. From a kinetic point of view, electrodes should exhibit primarily electronic conductivity generated by a small number of very mobile electrons or holes. In the following sections, the fundamental relations for the transport of the ions in the electrodes will be described.

8.2 Transport of ions and electrons in mixed conductors

Diffusion of the electroactive species within the electrode toward or away from the interface with the electrolyte is an irreversible process. The sum of the products of the forces and fluxes corresponds to the entropy production. In order to avoid space charge accumulation, the motion of at least two types of charged species has to be considered for charge compensation. Onsager's equations read in the isothermal case (neglecting energy fluxes)

$$j_1 = L_{11} \operatorname{grad} \tilde{\mu}_1 + L_{12} \operatorname{grad} \tilde{\mu}_2, \tag{8.7}$$

$$j_2 = L_{21} \operatorname{grad} \tilde{\mu}_1 + L_{22} \operatorname{grad} \tilde{\mu}_2, \tag{8.8}$$

where $\tilde{\mu}$ and L_{ij} are the electrochemical potential and Onsager's coefficients for which the relation

$$L_{ij} = L_{ji} \tag{8.9}$$

holds (Rickert, 1985). L_{12} and L_{21} indicate the coupling of the fluxes of the different species. In many cases the coupling may be neglected compared to the flux under the influence of the electrochemical potential gradient of the species under consideration.

On the other hand, from a phenomenological point of view, the flux is given by the product of the concentration and the average velocity of the species,

$$j = cv. \tag{8.10}$$

The average velocity is the result of the force $\mathbf{F} = -\text{grad}\,\tilde{\mu}$. The ratio of the velocity over the force is defined as the general mobility b

$$\mathrm{b} = \frac{v}{\mathbf{F}} = -\frac{v}{\text{grad}\,\tilde{\mu}} \tag{8.11}$$

and Eqn (8.10) yields the following general flux equation

$$j = -cb\,\text{grad}\,\tilde{\mu}. \tag{8.12}$$

This may be separated into two terms, one involving an activity gradient (diffusion) and the other an electrical field (migration)

$$j = -cb\,\text{grad}\,\mu - cbzq\,\text{grad}\,\phi. \tag{8.13}$$

The first term may be written in terms of a concentration gradient considering $\mu = \mu^\circ + kT \ln a$ and the expansion by $\mathrm{d} \ln a/\mathrm{d} \ln c$ since $c\,\mathrm{d} \ln c = \mathrm{d}c$

$$j_D = -bkT\frac{\mathrm{d} \ln a}{\mathrm{d} \ln c}\frac{\mathrm{d}c}{\mathrm{d}x}. \tag{8.14}$$

The expression $\mathrm{d} \ln a/\mathrm{d} \ln c$ is known as the thermodynamic factor and is a special case of the Wagner factor (or thermodynamic enhancement factor) which plays an important role for the *kinetic* properties of electrodes. This term indicates the deviation from ideality of the mobile component. For ideal systems this quantity becomes 1 and comparison with Fick's first law yields

$$D = bkT, \tag{8.15}$$

which is known as the Nernst–Einstein relation. D is called Fick's diffusion coefficient. In the general (non-ideal) case, this quantity $D = bkT$ is called diffusivity because it describes the mobility of the ions in the material without directed driving forces.

The second term of Eqn (8.13) may be compared with Ohm's law which holds at sufficiently low electrical fields as long as the product of the potential difference per atomic distance and the elementary charge is sufficiently low compared to kT (Rickert, 1985),

$$i = -\sigma\,\text{grad}\,\phi. \tag{8.16}$$

Considering $i = zqj$, comparison with the second term of Eqn (8.13) yields

$$j_E = -\frac{\sigma}{zq}\,\text{grad}\,\phi = -cbzq\,\text{grad}\,\phi = -\frac{cDzq}{kT}\,\text{grad}\,\phi. \tag{8.17}$$

Instead of the general mobility it is more common to use the electrical mobility u which is defined as the ratio of the average velocity over the electrical field. Since the electrical force is given by $\mathbf{F} = -zq \operatorname{grad} \phi$, comparison with Eqn (8.11) yields for the relation between the electrical and general mobility

$$u = |z|qb. \tag{8.18}$$

The fluxes due to the influence of both an electrical field and an activity gradient may be given by the following relations

$$\begin{aligned} j &= -cb \operatorname{grad} \tilde{\mu} \\ &= -\frac{cu}{|z|q} \operatorname{grad} \tilde{\mu} \\ &= -\frac{cD}{kT} \operatorname{grad} \tilde{\mu} \\ &= -\frac{\sigma}{z^2 q^2} \operatorname{grad} \tilde{\mu}. \end{aligned} \tag{8.19}$$

For the diffusion process in electrodes the interaction between all mobile species has to be taken into consideration. Overall charge neutrality holds

$$\sum_i z_i j_i = 0. \tag{8.20}$$

In general, only the two fastest species have to be taken into consideration and all other ones may be neglected. The second fastest species are rate controlling since these limit the motion of the fastest ones. In general, electrons (or holes) and the fastest moving ionic species have to be considered in electrodes. For the combined motion, the proportionality constant between the flux and the concentration gradient is called the chemical diffusion coefficient $\tilde{D}_i$

$$j_i = -\tilde{D}_i \frac{\partial c_i}{\partial x}, \qquad i = 1, 2. \tag{8.21}$$

By inserting this expression into Eqn (8.20) and considering that $z_1 c_1 = -z_2 c_2$ it can be shown that the chemical diffusion coefficients of both species are the same

$$\tilde{D}_1 = \tilde{D}_2 := \tilde{D}. \tag{8.22}$$

This means that the motions of the two species are coupled and show the same transport rate because of electroneutrality. The fastest species move ahead and generate an electrical field in such a way that the faster species are slowed down and the slower ones are accelerated.

The internal electrical field is not known. It may, however, be eliminated from the equations for the fluxes (Eqn (8.19)) by considering $\tilde{\mu} = \mu + zq\phi$, $\mu = \mu^\circ + kT \ln a$ and the electroneutrality condition (8.20). The result is in the general case of any number of mobile species

$$j_i = -D_i\left(\frac{\partial \ln a_i}{\partial \ln c_i} - \sum_j t_j \frac{z_i}{z_j} \frac{\partial \ln a_j}{\partial \ln c_i}\right)\frac{\mathrm{d}c_i}{\mathrm{d}x}, \tag{8.23}$$

where t is the transference number. Because of the ionisation equilibrium between the ions and the electrons

$$\mathrm{d} \ln a_i + z_i \,\mathrm{d} \ln a_\mathrm{e} = \mathrm{d} \ln a_i - z_i \,\mathrm{d} \ln a_\mathrm{h} = \mathrm{d} \ln a_{i\mathrm{x}}, \tag{8.24}$$

Eqn (8.23) may also be written in terms of the neutral component

$$j_i = -D_i\left(\frac{\partial \ln a_{i\mathrm{x}}}{\partial \ln c_{i\mathrm{x}}} - \sum_{j \neq \mathrm{e,h}} t_j \frac{z_i}{z_j} \frac{\partial \ln a_{j\mathrm{x}}}{\partial \ln c_{i\mathrm{x}}}\right)\frac{\mathrm{d}c_{i\mathrm{x}}}{\mathrm{d}x} \tag{8.25}$$

and comparison with Eqn (8.21) yields

$$\tilde{D}_i = D_i W_i \tag{8.26}$$

with

$$W_i = \left[(1 - t_i)\frac{\partial \ln a_{i\mathrm{x}}}{\partial \ln c_{i\mathrm{x}}} - \sum_{j \neq i,\mathrm{e,h}} \frac{z_i}{z_j} \frac{\partial \ln a_{j\mathrm{x}}}{\partial \ln c_{i\mathrm{x}}}\right]. \tag{8.27}$$

This expression is the general Wagner factor which includes the influence of all the motion of the other species on the motion of species i by the effect of the internal electric fields. W may be larger than 1 which indicates an enhancement of the motion by the simultaneous motions of other species, or W may be smaller than 1 which means that the species are slowed down because of the immobility of other species which are therefore unable to compensate for the electrical charges. The first situation is desirable for electrodes whereas the second one is required for electrolytes in which mobile species should not move except when electrons are provided through the external circuit. Since the transference numbers in Eqn (8.27) include the partial and total conductivities ($t_j = \sigma_j/\sum_k \sigma_k$) or the products of the diffusivities (or mobilities) and the concentrations, Eqn (8.27) shows that W depends both on kinetic parameters (diffusivities, mobilities) and thermodynamic properties (stoichiometries, activities).

If the activity coefficients γ_j for all ions and electrons are constant, i.e. Henry's or Raoult's law or the law of ideally diluted solutions holds, Eqn (8.27) reads

$$W_i = 1 - t_i - \sum_{j \neq i} t_j \frac{z_i c_i}{z_j c_j} \frac{\partial c_j}{\partial x} \left(\gamma_i = \frac{a_i}{c_i} = \text{const.} \right). \tag{8.28}$$

Electrode reactions are analogous to the growth of tarnishing (corrosion) layers (Weppner and Huggins, 1977). Assuming that bulk transport is the rate determining step, the growth rate of the reaction product is inversely proportional to the instantaneous thickness L

$$\frac{\mathrm{d}L}{\mathrm{d}x} = \frac{k_t}{L}, \tag{8.29}$$

where k_t is Tammann's parabolic or practical tarnishing constant which is generally dependent on the activities of the components. On the other hand, the growth of the electrode by the uptake of the electroactive species is given by the flux according to Fick's law (8.21)

$$\frac{\mathrm{d}L}{\mathrm{d}t} = \frac{\tilde{D}}{\bar{c}_i} \left| \frac{\mathrm{d}c_i}{\mathrm{d}x} \right|, \tag{8.30}$$

where $\bar{c}_i$ is the average concentration of the moving species in the product phase.

The chemical diffusion coefficient may be expressed by the diffusivity of the mobile species i, the transference number and the variation of the activity of the mobile component as a function of its concentration according to Eqns (8.26) and (8.27), if the transference numbers of other ionic species are negligibly small,

$$\frac{\mathrm{d}L}{\mathrm{d}t} = D_i t_e \left| \frac{\partial \ln a_{ix}}{\partial x} \right|. \tag{8.31}$$

Combining Eqn (8.29) with Eqn (8.30) or (8.31) and subsequent integration over the thickness L of the sample provides the following relationship

$$k_t = \frac{1}{\bar{c}_i} \int_0^L \tilde{D}\, \mathrm{d}c_i = \int_0^L D_i t_e \,\mathrm{d} \ln a_{ix}. \tag{8.32}$$

This equation relates the growth rate of the electrode to the chemical diffusion coefficient $\tilde{D}_i$ or the product of the diffusivity D_i and the transference number of the electrons t_e. If these basic parameters are known, the reaction rate is given or an optimum value may be chosen, if bulk transport is rate determining.

In view of what we have now learnt concerning the interplay between the motions of electronic and ionic species the features that an electrode must possess in order to provide optimum performance in a cell, will now be considered in the following section.

8.3 Kinetics of electrodes and the role of electrons in atomic transport

In electrodes the electronic species typically have the highest transference number. The motion of the most mobile ions generally determines the rate of the discharging and charging processes. But the ionic transport rate may be largely influenced by many orders of magnitude by the interaction with the electrons and holes. Eqn (8.23) reads in this specific case

$$j_i = -D_i t_e \left(\frac{\partial \ln a_i}{\partial \ln c_i} + z_i \frac{\partial \ln a_e}{\partial \ln c_i} \right) \frac{dc_i}{dx}$$

$$= D_i t_e \frac{\partial \ln a_{ix}}{\partial \ln c_{ix}} \frac{dc_{ix}}{dx}. \tag{8.33}$$

Compared to Fick's first law for ideal systems (which real systems only approach at the high dilution), two extra terms are introduced: (i) the transference number of the electrons and (ii) the Wagner factor. Their part in the electrode kinetics will be discussed in more detail.

t_e may be determined experimentally by transference or polarisation measurements. Its value should be close to 1 in order to produce fast equilibration of the electrodes.

The factor $\partial \ln a_{ix} / \partial \ln c_{ix}$ may be determined experimentally by measuring the emf (which is proportional to $\ln a_{ix}$) as a function of the composition (see Section 8.5). In order to give rise to a high Wagner factor W, the emf should change rapidly as the concentration of the mobile component varies (Fig. 8.1). Every phase accommodates a certain range of activities over which it is stable. Accordingly, phases which are formed rapidly have narrow ranges of stoichiometry. It is kinetically more advantageous for a galvanic cell reaction to form a new phase with a small stoichiometric width than to change the stoichiometry of a phase with a very wide stoichiometric range. Compounds with large stoichiometric ranges, such as intercalation compounds, have previously been considered to be good electrode materials, but this is not true from a kinetic point of view. It will be shown later that the requirement of narrow stoichiometric width corresponds to the requirement of semiconductivity rather than metallic conductivity of the electrode. The often stated desirability for metallic electrodes has to be revised in this respect.

Of course, the Wagner factor is only capable of enhancing the motion of the ions. In the case of very low diffusivities, W may enhance the diffusion by many orders of magnitude, but the chemical diffusion coefficient may still be low. In contrast one might think that a high diffusivity in a metallic system would not require any further enhancement. Based on our present state of knowledge, the ionic diffusivities in metals and semiconductors are rather similar and it is therefore helpful to make use of large Wagner factors. This is especially helpful because there are ways to control this factor by suitable doping of semiconductors.

In order to gain some insight into the ionic and electronic mobilities and concentrations that are required for a favourable electrode performance, the simplifying case of constant γ_i and γ_e (Henry's or Raoult's law) is considered. Under these conditions the Wagner factor reads according

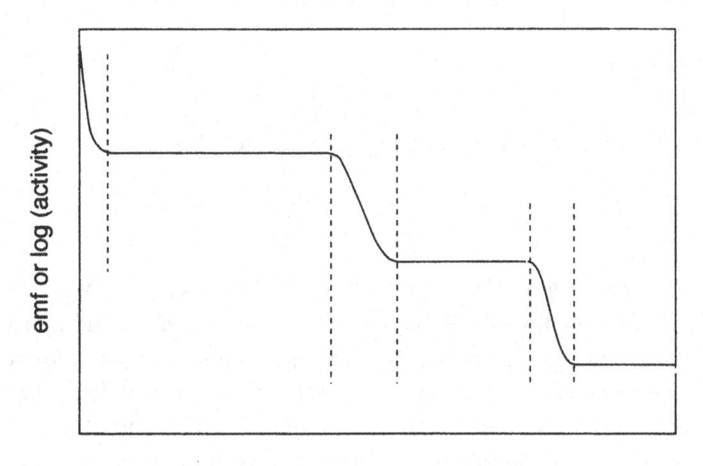

Fig. 8.1 Variation of the activity (logarithmic scale) or galvanic cell voltage as a function of the composition (schematic). The plateaux indicate multi-phase regions in which the activity is fixed according to Gibbs' phase rule.

to Eqn (8.33)

$$W = t_e\left(1 + z_i \frac{\partial \ln c_e}{\partial \ln c_i}\right). \qquad (8.34)$$

Since only the concentrations of ions i and electrons e are changing locally, the electroneutrality condition requires

$$dc_e = z_i \, dc_i \qquad (8.35)$$

and the Wagner factor becomes under the given assumptions

$$W = t_e\left(1 + z_i^2 \frac{c_i}{c_e}\right). \qquad (8.36)$$

In order to produce a large W factor, the concentration of the mobile ions should be high compared to the concentration of the electronic species. However, c_i should not be very much larger than c_e in order to keep the transference number of the electrons close to 1. These requirements are somewhat contradictory, but may best be fulfilled if the mobility of the small number of electrons is very large compared to the mobility of the ions. The conductivity of the electrons may in this way be kept larger than the conductivity of the ions. In order to come up with a quantitative relationship, the transference number of the electrons t_e is substituted by the product of the concentrations and the diffusivities or mobilities

$$t_e = \frac{\sigma_i}{\sigma_e + \sigma_i} = \frac{c_e u_e}{c_e u_e + z_i^2 c_i u_i}. \qquad (8.37)$$

Substitution of Eqn (8.37) into Eqn (8.36) results in

$$W = \frac{(c_e^2 + z_i^2 c_e c_i) u_e}{c_e^2 u_e + z_i^2 c_e c_i u_i}. \qquad (8.38)$$

Fig. 8.2 shows a graphical representation of the Wagner factor W as a function of the ratios of the electronic to ionic mobilities and concentrations, u_e/u_i and c_e/c_i, respectively, in a logarithmic form. It is found that large enhancement factors are obtained at sufficiently low electronic concentrations and high electronic mobilities. This situation is most favourable for semiconductors (in agreement with the previous consideration) which commonly fulfil both requirements. The mobility of the electrons is commonly much larger in semiconducting than in metallic materials. Large electronic concentrations and mobilities provide a typical value of $W = 1$. If the mobility of the electrons is low, values of $W \leq 1$ may be observed. This condition is required for solid electrolytes in which the motion of the ions is blocked because no charge compensating electrons are available. The requirement for electrolytes is very low electronic mobility rather than a low concentration of electrons. From the point of view of these electronic properties, electrodes and electrolytes are clearly distinct from each other. Electrodes should have highly mobile electronic species whereas electrolytes require low electronic mobilities. Studies of the electronic properties are therefore very useful in the search for electrode and electrolyte materials.

An example of a material (Li_3Sb) with a very large Wagner factor is shown in Fig. 8.3. The effective chemical diffusion coefficient is compared with the diffusivity as a function of non-stoichiometry. These data were determined by electrochemical techniques (see Section 8.5). An increase of the diffusion coefficient is observed at about the ideal stoichiometry which corresponds to a change in the mechanism from a predominantly vacancy to interstitial mechanism. The Wagner factor W is as large as 70 000 at the ideal stoichiometry. This gives an effective diffusion coefficient which is more typical of liquids than solids. It is a common

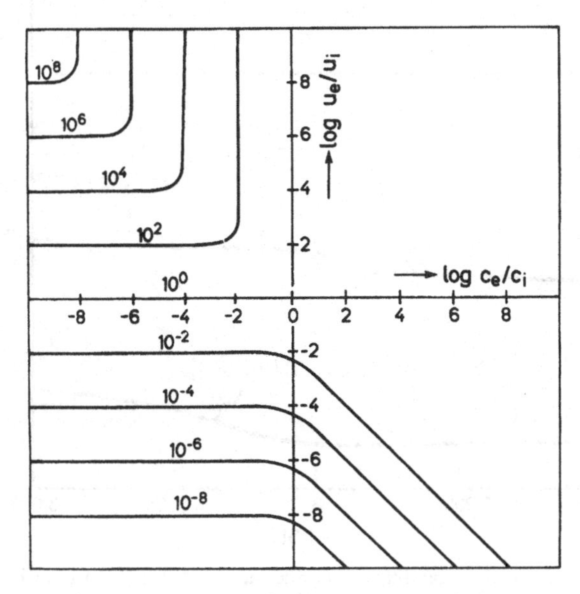

Fig. 8.2 The Wagner factor W as a function of the ratios of the mobilities u and concentrations c of the electronic and mobile ionic species (logarithmic scale).

phenomenon that the Wagner factor raises the diffusivity of semiconducting materials from typical solid-like to liquid-like behaviour. *W* shows a maximum at the ideal stoichiometry which corresponds to the lowest electronic concentration as required by Eqns (8.10) and (8.38). Other electrode materials which show fast chemical diffusion in the solid state are shown in Fig. 8.4. A large variety of lithium compounds have been studied because of the interest in electrodes for lithium cells. Their chemical diffusion coefficients are liquid-like and, for example, higher than the diffusivity of copper in liquid copper, in spite of the lower temperature. In some cases, the chemical diffusion coefficient of electrode materials is even higher or of the same order of magnitude as is typically found for gases. The most prominent example is Ag_2S which shows a faster silver ion diffusion than oxygen in air. It has been found that the kinetics is even faster at lower temperatures than at higher temperatures. This

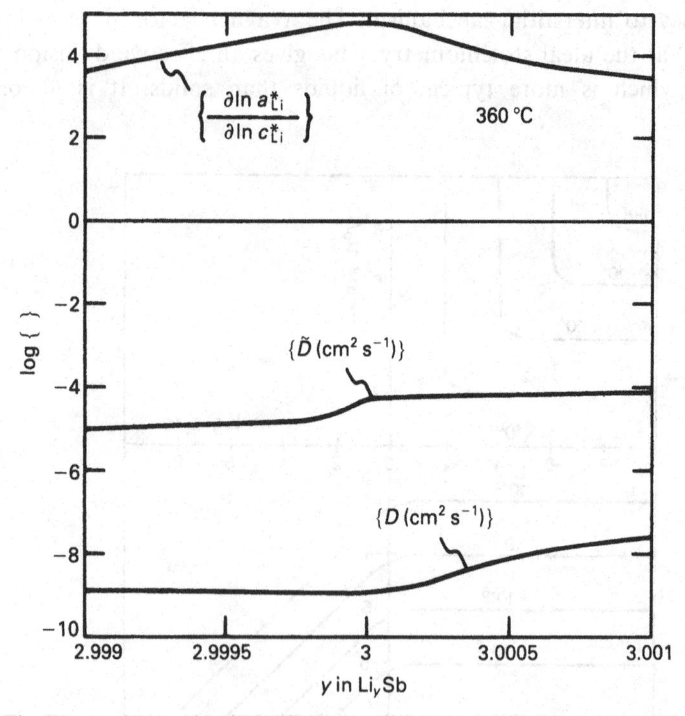

Fig. 8.3 The Wagner factor, chemical diffusion coefficient and diffusivity of Li in Li_3Sb as a function of stoichiometry at 400 °C. The Wagner factor is as large as 70 000 at ideal stoichiometry.

phenomenon is caused by the increasing number of electronic species with increasing temperature. The electrons increasingly shield the internal electrical field more and more which is responsible for the enhancement of the silver ion motion.

In the favourable case of low concentrations of highly mobile electronic species, both electrons and ions diffuse under the influence of concentration gradients. The electrons move ahead of the ions and in this way generate an internal electrical field in which the ions are accelerated and the electrons are slowed down in order to maintain electroneutrality. If

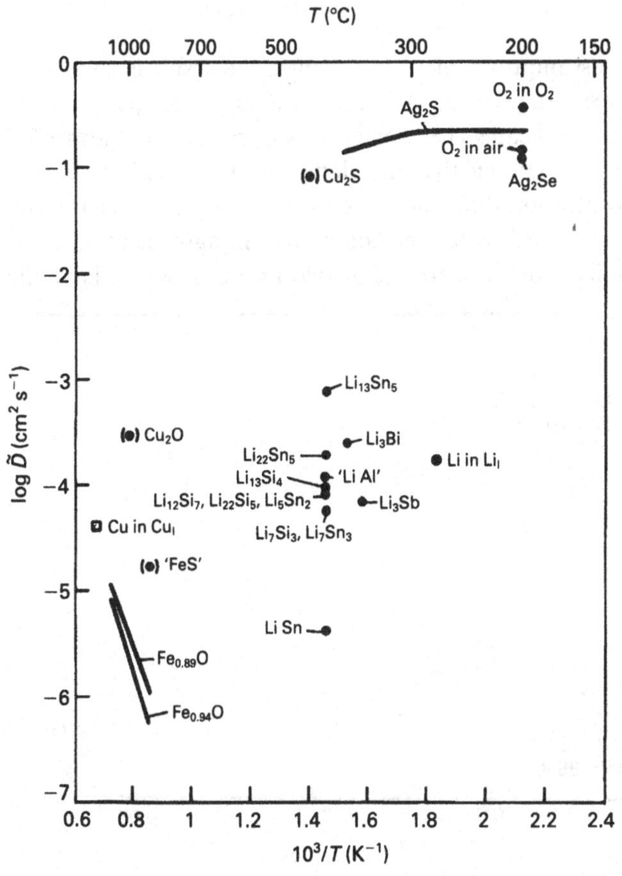

Fig. 8.4 Chemical diffusion coefficient as a function of temperature for various predominantly electronically conducting materials. For comparison the diffusion coefficients of Cu in liquid Cu and of O_2 in air are shown.

the concentration of electrons is large, a small backward diffusion of this large number would compensate the electrical field. In other words, the electronic species would shield the electrical field from the ions. The electrical field exerts the maximum enhancement of the motion of the ions if the concentration of electrons is small.

In order to estimate the order of magnitude of the internal electrical field, the two flux equations for the ions and electrons may be solved for the electrostatic potential gradient (rather than eliminating this quantity) as a function of the local difference in concentration (Weppner, 1985).

$$\frac{d\phi}{dx} = -q\left(\frac{z_i c_i D_i(\partial\mu_i/\partial x) - c_e D_e(\partial\mu_e/\partial x)}{zi^2 c_i D_i + c_e D_e}\right). \tag{8.39}$$

Under the assumptions of γ_i, γ_e = const., the result is shown in Fig. 8.5 for the three compounds Li_3Sb, β-Ag_2S and $Fe_{1-\delta}O$. The end of the straight lines indicates the maximum stoichiometric width of the compounds. It can be seen that the slope, i.e. the potential difference for a given concentration difference at two locations, is larger the smaller the stoichiometric width is for the compound, in agreement with the discussion of semiconductors above. For two locations which have the

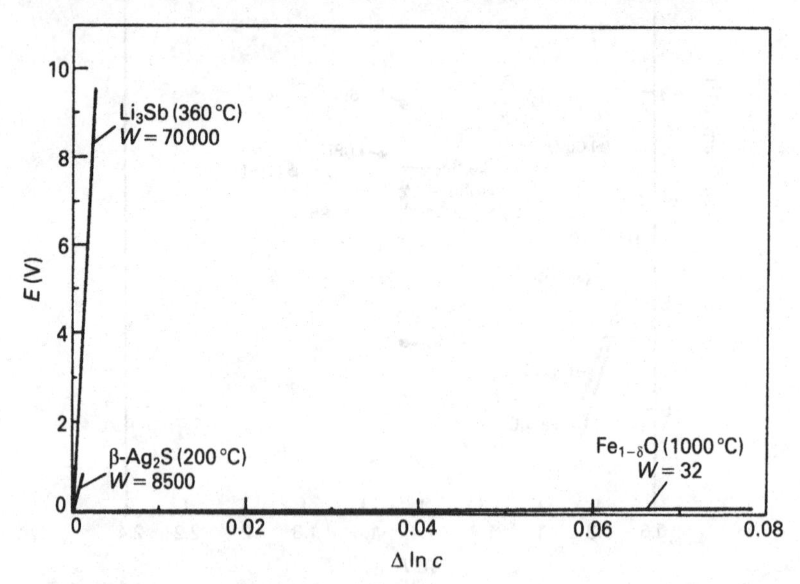

Fig. 8.5 Electrostatic potential differences between two locations as a function of the compositional difference for three different electrode materials. The end of the lines indicates the maximum width of the stoichiometry of the compounds.

maximum difference in the stoichiometric width, the voltage may be as large as 10 V in the case of Li_3Sb. For an initial period of variation of the stoichiometry of this compound when the total stoichiometric width changes over, for example, a distance of 1 μm, an electrical field as large as 10^5 V cm^{-1} is present.

The motion of the lithium ions is in this case to a large extent due to the electrical field rather than to a concentration gradient in spite of the fact that the motion originates from the concentration gradient as the driving force. The origin of the motion of ions in electrodes is a local difference in the concentration but in fast electrodes the electrical field becomes the real driving force.

The need for a small phase width for fast electrode kinetics requires the electrodes to undergo phase changes upon discharging and charging the galvanic cell. The phase change occurs at the interface with the electrolyte. Because of the fast kinetics, this phase is nearly always in equilibrium and its formation only requires a slightly lower (or higher) activity of the electroactive component than that of the starting electrode material, i.e. the polarisation of the electrode is only small because of the fast intermediate phase formation.

Semiconducting electrodes appear to be disadvantageous compared to metallic electrodes from the point of view of a large electrode resistance. This is, however, only a minor drawback since the larger IR drop causes only a comparatively small decrease of the voltage compared to the larger reduction of the concentration polarisation by fast diffusion. In addition, the resistance of a semiconducting electrode is commonly much smaller than the resistance of the electrolyte and is therefore usually negligible.

In addition to the electrons several other factors may influence the kinetics and the cell voltage. One factor is the formation of metastable phases: this has been found, for example, in the case of the Li_xFeS_2 electrode (Fig. 8.6). The voltage is 79 mV (at 400 °C) lower than expected for the formation of the thermodynamically most favourable compounds (Weppner and Schmidt, 1984). The addition of lithium by the discharge of the galvanic cell produces a higher lithium activity by the formation of a kinetically more favourable compound. Rather than changing the ratios of the three compounds FeS, FeS_2 and Li_2S which are thermodynamically stable with each other, the metastable phase Li_2S_2 forms which is thermodynamically unstable but kinetically more favourable. As a result there is a loss of practically useful electrical energy but the kinetics is more favourable and allows a higher current to be drawn.

Another kinetic aspect is observed if a component other than the electroactive species is predominantly mobile. The electroactive species are in this case made available to the electrolyte by the motion of the other components in the opposite direction. In a binary compound this does not make a difference to the electrode performance. But in the case of a compound with more than two components the composition is changed to values which are not expected from a thermodynamic point of view for the variation of the concentration of the electroactive species. Other phases are formed which may provide a lower or higher activity of the electroactive species than that expected thermodynamically. This has an influence both on the current and the cell voltage. Upon discharging and charging a galvanic cell, the composition of the electrode at the interface with the electrolyte may follow very different compositional pathways (Weppner, 1985).

8.4 Thermodynamics of electrodes

The electrodes store chemical energy. The amount depends on the thermodynamic quantity of the Gibbs energy of reaction or differences of

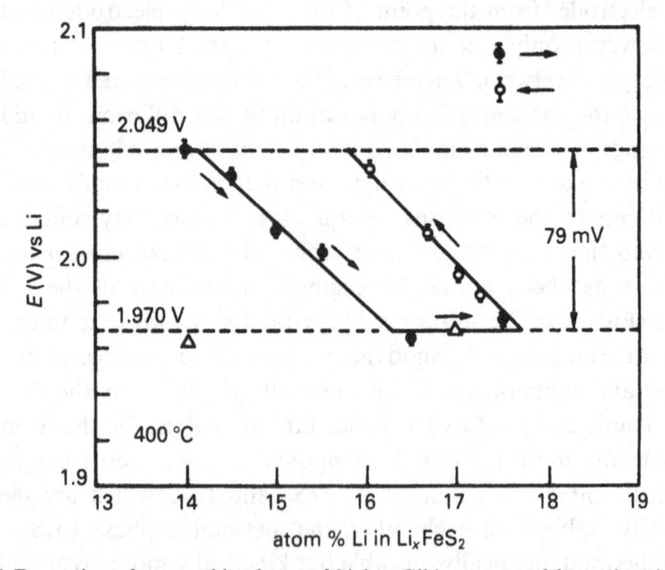

Fig. 8.6 Formation of metastable phases of higher Gibbs energy of formation during the discharge of Li_xFeS_2. Upon recharging the electrode returns to the thermodynamically more favourable phases.

the Gibbs energy of formation. Starting with an electrode composition $A_{y+\delta}B$, virtually $d\delta A^{z+}$ ions are passed through the electrolyte to change the composition into $A_{y+\delta+d\delta}B$. The corresponding Gibbs energy of reaction is given by the difference in the Gibbs energies of formation of the two compounds

$$\Delta G = \Delta G_f^\circ(A_{y+\delta+d\delta}B) - \Delta G_f^\circ(A_{y+\delta}B). \tag{8.40}$$

This Gibbs energy of reaction is converted into the electrical energy $zq\, d\delta E$. Then we have for the cell voltage

$$E = -\frac{1}{zq}\frac{d[\Delta G_f^\circ(A_{y+\delta}B)]}{d\delta}. \tag{8.41}$$

The cell voltage depends on the variation of the Gibbs energy of formation of the electrode compound with the stoichiometry of the electroactive species. Energy is accordingly only stored if chemical work is necessary to add or take away the electroactive component.

Even more important than the cell voltage is the energy that may be obtained from a galvanic cell as it discharges and the composition of the electrode changes. Integration of Eqn (8.41) provides the following expression

$$\Delta G_f^\circ(A_{y+\delta_0}B) - \Delta G_f^\circ(A_{y+\delta}B) = zq\int_{\delta_0}^{\delta} E\, d\delta. \tag{8.42}$$

The difference in the Gibbs energies of formation corresponds to the integral of the cell voltage as a function of the composition between the limits of the starting and the final compositions. The value for ΔS_f° is determined from the variation of the cell voltage as a function of the temperature

$$\Delta S_f^\circ(A_{y+\delta}B) = -\frac{\partial \Delta G_f^\circ(A_{y+\delta}B)}{\partial T} = zq\int_0^{\delta}\frac{\partial}{\partial T}E(\delta, T)\, d\delta \tag{8.43}$$

and the enthalpy of formation is given by

$$\begin{aligned}\Delta H_f^\circ(A_{y+\delta}B) &= \Delta G_f^\circ(A_{y+\delta}B) + T\Delta S_f^\circ(A_{y+\delta}B)\\ &= -zq\left[\int_0^{\delta} E\, d\delta - T\int_0^{\delta}\frac{\partial}{\partial T}E(\delta, T)\, d\delta\right].\end{aligned} \tag{8.44}$$

The activity a_A of the electroactive species A is obtained directly as a function of the composition from the cell voltage

$$\ln a_A(\delta) = -zqE(\delta)/kT. \tag{8.45}$$

In addition, the activity of the second component, a_B, of the binary compound $A_{y+\delta}B$ may be readily determined as a function of the composition. Since $\Delta G_f°(A_{y+\delta}B)$ may be separated into $(y+\delta)\mu_A + \mu_B$, the following relation holds

$$\ln a_B(\delta) = [(y+\delta)zqE(\delta) + \Delta G_f°(A_{y+\delta}B)]/kT \tag{8.46}$$

or, if $\Delta G_f°$ is written in terms of the cell voltage

$$\ln a_B(\delta) = \frac{zq}{kT}\left[(y+\delta)E(\delta) - \int_0^\delta E\,d\delta\right]. \tag{8.47}$$

Another convenient technique is the measurement of the equilibrium cell voltages of the various N-phase regions of an N-component system. Commonly, these regimes cover most parts of the phase diagram. The technique is also applicable if compositional changes cannot be easily induced by coulometric titration. The ΔG value of the virtual cell reaction is given by the Gibbs energies of formation of the N compounds multiplied by their amounts of change. These may be written in terms of determinants composed of the stoichiometric numbers of the N compounds (Weppner, Chen Li-chuan and Piekarczyk, 1980). Then, the cell voltage is given by

$$E = \frac{(-1)^k}{z_k q d} \sum_{i=1}^{N} (-1)^{i+k} d_{ik} \Delta G_f°(A_{1_{x_{1i}}} A_{2_{x_{2i}}} \ldots A_{N_{x_{Ni}}}) \tag{8.48}$$

with

$$d = \begin{vmatrix} x_{11} & x_{21} & \cdots & x_{N1} \\ x_{12} & x_{22} & \cdots & x_{N2} \\ \vdots & & & \\ x_{1N} & x_{2N} & \cdots & x_{NN} \end{vmatrix}$$

and d_{ik} being the minor of d which is formed by eliminating the ith row and kth column (electroactive component). Eqn (8.48) relates the cell voltage to the Gibbs energies of formation of the coexisting N phases. Application of this formula to the various N-phase regimes of the N-component system allows all the Gibbs energies of formation to be determined. The consistency of the data indicates that the assumed phase equilibria are correct.

8.5 Measurement of kinetic and thermodynamic electrode parameters

In-situ electrochemical techniques may be conveniently used to analyse the fundamental thermodynamic and kinetic parameters which are responsible for the performance of electrodes. The methods are non-destructive and may be applied to the actual galvanic cell. The data may be easily determined as a function of the discharge state.

Electrochemical techniques are based on the fact that galvanic cells translate thermodynamic and kinetic quantities directly into precisely measurable electrical parameters. The galvanic cell reaction may be readily controlled by applying voltages and currents. An important and convenient technique which combines thermodynamic and kinetic measurements is called the galvanostatic intermittent titration technique (GITT) (Weppner and Huggins, 1977). A major difference between this technique and the analysis of conventional discharge curves under load and voltametric techniques is the equilibration which is reached between incremental charging or discharging steps of the electrode. GITT combines coulometric titrations with electrochemical relaxation measurements.

A current through the galvanic cell changes the composition according to Faraday's law. The time integral of the current $\int I \, dt$ is a very precise measure of the variation of the concentration of the electroactive component

$$\Delta\delta = \frac{M}{zmF} \int_0^t I \, dt, \tag{8.49}$$

where M, m and F are the molecular weight of the sample, the original mass of the electrode and Faraday's constant, respectively. The resolution is extremely high compared to typical measurements using balances. Changes of the order of less than 10^{-10} g may be readily determined. After an appropriate amount of current has been passed through the galvanic cell, the open circuit equilibrium cell voltage is determined and plotted as a function of the composition. Integration of this curve allows determination of the ΔG value as a function of the composition. It is important to take the thermodynamic equilibrium cell voltage rather than the actual voltage while current is passing, which provides information about the energy loss due to various polarisations. With the temperature dependence of the Gibbs energy of formation, a large variety of further thermodynamic quantities may be derived as described in Section 8.4.

Also the activities of the electro-active and other components may be readily determined as a function of the stoichiometry. These values have to be within the stability range ('stability window') of the electrolyte.

The phase diagram of an electrode material may be determined from the slope of the coulometric titration curve. An electrode of N components shows activities which are independent of the composition as long as the maximum number of N phases are in equilibrium with each other. Relative changes in the amounts of the different phases do not change the activities of the components and therefore keep the cell voltage constant. This causes voltage plateaux for any region of the equilibrium of the maximum number of phases.

An example of equilibrium voltage measurements as a function of the composition is shown in Fig. 8.7 for adding lithium to Sb and Bi. In the Li–Sb system the first plateau corresponds to the equilibrium between Sb and Li_2Sb. Within the single phase regime of Li_2Sb a drop of the open circuit cell voltage occurs. At a higher lithium content, Li_3Sb will be formed and another plateau is observed due to the equilibrium between the two binary phases Li_2Sb and Li_3Sb before the voltage drops rapidly to 0 V after the Li_3Sb is exhausted. The curve for the addition of lithium to Bi is similar except for the large solubility of lithium in elemental

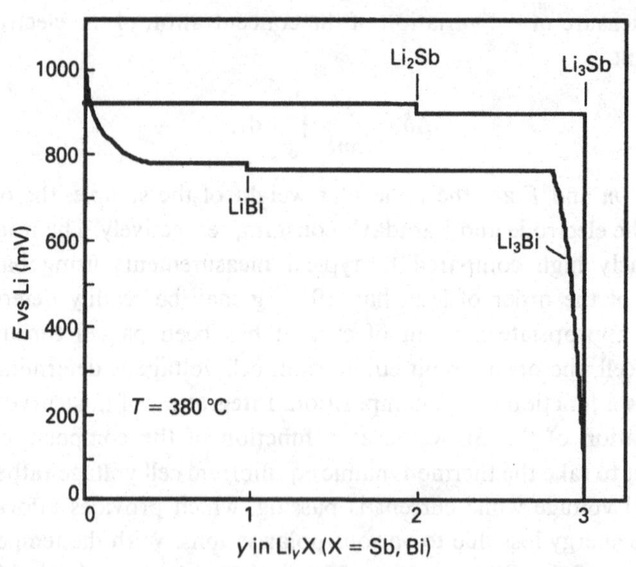

Fig. 8.7 Coulometric titration curve for the addition of Li to Sb and Bi. The plateaux indicate two-phase regions whereas the voltage drops correspond to single phase regions.

bismuth before the two-phase regimes Bi(+Li)/LiBi and LiBi/Li_3Bi occur. It is found that different intermediate phases, Li_2Sb and LiBi, are formed in spite of the similar chemistry of Bi and Sb. The small differences in the two plateaux in both cases are caused by similar Gibbs energies of formation of Li_3Sb and Li_3Bi due to the reaction of Sb and Bi with lithium, compared with the formation of Li_2Sb and LiBi related to the same number of lithium atoms. During passage of a current, the voltage steps of Li_2Sb and LiBi are not generally seen because of the sluggish kinetics of their formation.

The integration of the coulometric titration curve for the electrodes over single phase regimes allows us to determine the Gibbs energy of formation as a function of composition (see Fig. 8.8 for Li_3Sb). The resolution is extraordinarily high, i.e. of the order of 1 J mol^{-1}. The integration constant, i.e. the Gibbs energy of formation for a reference composition, is commonly known with much lower precision.

The high resolution for controlling the stoichiometry allows us to determine even very narrow ranges of stoichiometry including that of so-called line phases. Non-stoichiometries of the order of 10^{-8} or lower may be readily resolved. Knowledge of the variation of the cell voltage

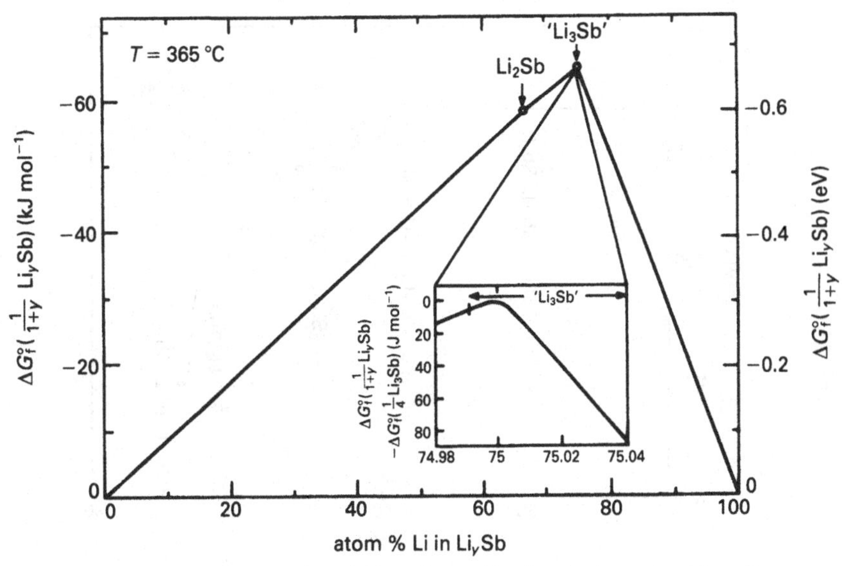

Fig. 8.8 Gibbs energy of formation of $Li_{3+\delta}Sb$ as a function of the composition. The curve is derived from the integration of the coulometric titration curves.

with the stoichiometry is important because it allows the Wagner factor to be determined. If it is assumed that the lattice parameters remain unchanged during the variation of the stoichiometry, we have

$$\frac{dE}{d \ln c_{i^\times}} = y \frac{dE}{dy}. \tag{8.50}$$

The voltage is related to the activity of the electroactive component by Nernst's law which then provides the following relation

$$\frac{d \ln a_{i^\times}}{d \ln c_{i^\times}} = -\frac{zqy}{kT} \frac{dE}{dy}. \tag{8.51}$$

The slope of the coulometric titration curve is accordingly proportional to the Wagner factor in the case of predominant electronic conductivity.

Also, the phases formed in the course of discharge of an electrode with three or more components may be readily detected by reading the equilibrium cell voltage. As an example, the determination of the quite complex ternary phase diagram of the system Li–In–Sb is shown in Fig. 8.9. In this case, plateaux are observed in the presence of three-phase equilibria. In order to obtain the complete phase diagram it is necessary

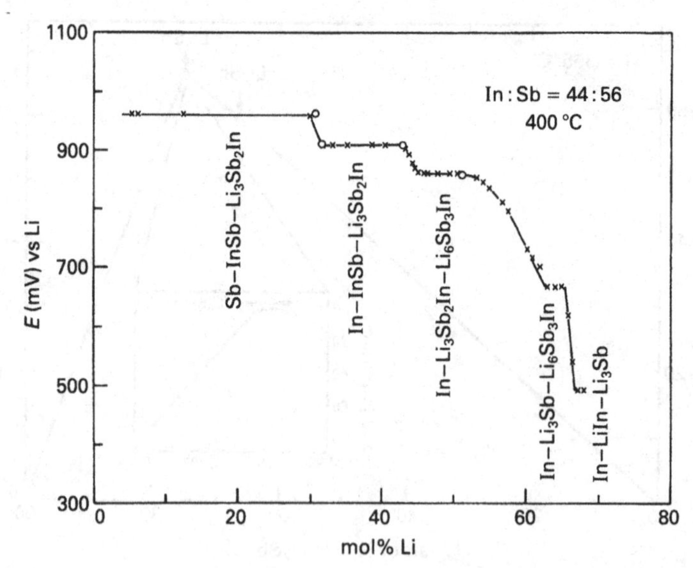

Fig. 8.9 Coulometric titration curve for the ternary system Li–In–Sb. Plateaux correspond to three-phase regions. The phases are indicated.

to employ several ratios of the non-electroactive components. Several thermodynamic considerations may be used to minimise the number of sample preparations. For example, the phase diagram has to be constructed in such a way that the cell voltage has to increase continuously along the path of composition with increasing distance from the electroactive component. Also, regions with the same cell voltages for different compositions with regard to any component belong to the same three-phase equilibrium. Compared to conventional techniques of phase diagram evaluation of electrodes, only a comparatively small number of samples has to be used, no quenching and destruction of the electrodes is required and thermodynamic information is obtained in addition.

GITT also provides very comprehensive information about the kinetic parameters of the electrode by analysis of the electrical current. The current I, which is driven through the galvanic cell by an external current or voltage source, determines the number of electroactive species added to (or taken away from) the electrode and discharged at the electrode/electrolyte interface. A chemical diffusion process occurs within the electrode and the current corresponds to the motion of mobile ionic species within the electrode just inside the phase boundary with the electrolyte (at $x = 0$)

$$I = -Sz_i q\tilde{D}\frac{\partial c_i}{\partial x} \qquad (x = 0), \tag{8.52}$$

where S is the area of the sample–electrolyte interface.

The principles of the procedure are illustrated in Fig. 8.10. Starting with a homogeneous composition throughout the electrode, corresponding to the cell voltage E_0, a constant current I_0 is applied to the cell at $t = 0$ with the help of a constant current source. According to Eqn (8.52), this produces a constant concentration gradient $\partial c_i/\partial x$ of the mobile ions i at the phase boundary with the electrolyte. The applied cell voltage increases or decreases (depending on the direction of the current) with time in order to maintain the constant concentration gradient. A voltage drop due to other polarisations within the cell is superimposed, but the ohmic resistance (IR drop) due to the current flux through the electrolyte and the interfaces remains constant with time and does not change the shape of the time dependence of the cell voltage. This behaviour makes galvanostatic processes advantageous compared to other relaxation techniques which often involve large changes of the cell current, for example, in the case of a potential step which requires an initially large current which finally becomes zero. After a constant current is applied for a certain time interval τ, the current flux is interrupted whereupon the composition has changed according to Eqn (8.49). During the following equilibrium process, the voltage is directly related to the interfacial composition of the electrode and approaches a new steady state value E_1. After the electrode has again reached an equilibrium, the procedure may be repeated, making use of E_1 as a new starting voltage. The process is continued until a phase change occurs in the electrode.

In order to relate the time dependence of the voltage E to the ionic transport in the electrode, Fick's second law is solved for the concentration of the mobile ions at the interface $x = 0$ which is observed experimentally by reading the cell voltage. With the appropriate initial and boundary conditions of a homogeneous concentration throughout the electrode at $t = 0$, a constant concentration gradient at $x = 0$ at any time (as given by Eqn (8.5.4)) and a zero concentration gradient at the opposite surface of the cathode (because of an assumed impermeability of the ions at this

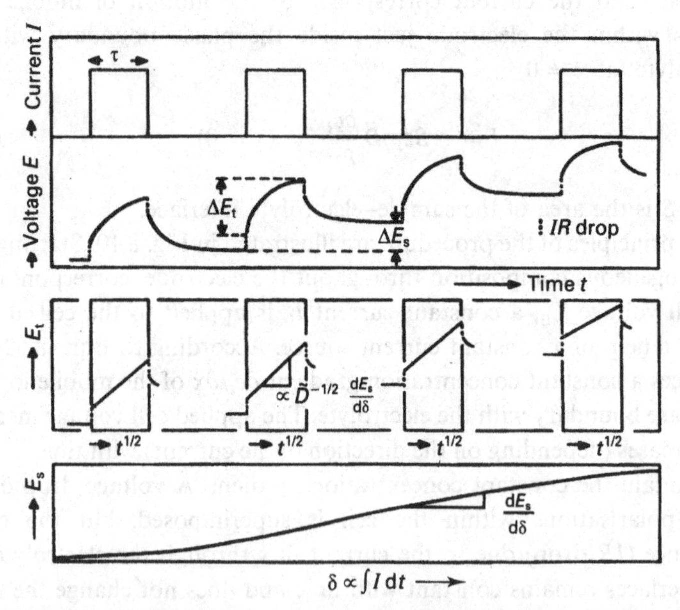

Fig. 8.10 Principles of GITT for the evaluation of thermodynamic and kinetic data of electrodes. A constant current I_0 is applied and interrupted after certain time intervals τ until an equilibrium cell voltage is reached. The combined analysis of the relaxation process and the variation of the steady state voltage results in a comprehensive picture of fundamental electrode properties.

location), the following solution for the concentration of the mobile species at the interface is derived as a function of time (Carslaw and Jaeger, 1967; Crank, 1967):

$$c_i(x=0,t) = c_0 + \frac{2I_0 t^{1/2}}{Sz_i q\tilde{D}^{1/2}} \sum_{n=0}^{\infty} \left[\mathrm{ierfc}\left(\frac{nL}{(\tilde{D}t)^{1/2}}\right) + \mathrm{ierfc}\left(\frac{(n+1)L}{(\tilde{D}t)^{1/2}}\right)\right], \tag{8.53}$$

where ierfc $(\lambda) = [\pi^{-1/2}\exp(-\lambda^2)] - \lambda + [\lambda\,\mathrm{erf}(\lambda)]$ is the error function.

For short times, $t \ll L^2/\tilde{D}$, the infinite sum can be approximated by the first term. In this case, the concentration c_i should change with the square root of time:

$$\frac{\mathrm{d}c_i(x=0,t)}{\mathrm{d}t^{1/2}} = \frac{2I_0}{Sz_i q(\tilde{D}\pi)^{1/2}} \qquad (t \ll L^2/\tilde{D}). \tag{8.54}$$

Now the voltage provides information on the activity according to Nernst's law whereas Eqs (8.53) and (8.54) indicate the time dependence of the concentration. This may be overcome by expanding Eqn (8.54) by $\mathrm{d}E$ and by introducing the relation between changes in the concentration and the stoichiometry, $\mathrm{d}c_i = (N_A/V_M)\,\mathrm{d}\delta$, where N_A is Avogadro's number and V_M is the molar volume of the electrode material:

$$\frac{\mathrm{d}E}{\mathrm{d}t^{1/2}} = \frac{2V_M I_0}{SFz_i(\tilde{D}\pi)^{1/2}} \frac{\mathrm{d}E}{\mathrm{d}\delta} \qquad (t \ll L^2/\tilde{D}). \tag{8.55}$$

$\mathrm{d}E/\mathrm{d}\delta$ is the slope of the coulometric titration curve at the given composition. With this information, the chemical diffusion coefficient may be calculated from the time dependence of the cell voltage during the application of the constant current.

If the time during which the current is passed through the cell is short compared to $L^2/\tilde{D}$ (i.e. Eqn (8.55) is valid for the entire period of time of the current flux), and the change of the voltage is sufficiently small to consider $\mathrm{d}E/\mathrm{d}\delta$ to be constant, $\mathrm{d}E/\mathrm{d}\delta$ may be replaced by the variation of the equilibrium steady state voltage ΔE_s over the variation of the stoichiometry $\Delta\delta$ that corresponds to the current flux I_0 for the period of time τ. In addition, if one ensures that the voltage E vs $t^{1/2}$ shows the expected straight line behaviour over the entire time period of the current flux, $\mathrm{d}E/\mathrm{d}t^{1/2}$ may be replaced by the variation of the transient voltage ΔE_t (disregarding the IR drop) over $\tau^{1/2}$. This provides the following simple expression under the indicated assumptions for $\tilde{D}$ as a function of the electrode composition:

$$\tilde{D} = \frac{4}{\pi\tau}\left(\frac{m_B V_M}{M_B S}\right)^2 \left(\frac{\Delta E_s}{\Delta E_t}\right)^2 \qquad (t \ll L^2/\tilde{D}), \tag{8.56}$$

where B refers to the non-electroactive components.

A variety of other kinetic quantities may be readily derived from this information and the coulometric titration curve. Only a few examples will be given.

As discussed in Section 8.2 the relation between the chemical diffusion coefficient and diffusivity (sometimes also called the component diffusion coefficient) is given by the Wagner factor (which is also known in metallurgy in the special case of predominant electronic conductivity as the thermodynamic factor) $W = \partial \ln a_A / \partial \ln c_A$ where A represents the electroactive component. W may be readily derived from the slope of the coulometric titration curve since the activity of A is related to the cell voltage E (Nernst's law) and the concentration is proportional to the stoichiometry of the electrode material:

$$W = \frac{\partial \ln a_A}{\partial \ln c_A} = -\frac{z_A q c_A V_M}{kTN_A} \frac{dE}{d\delta}. \tag{8.57}$$

In view of this relationship, the diffusivity D_A is given by

$$D_A = \left(\frac{\partial \ln a_A}{\partial \ln c_A}\right)^{-1} \tilde{D} = -\frac{4kTm_B V_M I_0}{\pi c_A z_A{}^2 q^2 M_B S^2} \frac{\Delta E_s}{(\Delta E_t)^2} \qquad (t \ll L^2/\tilde{D}). \tag{8.58}$$

The diffusivity is independent of the motion of any other species (e.g. electrons or holes) and is not influenced by internal electrical fields as in the case of chemical diffusion processes which require the simultaneous motion of electronic or other ionic species. The partial ionic conductivity of the mixed ionic and (predominantly) electronic conducting electrode is given by the product of the concentration and the diffusivity and may be related to the variations of the steady state and transient voltage:

$$\sigma_A = \left(\frac{\partial \ln a_A}{\partial \ln c_A}\right)^{-1} \frac{(z_A q)^2 c_A \tilde{D}}{t_e kT} = -\frac{4}{\pi} \frac{m_B V_M I_0 \Delta E_s}{t_e M_B S^2 (\Delta E_t)^2} \qquad (t \ll L^2/\tilde{D}), \tag{8.59}$$

where t_e is the transference number of the electrons. This method of determining partial ionic conductivities is especially valuable in the case of

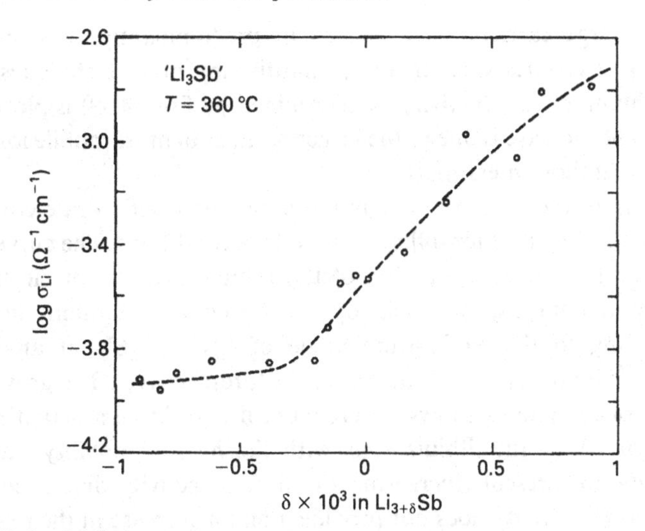

Fig. 8.11 Partial lithium ion conductivity of the predominantly electronically conducting compound $Li_{3+\delta}Sb$ as a function of stoichiometry. The variation in the conductivity is due to changes of the transport mechanism.

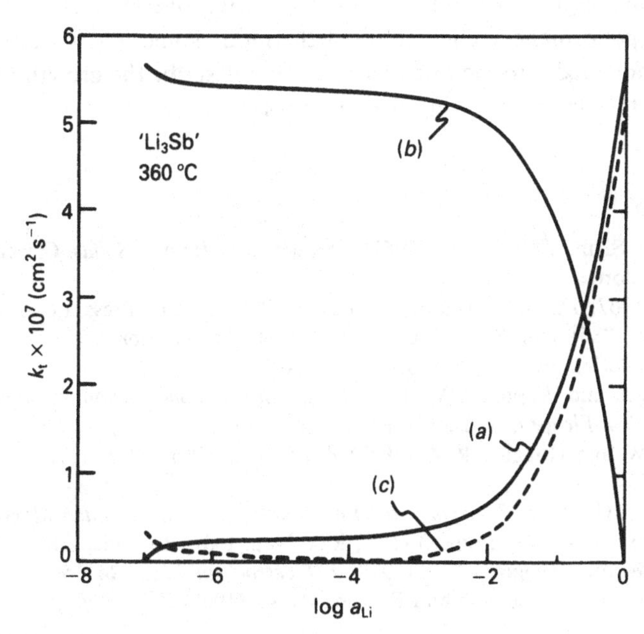

Fig. 8.12 Parabolic reaction rate constant for the formation of Li_3Sb under three different conditions at the opposite side: (*a*) a_{Li} corresponds to that of the equilibrium Li_2Sb/Li_3Sb, (*b*) $a_{Li} = 10^{-4}$, (*c*) $a_{Li} = 1$.

minority charge carriers such as ions in predominantly electronically conducting electrodes since there are hardly any other techniques. The partial lithium ion conductivity as determined by Eqn (8.59) is plotted in Fig. 8.11. The increase is due to the larger number of more mobile ions, i.e. most likely lithium interstitials.

Returning to the $Li_{3+\delta}Sb$ system; the parabolic reaction rate constant k_t is plotted in Fig. 8.12 for different boundary conditions. The curves (*a*), (*b*) and (*c*) show k_t as a function of the lithium activity for electrodes which are maintained on their opposite side at a lithium activity corresponding to the equilibrium value of Li_2Sb/Li_3Sb, at an intermediate activity of $a_{Li} = 10^{-4}$ and at $a_{Li} = 1$, respectively. The growth or shrinkage of the phases shows a maximum if high lithium activities are provided at which the lithium ions with the highest mobility (lithium interstitials) are present. Increasing the lithium activity difference over ranges of lower activity does not provide a major increase in the reaction rate since only lithium defects of lower mobilities are introduced.

In conclusion, the combination of thermodynamic measurements over single phase and multiphase regimes and kinetic measurements within single phase regions provides a comprehensive overall picture of the electrode performance. The extracted thermodynamic and kinetic data may be converted into the rate constants that describe the growth of new phases in the electrode during the course of discharge.

References

Carslaw, H. S. and Jaeger, J. C. (1967) *Conduction of Heat in Solids*, Clarendon Press, Oxford.

Crank, J. (1967) *The Mathematics of Diffusion*, Oxford Univ. Press, London.

Rickert, H. (1985) *Solid State Electrochemistry, An Introduction*, Springer-Verlag, Berlin, Heidelberg, New York.

Schmidt, J. A. and Weppner, W. (1982) *Proc. 7th Int. Conf. on Solid Compounds of Transition Elements, Grenoble, France*, p. IB 13a.

Weppner, W. and Huggins, R. A. (1977) *Z. Physikal. Chem. N. F. (Frankfurt)*, **108**, 105.

Weppner, W. (1985) in *Transport-Structure Relations in Fast Ion and Mixed Conductors*, Eds F. W. Poulsen *et al.*, Risø Natl. Lab., Roskilde, DK, p. 139.

Weppner, W. and Huggins, R. A. (1977) *J. Electrochem. Soc.*, **124**, 1569.

Weppner, W., Chen Li-chuan and Piekarczyk, W. (1980) *Z. Naturforsch.*, **35a**, 381.

9 Polymer electrodes

B. SCROSATI
Dept of Chemistry, University of Rome

9.1 Introduction

The discovery that certain classes of polymers may acquire high electronic conductivity following chemical or electrochemical treatment has opened a new and exciting area of research and development.

The concept of electric transport in polymers due to the availability of polymeric materials with characteristics similar to those of metals is certainly fascinating and, indeed, many studies have been directed towards the preparation and the characterisation of these new electroactive conductors. The final goal is their use as new components for the realisation of electronic and electrochemical devices with exotic designs and diverse applications.

The idea of exploiting these new conducting polymers for the development of flexible diodes and junction transistors, as well as for selective field effect transistor sensors, has been proposed and experimentally confirmed, and thus we may, perhaps optimistically, look forward to a time when popular electronic devices can be based on low cost, flexible and modular polymer components.

Even more interesting than all this is the fact that conducting polymers allow the fabrication of not only polymer-like electronic devices but also the battery which is necessary to power them. Indeed, such polymers are capable of acquiring high conductivity by reversible electrochemical processes, and thus they may be regarded as new electrode materials which can operate in the same way as the conventional battery electrodes, while still maintaining their unique mechanical characteristics. In this way, the realisation of thin-layer, flexible batteries, which may replace the heavy and polluting dry cells or even the low energy density lead acid cells, is now a feasible proposition.

Furthermore, since the electrochemical processes modify the optical properties as well as the electronic transport, selected polymer materials are now available for the development of optical displays with a multi-chromatic response controlled and switched by electrochemical pulses.

In this chapter we will attempt to provide a brief but illustrative description of the various aspects of the research and technology of conducting polymers. To appreciate fully the diverse range of operations that these materials may fulfil, it is crucial to understand their basic properties. Therefore, particular attention will be devoted here to the description of the mechanism of charge transport and to the characteristics of the electrodic processes in electrochemical cells.

9.2 The case of polyacetylene

The first and most important event in the history of conducting polymers occurred in 1978 when it was announced that the electrical properties of polyacetylene could be dramatically changed by chemical treatment (Chiang *et al.*, 1978).

Polyacetylene, $(CH)_x$, is a simple, conjugated polymer which may have either a *trans* or a *cis* configuration (Fig. 9.1). Free-standing films of polyacetylene can be easily obtained by catalytic polymerisation of gaseous acetylene, the most common procedure being the Shirikawa

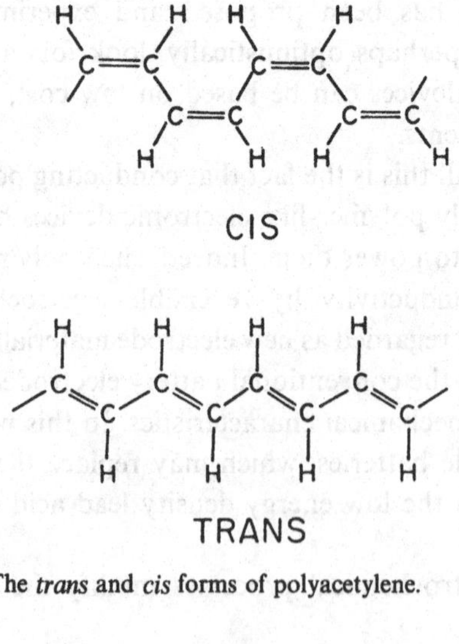

CIS

TRANS

Fig. 9.1 The *trans* and *cis* forms of polyacetylene.

method (Shirikawa and Ikeda, 1971). These films are crystalline with a structure and a morphology which depend on the temperature of synthesis and the concentration of the catalyst used. Most commonly, the films are in the *trans* configuration and have a fibrillar-type morphology (Fig. 9.2) with a typical fibril diameter between 200 Å and 800 Å.

The room temperature conductivity of polyacetylene is in the region of 10^{-5} S cm^{-1} for the *trans* form and 10^{-9} S cm^{-1} for the *cis* form (Shirikawa, Ito and Iteda, 1978). Therefore, the electrical properties of $(CH)_x$ were for a long time considered to be those typical of a poorly conductive material. However, a major breakthrough was achieved by the discovery that, upon exposure to oxidising or reducing agents, the conductivity of the polymer could be increased by several orders of magnitude. A typical example is shown in Fig. 9.3, which illustrates the effect on the conductivity of $(CH)_x$ upon exposure to halogens (Chiang *et al.*, 1978): the chemical reaction transforms polyacetylene from a poor conductor to a lustrous polymer with a conductivity approaching that of metals.

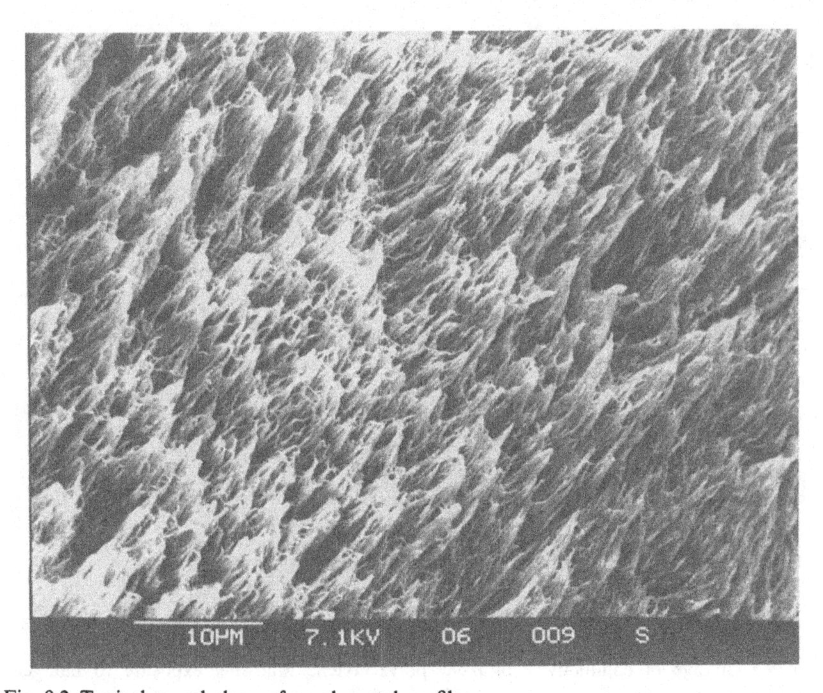

Fig. 9.2 Typical morphology of a polyacetylene film.

The reactions which induce enhancements in the conductivity of polyacetylene have been termed *p-doping* or *n-doping* processes, by direct analogy with classical semiconductor terminology. However, these processes may be more correctly described as redox reactions, which involve oxidising or reducing agents with the formation of polyanions or polycations whose charge is balanced by the counterions.

For instance, exposure to an oxidising agent, say X, leads to the formation of a positively charged polymer complex:

$$(CH)_x \rightarrow [(CH^{y+})]_x + (xy)e^- \tag{9.1}$$

accompanied by the reduction of X:

$$(xy)X + (xy)e^- \rightarrow (xy)X^- \tag{9.2}$$

with a total reaction of the type:

$$(CH)_x + (xy)X \rightarrow [(CH^{y+})]_x + (xy)X^- \rightarrow [(CH^{y+}(X_y{}^-]_x, \tag{9.3}$$

where $X^- = I^-, Br^-, \ldots$

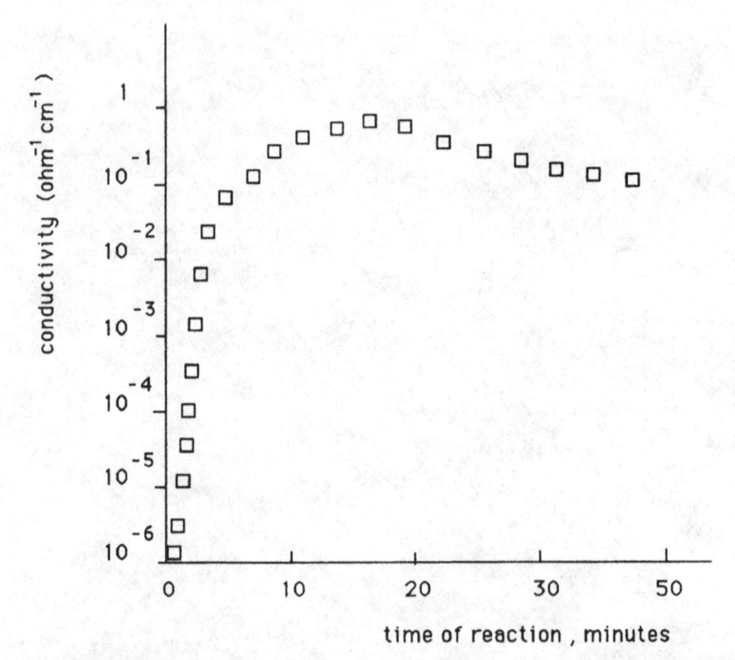

Fig. 9.3 Change of conductivity of *trans* polyacetylene following exposure to halogen oxidising agents (from Chiang *et al.* (1978)).

In line with the terminology adopted, X^- is commonly called the *dopant counter anion* and y, which represents the ratio between dopant ion and polymer repeating unit, is commonly called the *doping level.*

For example, the reaction following exposure to iodine may be illustrated as:

$$(CH)_x + \tfrac{1}{2}(xy)I_2 \rightarrow (CH^{y+})_x + (xy)I^- \tag{9.4}$$

$$(xy)I^- + (xy)I_2 \rightarrow (xy)I_3^- \tag{9.5}$$

$$(CH)_x + \tfrac{3}{2}(xy)I_2 \rightarrow (CH^{y+})_x + (xy)I_3^- \rightarrow [(CH^{y+})(I_3^-)_y]_x. \tag{9.6}$$

Similarly, we can describe the n-doping process due to exposure to a reducing agent, say M, as the formation of a negatively charged polymer complex:

$$(CH)_x + (xy)e^- \rightarrow [(CH^{y-})]_x \tag{9.7}$$

accompanied by the oxidation of M:

$$(xy)M \rightarrow (xy)M^+ + (xy)e^- \tag{9.8}$$

with a total reaction of the following type:

$$(CH)_x + (xy)M \rightarrow [(CH^{y-})]_x + (xy)M^+ \rightarrow [M_y{}^+(CH^{y-})]_x, \tag{9.9}$$

where $M^+ = Na^+, Li^+, \ldots$, is commonly called the *dopant counter cation.*

The redox reactions of polyacetylene – and of conducting polymers in general – lead to the formation of complexes which include the polyanion and the counterion in a combined structure which resembles that of the well-known intercalation compounds, such as the transition metal dichalcogenides or oxides, described in previous chapters of this book. In fact, in a similar fashion to these intercalation compounds, the redox reactions promote electron transport accompanied by the diffusion of the counterion into the polymer structure. For instance, it has been shown (Shacklette, Toth, Murthy and Baugham, 1985) that the process of reduction of polyacetylene by potassium is accompanied by the modification of the polymer structure upon K^+ ion insertion. In fact, to accommodate the guest dopant ion, the polymer chains undergo a structural rearrangement, the extent of which increases with increasing doping level. At high doping levels a first stage structure is formed, where the ratio of polyacetylene chains to columns of K^+ ions is 2, while at lower doping levels a 'second stage' structure is proposed with three polymer chains for each K^+ column (Shacklette *et al.*, 1985).

9.3 Electrochemical doping processes

A second major event in the saga of polymer conductors was the discovery that the doping processes of polyacetylene could be promoted and driven electrochemically in a reversible fashion by polarising the polymer film electrode in a suitable electrochemical cell (MacDiarmid and Maxfield, 1987). Typically, a three-electrode cell, containing the $(CH)_x$ film as the working electrode, a suitable electrolyte (e.g. a non-aqueous solution of lithium perchlorate in propylene carbonate, here abbreviated to $LiClO_4$-PC) and suitable counter (e.g. lithium metal) and reference (e.g. again Li) electrodes, can be used.

By anodically polarising the $(CH)_x$ electrode, the polymer may be oxidised (p-doping process), this being accompanied by the insertion of the anions (ClO_4^-) from the electrolytic solution:

$$(CH)_x + (xy)ClO_4^- \Leftrightarrow [(CH^{y+})(ClO_4^-)_y]_x + (xy)e^-, \qquad (9.10)$$

i.e. a reaction substantially similar to (9.3).

Similarly, by cathodic polarisation, one may promote reduction (n-doping process) accompanied by the insertion of the cations (Li^+) from the electrolytic solution:

$$(CH)_x + (xy)e^- + (xy)Li^+ \Leftrightarrow [(CH^{y-})(Li_y^+)]_x, \qquad (9.11)$$

i.e. a reaction substantially similar to (9.9).

The implication of the possibility of driving the doping processes electrochemically is that polyacetylene, and doped conducting polymers in general, can be exploited as innovative electrode materials. Indeed, this possibility has greatly stimulated research in the field. Shortly after it was found that $(CH)_x$ could be electrochemically doped to the highly conductive regime, it was discovered that other polymers could be treated in an analogous way. These are mainly polymers of the heterocyclic type, and of the polyaniline family.

9.4 Heterocyclic polymers

It has been shown that the chemical and electrochemical doping of polymers may be described as a redox reaction which involves the formation of charged polyions accompanied by the insertion of counter-ions. Consequently, the polymers which are likely to experience these processes are unsaturated materials with π electrons that can easily be removed or added and are thus delocalised in the polymer chain. This basic principle applies to conjugated polymers, such as polyacetylene described above, as well as to heterocyclic polymers, such as polypyrrole and polythiophene (Fig. 9.4).

An additional important feature of this class of polymers lies in the fact that their polymerisation and doping processes may be driven by a single electrochemical operation which, starting from the monomer, first forms the polymeric chain and then induces its oxidation and deposition in the doped form as a conductive film on a suitable substrate. The polymerisation reaction may be basically described as an electrophilic substitution which retains the aromatic structure and proceeds via a radical cation intermediate:

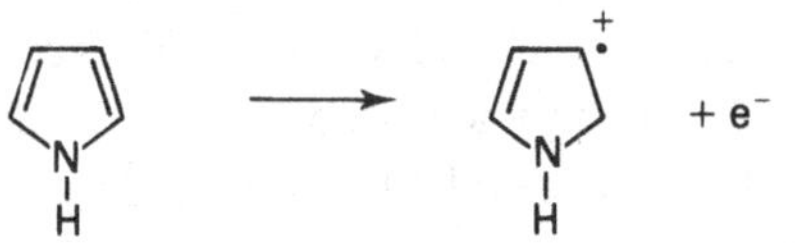

$$\text{pyrrole} \longrightarrow \text{pyrrole}^{+\bullet} + e^- \qquad (9.12)$$

The coupling occurs at the carbon atoms (generally via α–α' links) which are the most reactive to addition and substitution reactions:

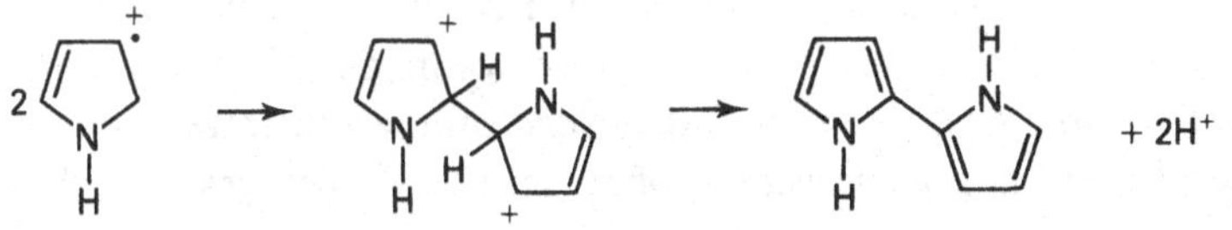

$$2\,\text{pyrrole}^{+\bullet} \longrightarrow \text{dimer dication} \longrightarrow \text{bipyrrole} + 2H^+ \qquad (9.13)$$

The polmerisation reaction proceeds between the radical cations of the monomer and those of the continuing forming oligomers since the latter, following the growth of the chain length, acquire an oxidation potential

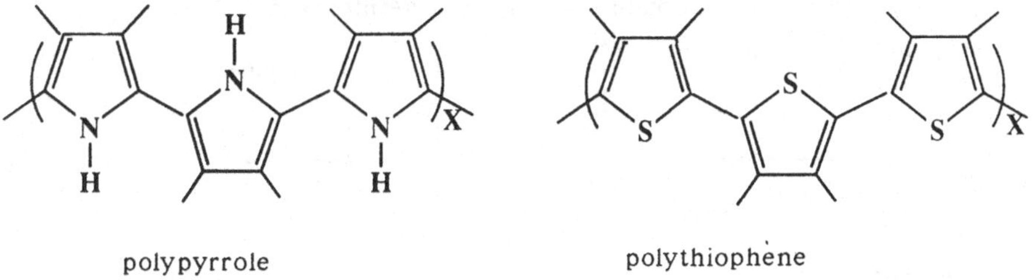

polypyrrole polythiophene

Fig. 9.4 Polypyrrole and polythiophene as typical examples of heterocyclic polymers.

progressively lower than that of the monomer (Fig. 9.5):

$$\text{H–N}\left(\text{pyrrole}\right)_{x-1}^{+\bullet} + \text{H–N(pyrrole)}^{+\bullet} \longrightarrow \text{H–N}\left(\text{pyrrole}\right)_{x} + 2H^{+} \qquad (9.14)$$

The doping process (here of the p-type) takes place with formation of the charged polycation accompanied by diffusion of the electrolyte counterion X^-:

$$\text{H–N}\left(\text{pyrrole}\right)_{x} + xyX^{-} \longrightarrow \left[\text{H–N}\left(\text{pyrrole}\right)^{y+} (X)^{y-}\right]_{x} + xye \qquad (9.15)$$

The entire operation can be achieved in a simple electrochemical cell consisting of two (usually planar) electrodes immersed in a solution (generally non-aqueous, e.g. an acetonitrile (CH_3CN) solution) containing the monomer (e.g. pyrrole, C_4H_5N) and a supporting electrolyte (e.g. $LiClO_4$ or alkylammonium salts, both of which are soluble and highly dissociated in aprotic solvents).

The electrodeposition voltage is specific for any given electropolymerisation process. The polymers are obtained directly in the oxidised conducting form. Consequently, upon applied polarisation, polymerisation takes place, the p-doping of the polymer occurs and, finally, a film of the selected conductive polymer is deposited on the substrate.

The electrochemical polymerisation offers several advantages over the

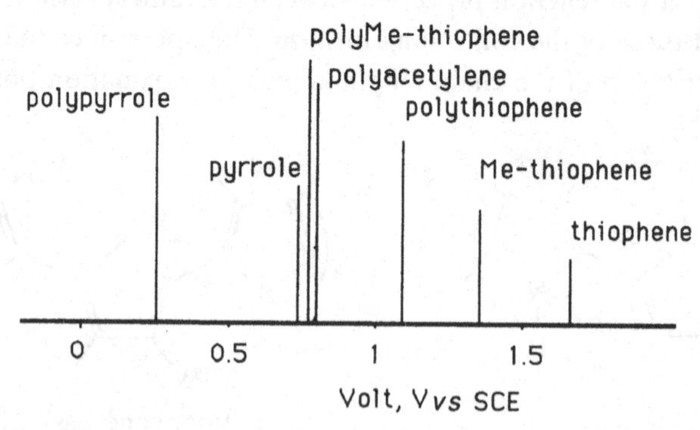

Fig. 9.5 Oxidation potentials of some selected examples of monomers and of their corresponding polymers. (SCE is a saturated calomel electrode.)

chemical one, particularly because of the possibility of optimising the characteristics of the polymer films. In fact, by controlling the charge involved in the electrodeposition process, the thickness of the polymer film may be varied from a few Å to many μm or even to mm.

By changing the nature of the counterion in solution, the electrical and the physicochemical properties of the final polymer can be tailored and by varying the value of the electropolymerisation current density, the morphology of the polymer can easily be controlled. Furthermore, different substrates can be used for the electrodeposition process, provided that they have sufficient conductivity to allow the passage of current without inducing substantial ohmic drop; thus substrates varying from platinum to silicon and indium–tin-oxide(ITO)-coated glasses, can be successfully used, and this, in turn, favours a large variety of applications.

9.5 The electrochemical doping of heterocyclic polymers

Once deposited as conductive films, the heterocyclic polymers can be repeatedly cycled from the undoped to the doped forms (and vice versa) in electrochemical cells substantially similar to those used for the electropolymerisation reactions.

For example, the p-doping process of a typical heterocyclic polymer, say polypyrrole, can be reversibly driven in an electrochemical cell by polarising the polymer electrode vs a counterelectrode (say Li) in a suitable electrolyte (say $LiClO_4$-PC). Under these circumstances the p-doping redox reaction (9.15) can be described by the scheme:

$$(C_4H_5N)_x + xyClO_4^- \Leftrightarrow [(C_4H_5N^{y+})(ClO_4^-)_y]_x + xy\,e^-, \quad (9.16)$$

which, in analogy with the doping process of polyacetylene and in line with the general doping mechanism of conducting polymers, involves the oxidation of polypyrrole with the formation of a positive polycation whose charge is counterbalanced by the electrolyte anion (ClO_4^- in this case) diffusing into the polymer matrix. Again, y is the doping level. Other heterocyclic polymers, such as polythiophene, can be grown and doped electrochemically by a process of the type:

$$(C_4H_5S)_x + xyX^- \Leftrightarrow [(C_4H_5S^{y+})(X^-)_y]_x + xy\,e^-. \quad (9.17)$$

It is worth noting that the doping level y acquired by the electrochemical processes may be expressed as the percentage of moles of the doping anion (X^-) over the moles of the monomeric repeating unit (e.g.

C_4H_5N). The numerical value of y is obtained as follows. The number of moles of X^- are expressed by the cyclable charge, Q_{cycl}, which, in turn, is given by the number of cyclable Faradays F_{cycl}, over the number of electrons involved in the electrochemical doping, ne_{dop} ($ne_{dop} = 1$ in the case of X^-):

$$\text{moles of doping anion} = Q_{cycl}\,\frac{F_{cycl}}{ne_{dop}}. \tag{9.18}$$

Accordingly, the moles of the monomeric unit are given by the ratio of the difference between the number of Faradays used for the entire process, F_{tot}, and the number of cyclable Faradays, F_{cycl}, over the number of electrons for the entire process ne_{tot} ($ne_{tot} = 2$ in the case of (9.16)):

$$\text{moles of monomeric unit} = \frac{F_{tot} - F_{cycl}}{ne_{tot}} = \frac{Q_{tot} - Q_{cycl}}{ne_{tot}} \tag{9.19}$$

and thus:

$$y = \frac{\text{moles of doping anion}}{\text{moles of monomeric unit}}\,\% = \frac{ne_{tot}}{ne_{dop}} \times \frac{Q_{cycl}}{Q_{tot} - Q_{cycl}}. \tag{9.20}$$

In the case of (9.16):

$$y = \frac{2Q_{cycl}}{Q_{tot} - Q_{cycl}}. \tag{9.21}$$

9.6 Polyaniline

Another conducting polymer of interest for electrochemical applications is polyaniline:

$$\left(\!\!-\!\!\bigcirc\!\!-\overset{\text{H}}{\underset{}{\text{N}}}-\right)_x \tag{9.22}$$

and its derivatives.

Polyaniline can also be prepared electrochemically, e.g. by oxidation of aniline in acid media (MacDiarmid and Maxfield, 1987). An interesting aspect of this polymer is that by doping, the oxidation state of the polyaniline may be continuously varied from the totally reduced insulator, *leucoemeraldine* (oxidation state $y = 1$):

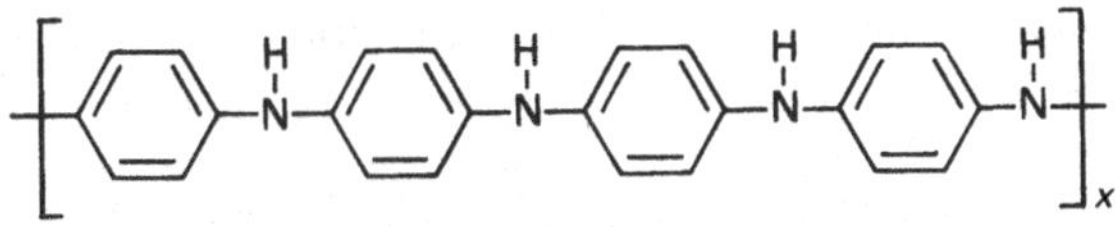

(9.23)

to the totally oxidised, insulator, *pernigraniline* (oxidation state $y = 0$):

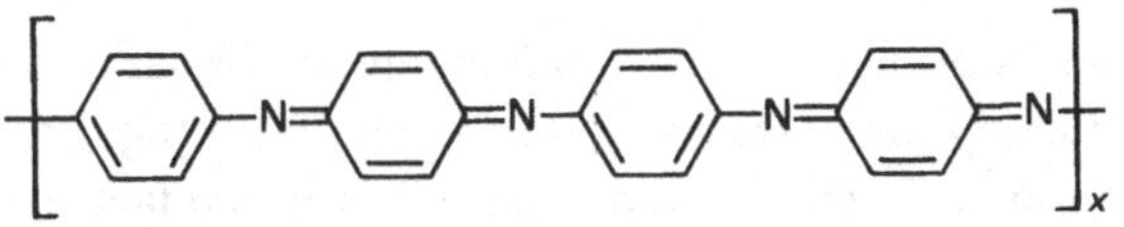

(9.24)

where the most stable form corresponds to a state consisting of an equal number of reduced and oxidised groups, i.e. the *emeraldine base* form (oxidation state $y = 0.5$):

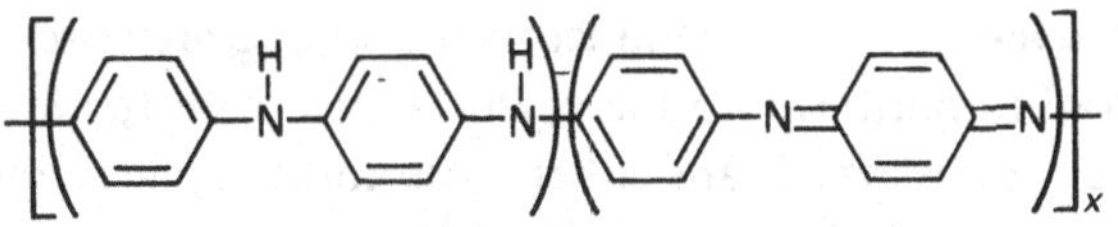

(9.25)

which again is an insulator. However, this polymer can react with dilute acids (e.g. HCl) to give the corresponding salts, with protonation which takes place preferentially on the —N= atoms:

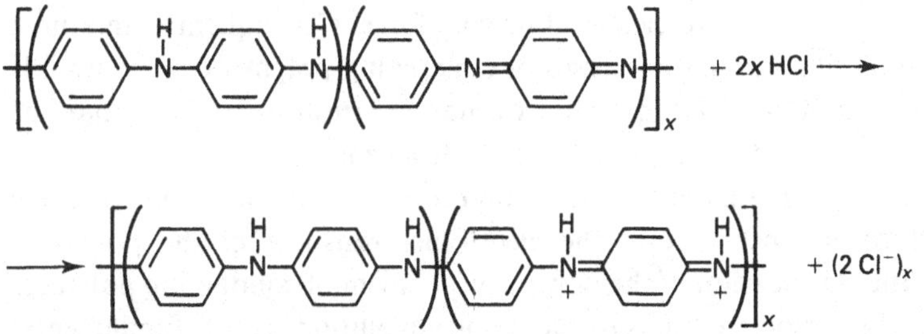

$+ 2x\,HCl \longrightarrow$

$\longrightarrow \quad + (2\,Cl^-)_x$ (9.26)

It is important to note that the protonation reaction induces a large increase in the polymer conductivity by a mechanism which again is believed to be a p-doping process.

The emeraldine base form of polyaniline may also react in non-aqueous electrolytes, such as a $LiClO_4$-propylene carbonate solution, with the formation of the conductive emeraldine hydroperchlorate salt:

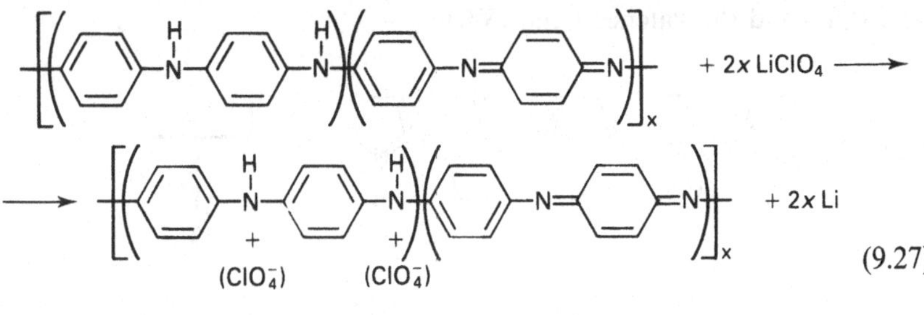

$+ 2x\,LiClO_4 \longrightarrow$

$\longrightarrow \quad + 2x\,Li$

(ClO_4^-) (ClO_4^-) (9.27)

i.e. by a reaction which is relevant in view of the practical applications of polyaniline.

9.7 Mechanism of the doping processes in conducting polymers

As already pointed out, even if the terminology is similar the doping processes of conducting polymers are quite different from those of classic, inorganic semiconductors. The basic difference lies in the fact that while semiconductors have a rigid lattice and the evolution of their electronic structure upon doping is well described by band models, polymers consist of flexible chains which, in turn, favour localised chain deformations. Therefore, while in inorganic semiconductors the doping processes involve the introduction of impurities into the crystal lattice with the formation of intergap energy levels which are either close to the conduction band (for donor impurities) or to the valence band (for acceptor impurities), in the case of conducting polymers the processes are quite different. The impurities or doping agents do not become part of the structure but, rather, are inserted within the polymer chains and can be easily removed by applying an opposite electrical driving force. Consequently and most significantly, the doping processes of conducting polymers are reversible and may be monitored by an external polarisation, which may be promoted and controlled in suitable electrochemical cells.

However, it remains to be understood how this type of doping process can induce enhancement in the electronic transport of the polymers. Indeed, the clarification of the doping mechanism of conducting polymers and of the associated electronic band evolution is of fundamental importance for the comprehension of the operational behaviour of these compounds as novel electrode materials.

In an attempt to illustrate in a simple way the general concept of the doping process in polymers, let us consider the p-doping (oxidation) process of polypyrrole. In the undoped state, polypyrrole is a poor electronic conductor with an energy gap of 3.2 eV between the conduction band (CB) and the valence band (VB):

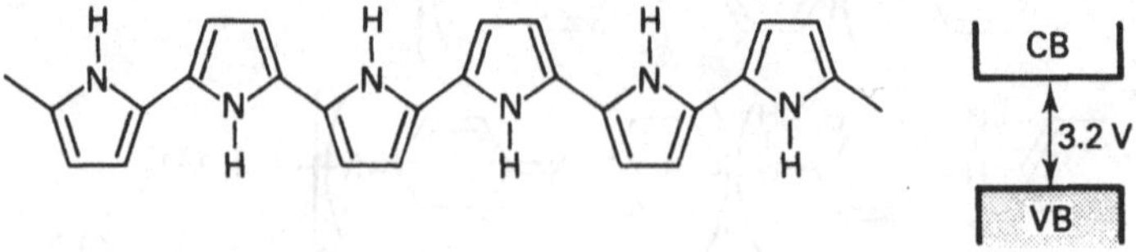

CB

3.2 V

VB

(9.28)

The initial removal of electrons (following the oxidation, p-doping process) leads to the formation of a positive charge localised in the polymer chain (radical cation), accompanied by a lattice distortion which is associated with a relaxation of the aromatic structural geometry of the polymer chain towards a quinoid form. This form extends over four pyrrolic rings:

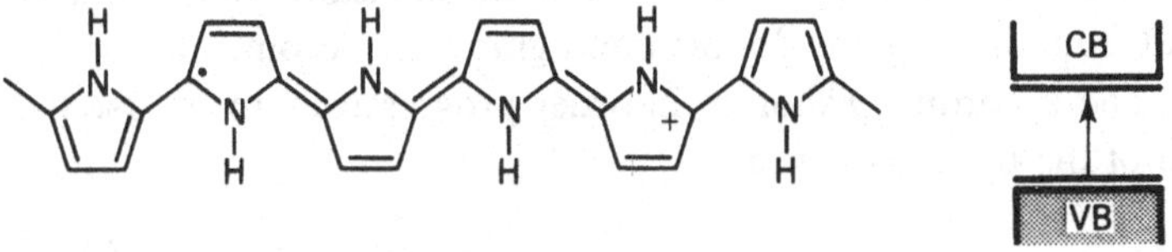

(9.29)

This radical cation (which is partially delocalised over the polymer segment and is stabilised by polarising the surrounding medium) is called a *polaron*. It is energetically described as a half-filled *polaron level* situated at approximately 0.5 eV from the band edges (for more detailed information on the mechanism of doping processes, see, for example, Bredas and Street (1985), Skotheim (1986) and Scrosati (1988)). When further electrons are removed from the chain, dications are formed, namely two positive charges localised in the same 'defect site'. This defect has been referred as a *bipolaron*, defined as a pair of like charges associated with a strong localised lattice distortion, which again extends over four pyrrolic rings:

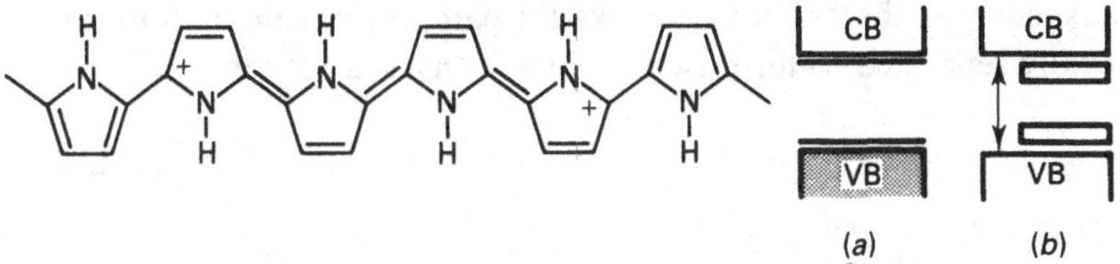

(9.30)

The bipolarons are energetically described as spinless *bipolaron levels* (scheme (9.30a)) which are empty and which, at high doping levels, may overlap with the formation of bipolaronic bands (9.30b). Finally, for polymers with band gap, E_g values smaller than that of polypyrrole – such as polythiophene – the bipolaronic bands may also overlap with the valence and conduction bands, thus approaching the metallic regime.

The band structure illustrated above applies to heterocyclic polymers, the cases of polyacetylene and polyaniline being somewhat different.

Polyacetylene is a polymer which has a degenerate ground state; it possesses two geometric structures having exactly the same energy and differing only in the sequence of the carbon–carbon single and double bonds (Fig. 9.6).

Since the two structures are chemically and energetically equivalent, it is possible that in long chains one part tends to assume one version of the structure and another part the other version with a transition zone in between. This zone may extend over a few carbon atoms and one may assume that from one side the double bond gradually become longer and the single bond accordingly shorter, in order progressively to reverse the alternation of the bond sequence:

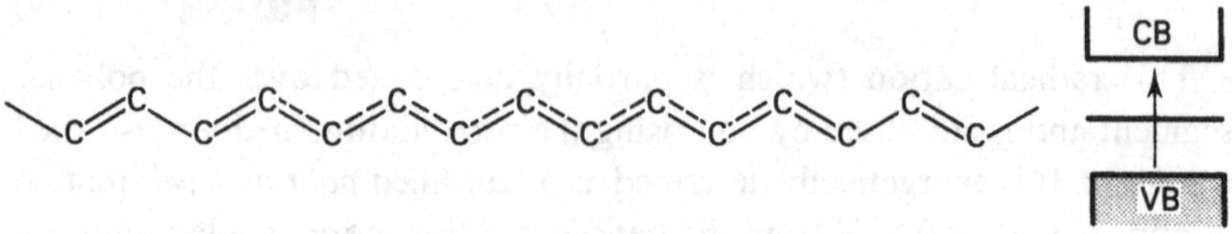

CB

VB

(9.31)

This is accompanied by a structural distortion and thus, according to the above model, to an electronic state which, due to the symmetry of the electronic distribution lies precisely at midgap. This single state, having half occupied levels in the gap, has been termed *native solitons*.

The degeneracy of the ground state of polyacetylene influences its charge distribution. In fact, upon doping the charges, which in other polymers, such as the heterocyclics, would pair to form bipolarons, are here readily separated to form two positively charged solitons:

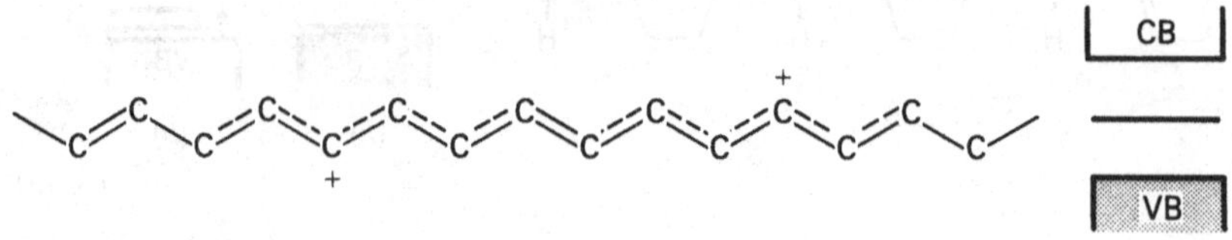

CB

VB

(9.32)

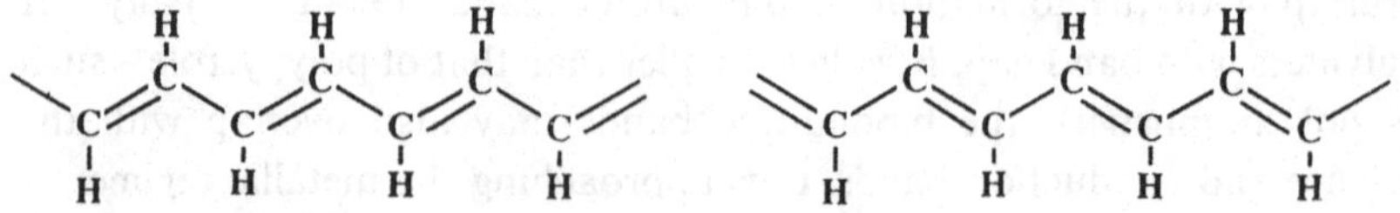

Fig. 9.6 The two different bond sequences which are possible in the *trans* configuration of polyacetylene.

Therefore, the p-doped polyacetylene chain acquires positive charges which are partly delocalised and which may be represented as non-occupied midgap states. Similarly, upon n-doping the formation of negatively charged solitons, which may be represented as double-occupied midgap states, takes place and the evolution of the electronic structure upon doping may again be described as the formation of solitonic bands.

The solitonic charges are mobile along the polymer chain by rearrangement of double and single bonds:

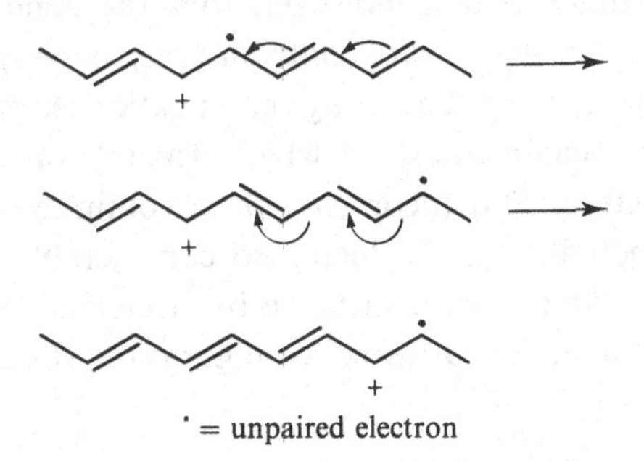

and this accounts for the conductivity mechanism of this, as well as of other similar systems (Kanatzidis, 1990).

Also the case of polyaniline is somewhat different from that of heterocyclic polymers. It has been proposed (MacDiarmid and Maxfield, 1987) that the doping process does not induce changes in the number of electrons associated with the polymer chain but that the high conductivity of the emeraldine salt polymers is related to a highly symmetrical π-delocalized structure.

In summary, although exhibiting some differences, the electronic properties of conducting polymers are generally associated with a band structure that accounts for the unique transport mechanism of this new class of conductors. Indeed, this mechanism lies between that of metals and that of semiconductors. Like metals, the conducting polymers have a high conductivity; however, like semiconductors they require doping processes to acquire this high conductivity. The nature of the charge carriers is different from that of both classes of materials: while in metals and in semiconductors the conduction proceeds via a coherent propagation of electrons and/or holes across the lattice, in conducting polymers the spinless bipolarons or solitons transport the current.

At high doping levels, when the coulombic attractions to the counterions are screened, the bipolarons or the solitons may become highly mobile and the conductivity of the polymer approaches that of metals. This latter situation is well illustrated in Fig. 9.7 which shows the temperature dependence of highly doped polyacetylene in comparison with that of silver: upon increasing temperature, the value of the conductivity of the polymer almost reaches that of metals, however, with a thermal trend which is more reminiscent of that of semiconductors.

One aspect of the conductivity of polymeric materials which has still not been fully clarified is the role of electron exchange between polymer chains. It seems quite certain, however, that the conductivity may be considered as the result of two contributions, namely the intrachain transport which depends on the average conjugation length of the polymer chain, and the interchain transport which depends on the regularity of the polymer structure. This theory has been confirmed by the fact that the overall conductivity may be increased considerably by ordering the structure of the polymer, as for instance by stretching the polyacetylene chains to obtain a highly oriented configuration (Lugli, Pedretti and Perego, 1985).

9.8 Methods for monitoring the doping processes of conducting polymers

The proposed mechanism of the doping processes in conducting polymers implies oxidation (p-doping) or reduction (n-doping) of the polymer with

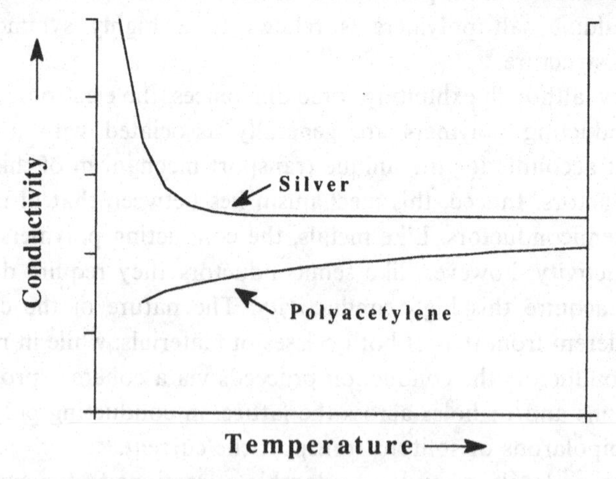

Fig. 9.7 Thermal behaviour of the conductivity of polyacetylene in comparison with that of silver (from Kanatzidis (1990)).

the diffusion of oppositely charged counterions from the electrolyte solution. Therefore, upon electrochemical doping, the polymers are expected to experience changes in their electronic structures as well as changes in their masses. It follows that techniques which are able to detect these changes may be employed both to control and evaluate the doping processes. Some examples are described below.

9.8.1 Optical absorption

The occurrence of bipolaronic states in the polymer chains promotes optical absorption prior to the π–π^* gap transitions. In fact, referring to the example (9.30) of the band structure of doped heterocyclic polymers, transitions may occur from the valence band to the bipolaronic levels. These intergap transitions are revealed by changes in the optical absorptions, as shown by Fig. 9.8 which illustrates the typical case of the spectral evolution of polydithienothiophene upon electrochemical doping (Danieli *et al.*, 1985).

Here curve (*a*) reflects the situation of the undoped state and it is characteristic of the π–π^* band, while curves (*b*)–(*e*) are the optical responses at progressively higher doping levels. It is evident that these curves show two new bands of increasing intensity, which, in turn, are related to transitions between the valence band and the two bipolaronic

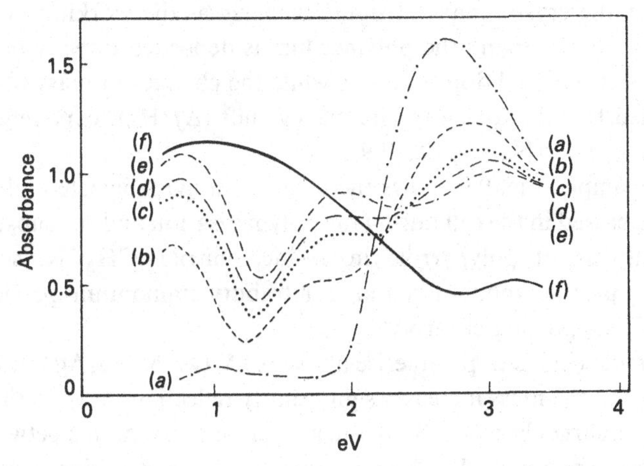

Fig. 9.8 Spectral evolution of polydithienothiophene upon electrochemical evolution (from Danieli *et al.* (1985)).

bands. As expected, the evolution of the two intergap bands is accompanied by a progressive decrease of the π–π^* band. Finally, at high doping levels, curve (*f*) reflects the quasi-metallic regime due to the overlapping of the bipolaronic bands with the conduction and/or the valence band.

The evolution of the band structure – and thus of the doping process – may be conveniently monitored by detecting *in situ* the optical absorption during the electrochemical process, by placing the cell directly into the spectrophotometer (Danieli *et al.*, 1985).

9.8.2 Microbalance studies

As repeatedly stressed, the doping processes imply the diffusion of electrolyte counterions to compensate for the electric charge assumed by the polymeric chain and thus polymers are expected to experience changes of mass upon doping. Consequently, by monitoring these changes it is possible to control the nature and the extent of the doping processes.

This may be conveniently carried out with the use of a quartz crystal microbalance (QCM), i.e. an apparatus capable of detecting mass changes in the nanogram (ng) range. The operation is performed by placing a gold-coated quartz crystal connected to both the potentiostat and the oscillator directly in the electrochemical cell (Naoi, Lien and Smyrl, 1991). In this way the crystal may act simultaneously as the working electrode and as the balance 'pan': the polymer film is deposited directly onto the working electrode and doped *in situ*, while the changes in mass (Δm(ng)) are detected by monitoring the frequency shift (Δf(Hz)) experienced by the quartz crystal (Naoi *et al.*, 1991).

As an example, Fig. 9.9 shows the changes in mass detected following the charge passed in the cell during the polymerisation and the subsequent doping process of polypyrrole in an acetonitrile, CH_3CN, solution containing pyrrole monomer and a tetrabutylammonium perchlorate ($TBAClO_4$) supporting electrolyte.

Polypyrrole begins to polymerise above 0.3 V (vs the Ag/Ag^+ reference electrode) and the mass increases significantly reflecting the growth of the film on the quartz electrode. Subsequent cycling at potentials between 0.3 and -1.5 V promotes the doping process which is accompanied by charge injection(doping)/release(undoping) to/from the polypyrrole film.

9.9 *Kinetics of the electrochemical doping processes*

In order to evaluate conducting polymers as possible electrode materials for novel devices, it is essential to investigate the kinetics of the doping processes.

As is well known in the field of electrochemistry in general, electrode kinetics may be conveniently examined by cyclic voltammetry (CV) and by frequency response analysis (ac impedance). The kinetics of the various polymer electrodes considered so far in this chapter will be discussed in terms of results obtained by these two experimental techniques.

9.9.1 *Kinetics of polyacetylene electrodes*

Fig. 9.10 shows a typical CV of a $(CH)_x$ film in a $LiClO_4$–propylene carbonate electrolyte. The voltammogram presents well-defined peaks both in the anodic (doping) and in the following cathodic (undoping) scans; this confirms that the doping process of polyacetylene, as suggested by (9.10), can indeed be driven electrochemically and in a reversible way.

Furthermore, the fact that the scan exhibits two peaks suggests the presence in the basic polyacetylene structure of at least two (if not more) different structural sites for the doping process, as effectively confirmed by independent structural studies (Shacklette *et al.*, 1985). Finally, the

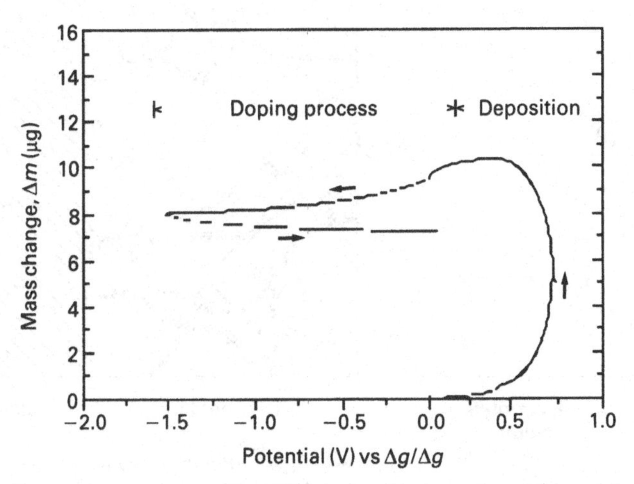

Fig. 9.9 Changes in mass detected by QCM during the electrodeposition and the subsequent doping process of polypyrrole in a $TBAClO_4$–CH_3CN solution (from Naoi *et al.*, 1991).

long tail at the end of the cycle suggests that the kinetics of the process are controlled by diffusion phenomena.

As schematically illustrated in Fig. 9.11, one may assume that the doping process of (9.10) proceeds via the following main steps:

(i) transport of the ClO_4^- anions from the bulk of the $LiClO_4$–propylene carbonate electrolyte to the electrode interface;

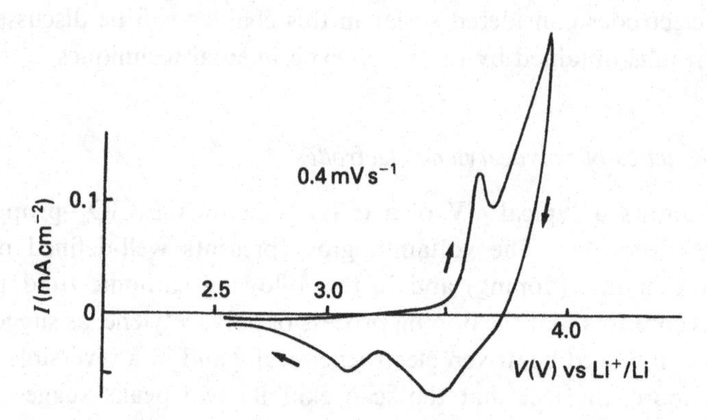

Fig. 9.10 Cyclic voltammetry of a polyacetylene film electrode in the $LiClO_4$–propylene carbonate electrolyte. Scan rate 0.4 mV s^{-1}.

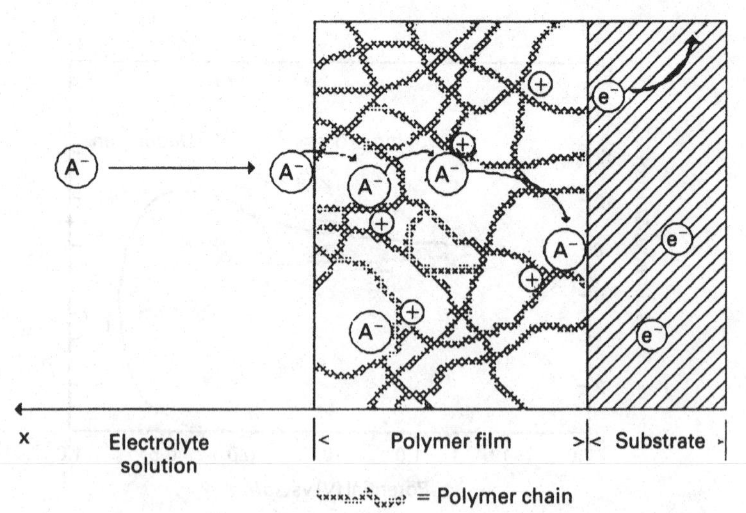

Fig. 9.11 Schematic of the electrochemical p-doping process of polymer film electrodes, involving transfer of electrolyte anions A^-.

(ii) electron removal from the $(CH)_x$ structure with the formation of the charged polycation;
(iii) diffusion of the ClO_4^- ions into the polymer matrix.

Since perchlorate ions, and more generally the majority of anions used in common electrolyte systems, are known to move rapidly in liquid solutions, it is reasonable to assume that the rate determining step in controlling the kinetics of the overall process is the ion diffusion throughout the polymer fibrils. This conclusion has been experimentally confirmed. For example, the diffusion coefficient of electrolyte counterions in bulk polyacetylene has been determined (Will, 1985) to be seven orders of magnitude lower than in liquid electrolytes, namely about 10^{-12} cm^2 s^{-1} vs 10^{-5} cm^2 s^{-1}.

Such a slow diffusion greatly affects the current which a polyacetylene electrode can sustain. This fact, combined with its poor chemical stability, makes polyacetylene of limited practical interest. Indeed, the promotion of polyacetylene as an electrode material for devices which led to such enthusiasm in the late 1970s (Nigrey, MacDiarmid and Heeger, 1979) is now regarded as hyperbole. However, the general idea of exploiting plastic-like electrode materials remains technologically very appealing and, therefore, it is not surprising that other polymers having faster response and higher chemical stability than polyacetylene are currently of substantial interest.

9.9.2 *Kinetics of heterocyclic polymers*

Considerable attention is presently devoted to heterocyclic polymers, such as polypyrrole, polythiophene and their derivatives. The kinetics of the electrochemical doping processes of these polymers has been extensively studied in electrochemical cells using non-aqueous electrolytes.

These studies have been mainly carried out using cyclic voltammetry and frequency response analysis as experimental tools. As a typical example, Fig. 9.12 illustrates the voltammogram related to the p-doping process of a polypyrrole film electrode in the $LiClO_4$–propylene carbonate electrolyte, i.e. the reaction already indicated by (9.16).

Two main features may be identified by examining the trend of this typical voltammogram; namely:

(i) The values of peak voltages associated with the oxidation and reduction waves are very close to each other and the integrated charge under the reduction (undoping) wave Q_{red} approaches that under the oxidation (doping) wave, Q_{ox}. These results may be interpreted as an indication that the electrochemical doping of polypyrrole has a good degree of reversibility, certainly higher than that of polyacetylene (compare Fig. 9.10).

(ii) A large, non-faradaic residual current is observed at the end of the anodic wave. This effect may be associated with a limiting capacitance, C_L, whose value may be evaluated from the expression:

$$C_L = i/v \tag{9.33}$$

where i is the current and v the voltage scan rate. Values of C_L of the order of 10–20 mF cm^{-2} are typically obtained (Panero, Prosperi, Passerini, Scrosati and Perlmutter, 1989). Such a high capacitance has been tentatively explained by charge saturation models which basically assume that when the process is driven to high doping levels, double layers of static charge, which involve molecular distances between weakly trapped ions and the surface

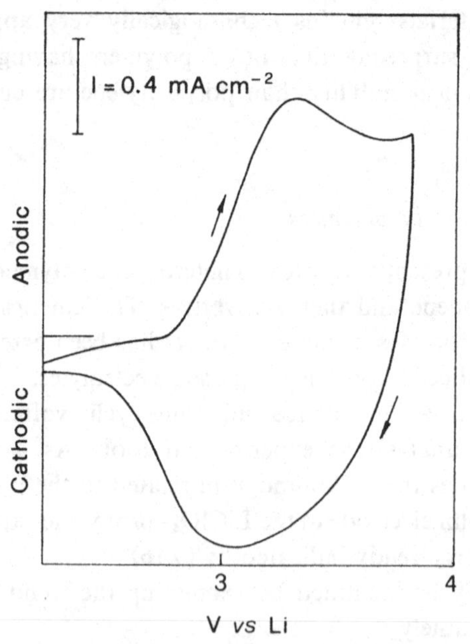

Fig. 9.12 Cyclic voltammetry of the p-doping(anodic)–undoping(cathodic) process of a polypyrrole electrode in $LiClO_4$–PC solution. Pt substrate, Li reference electrode, scan rate: 50 mV s^{-1}.

of molecular chains, are formed in the polymer matrix (Mermillod, Tanguy and Petiot, 1986; Tanguy, Mermillod and Hoclet, 1987). This phenomenon is certainly not common and suggests that the mechanism of the electrochemical doping process of these polymer electrodes is far from being fully clarified.

Further information on this subject can be obtained by frequency response analysis and this technique has proved to be very valuable for studying the kinetics of polymer electrodes. Initially, it has been shown that the overall impedance response of polymer electrodes generally resembles that of intercalation electrodes, such as TiS_2 and WO_3 (Ho, Raistrick and Huggins, 1980; Naoi, Ueyama, Osaka and Smyrl, 1990). On the other hand this was to be expected since polymer and intercalation electrodes both undergo somewhat similar electrochemical redox reactions, which include the diffusion of ions in the bulk of the host structures. One aspect of this conclusion is that the impedance response of polymer electrodes may be interpreted on the basis of electrical circuits which are representative of the intercalation electrodes, such as the Randles circuit illustrated in Fig. 9.13. The figure also illustrates the idealised response of this circuit in the complex impedance (jZ''–Z') plane.

The components of the circuit and the various frequency-limited regions of the response reflect the characteristics of the electrochemical processes. Referring to the model of electrochemical doping given in Fig. 9.11 – which basically holds for the doping processes of most polymer electrodes – the high frequency region is associated with the resistance, R_e, of the electrolyte solution, across which the ions are transported to reach the electrode interface. The medium frequency region is associated with the charge transfer at the interface (in the case under study to the complicated process which involves transfer of ions from their environment in solution to that in the solid matrix). The related relaxation effect is displayed in the ($-jZ''$–Z') plane as a semicircle, whose time constant is given by the product of the charge transfer resistance and the double-layer capacitance, $R_{ct}C_{dl}$. The two parameters may be obtained from the diameter of the semicircle on the real Z' axis and by the characteristic relaxation frequency f_c at the maximum of the semicircle, respectively (MacDonald, 1987).

At low frequencies the impedance is dominated by diffusion. Two regions may be identified in the complex impedance, a linear region with a phase angle of $\pi/4$ corresponding to semiinfinite diffusion and represented by the Warburg impedance Z_w, and a second linear region at yet lower frequencies with phase angle of $\pi/2$. Considering that ion diffusion is expected to be much faster in the liquid solution than in the solid polymer matrix, it is reasonable to assume that the mass transport process is dominated by ion transfer throughout the solid. The very low frequency region with phase angle $\pi/2$ is associated with a purely capacitive response. In this condition the diffusion process is progressively limited by charge accumulation into the polymer host, resulting in a limiting capacitance C_L, already discussed in relation to the cyclic voltammetry curve. The value of C_L can be obtained from the equation:

$$C_L = L^2/3DR_L, \tag{9.34}$$

where L is the thickness of the polymer electrode, D is the diffusion coefficient of the electrolyte counterion throughout the solid polymer host

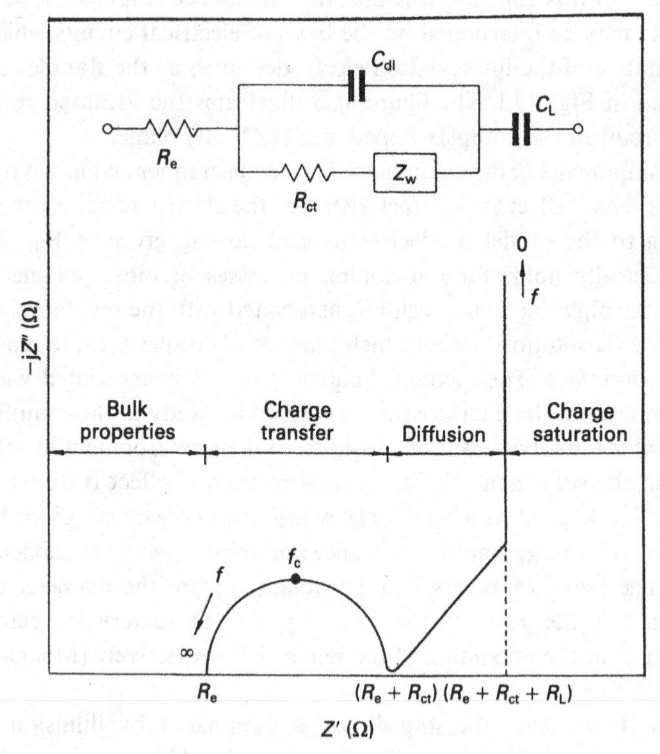

Fig. 9.13 Randles-type circuit (top) and its idealised response in the impedance plane. f = frequency of the ac signal.

structure and R_L is a limiting resistance determined from the intercept to the real axis.

As expected, the impedance responses obtained in practice do not fully match that of the model of Fig. 9.13. However, as shown by the typical case of Fig. 9.14 which illustrates the response obtained for a 5 mol% ClO_4-doped polypyrrole electrode in contact with a $LiClO_4$-propylene carbonate solution (Panero *et al.*, 1989), the trend is still reasonably close enough to the idealised one to allow (possibly with the help of fitting programmes) the determination of the relevant kinetics parameters, such as the charge transfer resistance, the double-layer capacitance and the diffusion coefficient.

In particular, from the analysis in the very low frequency region, the limiting capacitance C_L can be evaluated (via the limiting resistance R_L). Practical cases have given values of C_L of about 20 mF cm^{-2} (Panero *et al.*, 1989), i.e. values of the same order as those obtained by cyclic voltammetry, thus confirming the validity of the charge saturation model.

Furthermore, using Eqn (9.34), the diffusion coefficient of ClO_4^- in polypyrrole was estimated to be 1.3×10^{-9} cm^2 s^{-1} (Panero *et al.*, 1989). This value is three orders of magnitude greater than that found for the diffusion of the same anion in polyacetylene and this confirms that

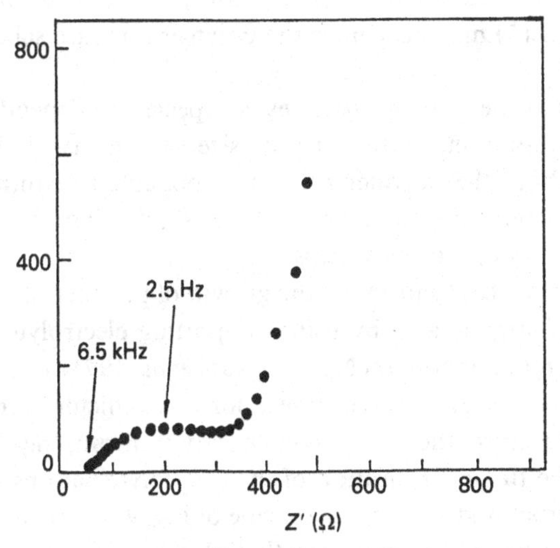

Fig. 9.14 The ac impedance response of a ClO_4-doped polypyrrole electrode over a frequency range extending from 0.006 Hz to 6.5 kHz.

the doping process in polypyrrole, and quite likely in several of the heterocyclic polymers, is kinetically faster than that of polyacetylene. However, a D value in the region of 10^{-9} cm^2 s^{-1} is still an indication of relatively slow diffusion and therefore that the rate of the overall electrochemical processes for heterocyclic polymer electrodes may ultimately be diffusion controlled. Consequently, an efficient application of these electrodes in practical devices would require some structural modifications capable of promoting enhancements in the ionic diffusion.

9.10 Methods for enhancing diffusion processes in polymer electrodes

There is clear evidence based on quartz microbalance measurements (Kaufman, Kanazawa and Street, 1979; Naoi *et al.*, 1991) and on secondary ion mass spectrometry studies (Chao, Baudoin, Costa and Lang, 1987) that the doping–undoping process of polymers – say of polypyrrole – are more complicated than simply involving anions – say ClO_4^- – entering the polymer as the doping (oxidation) proceeds and leaving as the process is inverted to promote undoping (reduction), as Eqn (9.16) would indicate. In fact, while the doping process may mainly involve the diffusion of anions, the undoping process is more likely characterised by multiple ion movements with the formation of cation– anion pairs and the release of these pairs plus the anions in excess.

During synthesis, the anions (e.g. ClO_4^-) are not the sole diffusing species: cations (e.g. Li^+) may also enter the polymer film (see scheme in Fig. 9.15(*a*)).

Since the kinetics of the doping processes is expected to depend upon the nature of the counterion, particularly its size (which may influence the mobility throughout the polymer host), it is possible to control the diffusion kinetics by selecting the nature of the supporting electrolyte employed in the electrodeposition process.

For instance, it is possible to promote the growth of polymer films with the incorporation of large anions by using supporting electrolytes such as sodium poly(4-styrenesulphonate) NaPSS (Naoi *et al.*, 1991), or sodium dodecylsulphonate, NaDS. Since these large anions are difficult to remove from the polymer structure, the charge compensation during the doping process is likely to be the incorporation of the electrolyte cations rather than the release of structured anions (see scheme of Fig. 9.15(*b*)). Since the diffusion of cations is generally faster than that of anions, the promotion of doping processes which mainly involve cation transport ultimately enhances the kinetics of the overall electrochemical process, as has indeed been experimentally verified for the PSS and the DS functionalised polypyrrole electrodes (Naoi *et al.*, 1991; De Paoli, Panero, Prosperi and Scrosati, 1990).

9.11 Applications of polymer electrodes

Although conjugated polymers can be both n-doped and p-doped – and thus, in principle, be capable of behaving either as negative or as positive electrodes – the majority of applications have been confined to the p-doping, positive side. Conductive polymers have been proposed and tested in a variety of advanced electrochemical devices. Due to lack of space, we will confine our attention to the description of the most illustrative examples which are rechargeable lithium batteries and multichromic optical displays.

9.12 Lithium rechargeable batteries

Many industrial and academic laboratories have investigated doped polymers as improved positive electrodes in rechargeable lithium batteries. A common example is the battery formed by a lithium anode, a liquid organic electrolyte (e.g. $LiClO_4$-PC solution) and a polypyrrole film

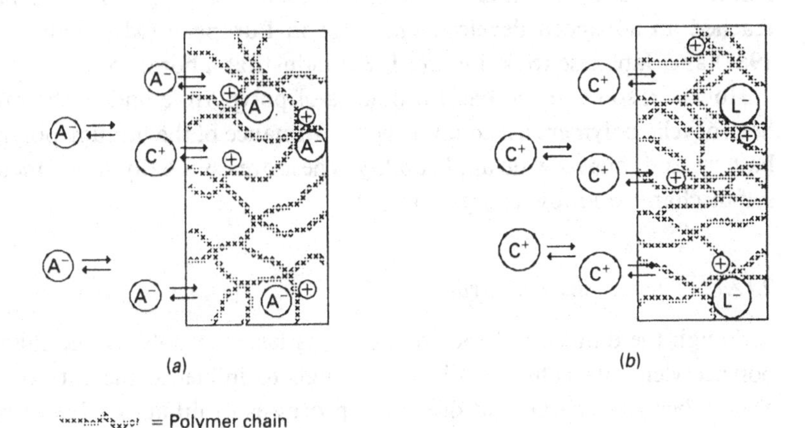

Fig. 9.15 Schematic models for ion insertion during redox processes of polymer electrodes electropolymerised in the presence of small (case (*a*)) and large (case (*b*)) anions. A^- = electrolyte anion; C^+ = electrolyte cation; L^- = large anion.

cathode:

$$Li/LiClO_4\text{-}PC/(C_4H_5N)_x, \tag{9.35}$$

characterised by the following electrochemical process:

$$(C_4H_5N)_x + (xy)LiClO_4 \underset{\text{discharge}}{\overset{\text{charge}}{\rightleftharpoons}} [(C_4H_5N^{y+})(ClO_4^-)_y]_x + (xy)Li. \tag{9.36}$$

The charging process implies the oxidation of the cathode polymer with the concurrent insertion of the ClO_4^- anions from the electrolyte and the deposition of lithium at the anode. In the discharging process the electroactive cathode material releases the anion and the lithium ions are stripped from the metal anode to restore the initial electrolyte concentration. Therefore, the electrochemical process involves the participation of the electrolyte salt to an extent which is defined by the doping level y.

Since y is proportional to the amount of charge passed, it is also related to the capacity (measured in ampere hours, A h) of the battery. Although oversimplified, we may describe this rechargeable battery by saying that during the charge cycle energy is furnished to activate electrochemically the polymer electrode by promoting the doping process up to a level y, while during discharge the energy is released by the undoping process and that this cycle can be repeated several times.

Prototypes of lithium/polypyrrole batteries are under study in various laboratories and low rate, small size versions of these batteries have reached an advanced development stage in European (Munstedt *et al.*, 1987) and Japanese (Sakai *et al.*, 1986) industrial laboratories.

However, some of the basic problems of polypyrrole and of the other heterocyclic polymers act to limit the performance of the lithium/polymer battery, and thus its wide applicability. These are essentially slow kinetics, self-discharge and low energy content.

9.12.1 Charge–discharge rate

Although the diffusion of the counterion is faster in polypyrrole than in polyacetylene, its value is still low enough to influence the rate of the electrochemical charge and discharge processes of lithium/polymer batteries. Indeed the current output of these batteries is generally confined to a few mA cm^{-2}. Possibly, improvements in the electrode kinetics, and thus in the battery rates, may be obtained by the replacement of 'standard' electrodes with large-size-ion-functionalised electrodes, supposedly characterised by faster ion diffusion (see Section 9.5 and Fig. 9.15).

There is some experimental evidence that the use of these 'modified' polymer electrodes is effectively beneficial in terms of battery performance. Fig. 9.16 compares the behaviour – in terms of per cent of theoretical capacity (evaluated on the basis of 33% maximum doping level) delivered upon cycling in the $LiClO_4$-PC electrolyte – of a lithium cell using a standard polypyrrole electrode (here indicated as pPy(ClO_4)) with that of a cell using a polypyrrole electrode synthesised in the presence of sodium dodecylsulphate, a large anion salt (here indicated as pPy(DS)). The figure clearly shows that the latter can be extensively cycled at a much higher capacity than the former (Scrosati, 1989).

9.12.2 *Self-discharge*

Another problem still to be solved in polymer batteries is the self-discharge of the polymer electrode in common electrolyte media. Effectively, the majority of the polymer electrodes show a poor charge retention in organic electrolytes. *In situ* spectroscopic measurements (Scrosati *et al.*, 1987) have clearly demonstrated the occurrence of spontaneous undoping processes. A typical example is illustrated in Fig. 9.17 which is related to the change of the absorbance of doped polypyrrole upon contact with the electrolyte.

The self-discharge phenomena are revealed by the continuous decrease of the low energy bands characteristic of the doped states and by the corresponding increase of those characteristic of the undoped state. These phenomena are not easy to explain. One important fact is that the self-undoping processes do not induce irreversible degradation of the

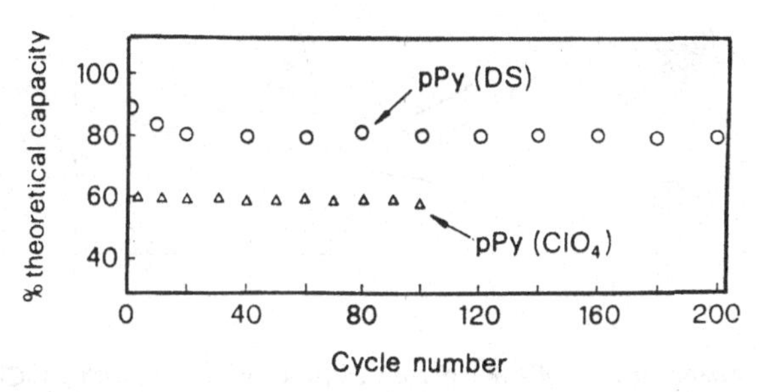

Fig. 9.16 Cyclic behaviour of lithium batteries using standard pPy(ClO_4) polypyrrole electrodes and modified pPy(DS) electrodes.

electrodes which can always be electrochemically regenerated to their initial doped states (see spectrum (5) of Fig. 9.17). Another plausible assumption is that the undoping (reduction) process of the polymer electrodes must be accompanied by a concurrent (oxidation) reaction. The nature of this reaction is still unclear. The participation of the electrolyte, with oxidation of the solvent and/or of impurities, is a reasonable hypothesis. Therefore, a suitable choice of the electrolyte medium appears crucial in ensuring the electrochemical stability of the polymer electrodes and thus the shelf life of lithium/polymer batteries.

9.12.3 Energy content

The *capacity* (measured in A h) of the lithium/polymer battery is controlled by the amount of cyclable charge, i.e. by the doping level y exchanged during charge and discharge. Consequently, the *energy content* (measured in W h) is, in turn, influenced by the capacity, as well as by the *average discharge voltage* (measured in V). If this parameter is then reported in relation to the *weight* or to the *volume* of the battery, one can characterise the system in terms of *specific energy* (measured in W h kg^{-1}) and of *energy density* (measured in W h cm^{-3}).

The majority of polymer electrodes cannot be doped to very high levels. For instance, polypyrrole may reach doping levels of the order of 33%. This inherent limitation combined with the fact that the operation of the lithium/polymer battery requires an excess of electrolyte (to ensure

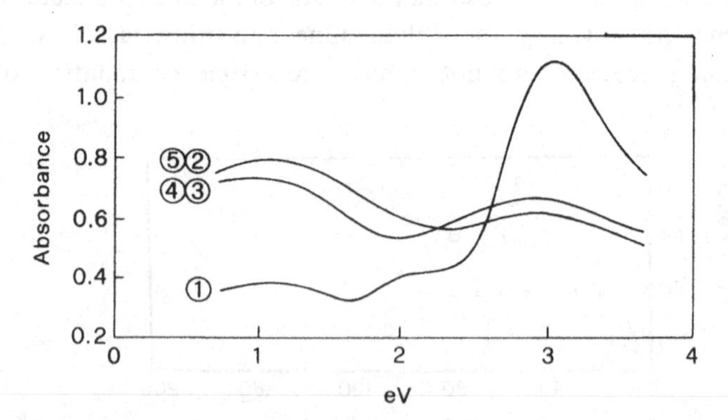

Fig. 9.17 *In-situ* absorbance spectra of a polypyrrole electrode in contact with a $LiClO_4$-PC electrolyte: (1) poorly doped; (2) highly doped immediately after charge; (3) after 5.5 hours and (4) after 17 hours of storage; (5) electrochemically regenerated (from Scrosati (1989)).

sufficiently low internal resistance also at the end of the charging (doping) process) confines the achievable specific energy and energy densities to modest values.

9.12.4 The future of the lithium/polymer battery

All the above factors, namely slow kinetics, self-discharge and low energy content, combine to limit the range of applications for polymer batteries. In fact, this battery appears to be most suitable for small size, low rate prototypes designed for the microelectronic market. However, the technology here is at a very preliminary stage and there is still room for substantial improvement. It is hoped that the polymer battery may reach a stage of development in terms of performance and cost, such that it may replace the common dry cells and nickel–cadmium batteries. This would be a great achievement, not only from the point of view of progress in advanced technology, but particularly because it would reduce the environmental pollution caused by the dumping of heavy metals, resulting from the disposal of common dry cells and batteries.

9.13 Optical displays

Since the band structure which develops upon doping induces changes not only in the conductivity but also in the optical absorption (see Fig. 9.8), conducting polymers may be exploited for *electrochromic displays*, which are optical devices with marked colour transitions. An example is illustrated diagramatically in Fig. 9.18.

Upon application of the anodic pulse, the polymer assumes the colour typical of the doped state and then returns to that typical of the neutral state when switched to the opposite cathodic pulse. These electrochromic displays (ECDs) offer several advantages over other non-emissive optical switches, such as liquid crystal displays (LCDs). The most significant are the unlimited visual angle, the feasibility of being constructed in large dimensions and the optical memory, i.e. the possibility of offering a colouration which can be extended over a wide panel and which does not vanish after the driving voltage pulse is removed.

ECDs operate in the diffuse reflectance mode and the basic requirements for their functionality are: (i) a primary electrochromic electrode (e.g. a polymer electrode) deposited on a substrate which is both optically transparent and electrically conducting (generally indium–tin-oxide(ITO)-

coated glasses are suitable for this purpose); (ii) an electrolyte having good conductivity by transport of the X^- counterion involved in the polymer doping process; and (iii) a counterelectrode capable of providing the electrochemical balance.

In the case of the scheme of Fig. 9.18 the primary electrode could be a thin film of polymethylthiophene ($[C_5H_7S]_x$) deposited on an ITO-coated glass, the electrolyte can be the usual $LiClO_4$-PC solution and the counterelectrode lithium metal (Li), to obtain the following structure:

$$Li/LiClO_4\text{-PC}/[C_5H_7S]_x,\ \text{ITO, glass.} \tag{9.37}$$

Essentially, a display is a battery with a colour change. The electrochemical process:

$$\underset{\text{(red)}}{[C_5H_7S]_x} + (xy)LiClO_4 \underset{\text{undoping}}{\overset{\text{doping}}{\rightleftharpoons}} \underset{\text{(blue)}}{[(C_5H_7S^{y+})(ClO_4{}^-)_y]_x} + (xy)Li \tag{9.38}$$

is, in fact, accompanied by a sharp change of colour which makes it suitable as the basis of an optical device.

There is increasing interest in optical devices commonly called *electrochromic* (*smart*) *windows* (EWs), i.e. ECDs which allow electrochemically

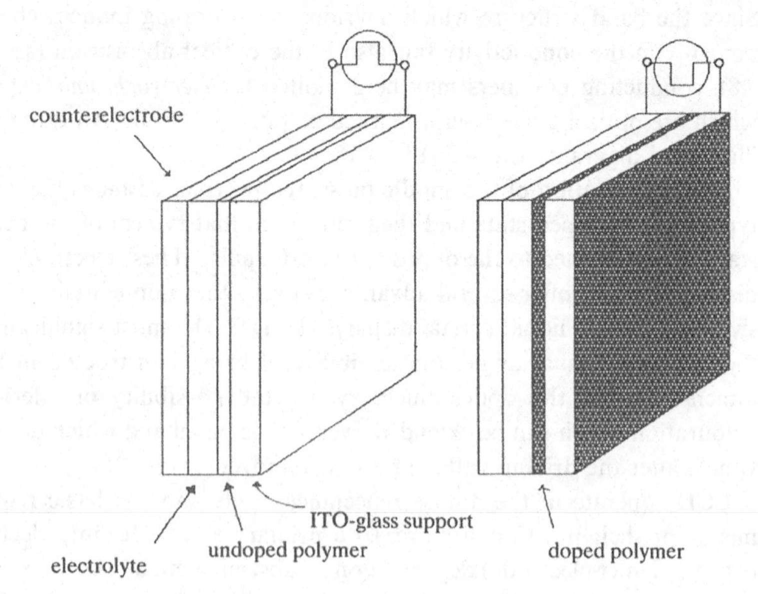

Fig. 9.18 Pictorial illustration of an electrochromic display.

driven modulations of light transmission and reflection. The basic difference between common ECDs and EWs is that in the latter case the entire system is in the optical path, thus requiring a transparent electrolyte and a counterelectrode which is either optically passive (i.e. colourless in both oxidised and reduced states) or electrochromic in a complementary mode with respect to the primary electrochromic electrode (i.e. if the latter is coloured cathodically, the former must be coloured anodically, and vice versa).

A promising EW may be obtained by combining tungsten trioxide, WO_3, a well-known primary electrochromic electrode, which is coloured by the following lithium insertion–deinsertion process:

$$\underset{\text{(light yellow)}}{x\text{Li} + \text{WO}_3 + xe^-} \rightleftharpoons \underset{\text{(dark blue)}}{\text{Li}_x\text{WO}_3} \tag{9.39}$$

with polyaniline PANI, $[CHN_4]_x$, which, in the presence of perchlorate ions, may be doped by a process:

$$\underset{\text{(light yellow)}}{[\text{CHN}_4]_x + (xy)\text{ClO}_4^-} \rightleftharpoons \underset{\text{(green)}}{[(\text{CHN}_4^{y+})(\text{ClO}_4)_y^-]_x}. \tag{9.40}$$

The colour change is, however, complementary to that of tungsten trioxide. Consequently, one may consider a window formed by the combination of the two electrodes in the $LiClO_4$-PC electrolyte, i.e.:

$$\text{glass/ITO/WO}_3\text{/LiClO}_4\text{–PC/[CHN}_4]_x\text{/ITO/glass} \tag{9.41}$$

Upon application of an external voltage pulse, lithium is intercalated in WO_3, promoting the dark blue colour configuration and $[CHN_4]_x$ is doped with perchlorate promoting the green colour configuration, so that the window is shifted to the full reflecting condition; by switching the polarity of the external pulse, the electrodes are restored to their initial pale yellow configuration and the window assumes the fully transparent condition, the cycle can be repeated many times.

9.13.1 *The future of polymer displays*

A possible limitation of ECDs and EWs is the relatively slow response time, which is typically in the range of a few seconds. This makes ECDs uncompetitive with LCDs for fast devices, such as electronic watches. However, if the applications are directed to realisation of large panels for optical information or for energy control, the time of response becomes of secondary importance with respect to colour contrast and optical memory. Therefore, in this field ECDs assume a prominent if not a unique role in the area of display technology. Their widespread use in important devices, such as automobile rear-view mirrors and energy-saving windows, is expected to increase significantly in the near future.

Acknowledgements

I would like to thank my coworkers and graduate students for their valuable experimental work and for the helpful discussions in the preparation of this chapter. In particular, I express may gratitude to Patrizia Morghen, Stefania Panero, Paola Prosperi and Daniela Zane.

References

Bredas, J. L. and Street, G. B. (1985) *Acc. Chem. Res.*, **18**, 309.
Chao, F., Baudoin, J. L., Costa, M. and Lang, P. (1987) *Makromol. Chem. Makromol. Symp.*, **8**, 173.
Chiang, C. K., Park, Y. W., Heeger, A. J., Shirakawa, H., Louis, E. J. and MacDiarmid, A. G. (1978) *J. Chem. Phys.*, **69** (**11**), 5098.
Danieli, R., Taliani, C., Zamboni, R., Giro, G., Biserni, M., Mastragostinio, M. and Testoni, A. (1985) *Synth. Met.*, **13**, 325.
De Paoli, M., Panero, S., Prosperi, P. and Scrosati, B. (1990) *Electrochim. Acta*, **55**, 1145.
Kanatzidis, M. G. (1990) *Chem. & Eng. News*, **68** (49), 36.
Kaufman, J. H., Kanazawa, K. K. and Street, G. B. (1979) *Phys. Rev. Letters*, **53** (26), 2461.
Lugli, G., Pedretti, U. and Perego, G. (1985) *J. Polym. Sci. Polym. Lett. Ed.*, **23**, 129.
Ho, C., Raistrick, I. D. and Huggins, R. A. (1980) *J. Electrochem. Soc.*, **127**, 343.
MacDiarmid, A. G. and Maxfield, M. R. (1987) *Organic polymers as electroactive materials*, in '*Electrochemical Science and Technology of Polymers*', Ed. Linford, R. G., Elsevier Applied Science, London, 67.
MacDonald, J. R. (1987) *Impedance Spectroscopy*, John Wiley, London.
Mermillod, N., Tanguy, J. and Petiot, F. (1986) *J. Electrochem. Soc.*, **133**, 1073.
Munstedt, H., Kohler, G., Mohwald, H., Neagle, D., Bittin, R., Ely, G. and Meissner, E. (1987) *Synth. Metals*, **18**, 259.
Naoi, K., Ueyama, K., Osaka, T. and Smyrl, W. H. (1990) *J. Electrochem. Soc.*, **137**, 494.
Naoi, K., Lien, M. and Smyrl, W. H. (1991) *J. Electrochem. Soc.*, **138**, 440.
Nigrey, P. J., MacDiarmid, A. G. and Heeger, A. J. (1979) *J. Chem. Soc. Chem. Comm.*, 594.
Panero, S., Prosperi, P., Passerini, S., Scrosati, B. and Perlmutter, D. D. (1989) *J. Electrochem. Soc.*, **136**, 3729.

References

Sakai, T., Furukawa, N., Nishio, K., Suzuki, T., Hasegawa, K. and Ando, O. (1986) *The 27th Battery Symposium in Japan*, Nov. 25–27, 189.

Scrosati, B., Panero, S., Prosperi, P., Corradini, A. and Mastragostino, M. (1987) *J. Power Sources*, **19**, 27.

Scrosati, B. (1988) *Progress Solid State Chem.*, **18**, 1.

Scrosati, B. (1989) *J. Electrochem. Soc.*, **136**, 2774.

Shacklette, L. W., Toth, J. E., Murthy, N. S. and Baugham, R. H. (1985) *J. Electrochem. Soc.*, **132**, 1529.

Shirikawa, H. and Ikeda, S. (1971) *Polymer J.*, **2**, 231.

Shirikawa, H., Ito, T. and Ikeda, S. (1978) *Makromol. Chemie*, **179**, 1565.

Skotheim, T. A. Ed. (1986) *Handbook of Conducting Polymers*, Vol. 1 & 2, Marcel Dekker Inc., New York.

Tanguy, J., Mermillod, N. and Hoclet, M. (1987) *J. Electrochem. Soc.*, **134**, 795.

Will, F. G. (1985) *J. Electrochem. Soc.*, **132**, 2093, 2351.

10 Interfacial electrochemistry

R. D. ARMSTRONG and M. TODD
Department of Chemistry, University of Newcastle upon Tyne

In this chapter we shall discuss both the theoretical and experimental aspects of the following types of interfaces:

(*a*) metal/polymer electrolyte, e.g. Li/PEO–$LiCF_3SO_3$ or Pt/PEO–$LiCF_3SO_3$;
(*b*) metal/crystalline ionically conducting solid, e.g. Ag/Ag_4RbI_5 or C/Ag_4RbI_5;
(*c*) aqueous salt solution/polymer electrolyte, e.g. $AsPh_4Cl$–H_2O/$AsPh_4BPh_4$–PVC or KCl–H_2O/$AsPh_4BPh_4$–PVC;
(*d*) polymer electrolyte/electronically and ionically conducting solid, e.g. PEO–$LiCF_3SO_3/Li_xV_6O_{13}$;
(*e*) solid electrolyte/solid electrolyte, e.g. Ag_4RbI_5/Ag-β-Al_2O_3.

We shall not attempt to review exhaustively the literature on interfacial electrochemistry in solid state systems. Instead we shall indicate the appropriate theoretical models for different situations. Most of the models and the related equations were developed some time ago in relation to the electrochemistry of aqueous systems. However, we will not assume a knowledge of these models on the part of the reader. It is important to realise that a direct transposition of models from one situation to another is fraught with difficulty, particularly since in aqueous electrochemistry a supporting electrolyte is generally present.

It is convenient for the purposes of this chapter initially to make a number of simplifying assumptions about the nature of the electrolytes under discussion. For example, Ag_4RbI_5 will be assumed to be an ionic conductor with negligible electronic conductivity and with only the silver ion mobile. Likewise Na-β-Al_2O_3 will be assumed to be a substance in which the only mobile charge is Na^+. Another simplifying assumption of a different sort which we will make is that the interfaces when formed do not, in general, have a third phase, e.g. an oxide film between the two bulk phases. Similarly the complications which arise from the presence of more than one phase in a particular solid (e.g. grain boundary material) will also be ignored. Later in the chapter the differences which arise when some of these assumptions are no longer valid will be discussed.

It is important to distinguish between interfaces across which charged species tend to equilibrate (once contact has been made between the bulk phases) which are called *non-blocking* interfaces and those where there is no immediate equilibration of charged species which are called *blocking* interfaces. An example of a non-blocking interface is Ag/Ag_4RbI_5 where the Ag^+ tends to equilibrate between the silver metal and the Ag_4RbI_5. In fact when a potential difference of $\Delta\phi_e$ exists, equilibrium is established between the two phases and no net current flows across the interface so the rate of Ag^+ transfer from the metal to the electrolyte is exactly balanced by the rate of transfer in the reverse direction (Fig. 10.1). The flux of ions across the interface at equilibrium is generally given by an exchange current (i_0) expressed in A cm^{-2}. If a potential difference between the bulk phases, different from $\Delta\phi_e$, is imposed on the interface, a net current of Ag^+ will flow (Fig. 10.2).

We can contrast this interface with the C/Ag_4RbI_5 interface where no charged species start to equilibrate once the bulk phases have been brought into contact. For a range of interfacial potential differences extending to 0.7 V there is an electrostatic equilibrium whereby the charge on the surface of the carbon is balanced by an equal and opposite charge

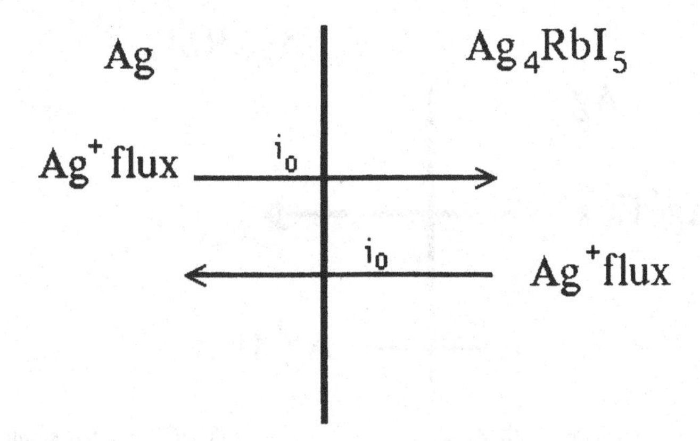

Fig. 10.1 Silver in contact with Ag_4RbI_5 at equilibrium. The fluxes of Ag^+ across the interface are equal and opposite so that no net current flows.

on the surface of the electrolyte. Under these circumstances this interface is blocking, i.e. no steady state current flows. However, if a sufficiently negative potential is applied to the carbon, charge will cross the interface and cause the deposition of silver metal onto the carbon;

$$e + Ag^+ = Ag.$$

If a sufficiently positive potential is applied to the carbon a different electrode reaction occurs;

$$2I^- = I_2 + 2e.$$

This demonstrates one general feature of a real blocking electrode, i.e. that it is only blocking provided that positive and negative extremes of potential difference are avoided. It should be noted that when an exchange current is very low at a non-blocking interface (perhaps below 10^{-10} A cm^{-2}) it will behave like a blocking interface, so that the distinction between the two types of interface is not as sharp as may at first sight appear.

Electrical measurements are the most commonly used method of investigating interfaces. It is not possible to give a full account of them in this chapter. For more detailed accounts the reader should consult the following references: Bruce (1987), MacDonald (1987), Armstrong and Archer (1980).

Electrical measurements which are made on interfaces generally involve changing the potential difference between two points in the two different

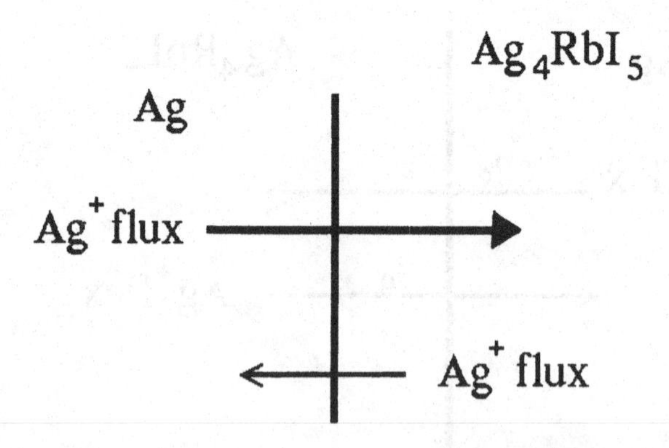

Fig. 10.2 Silver in contact with Ag_4RbI_5 at a potential different from the equilibrium potential (a positive overpotential). The Ag^+ flux from the metal exceeds that from the electrolyte and a net current flows.

phases concerned in one of three different ways:

(1) The potential difference is rapidly changed from one fixed value to another, perhaps in a time interval as short as a few microseconds by the use of a potentiostat, and the current which flows is recorded as a function of time. For a blocking interface there will be no current flow either before or after the change in potential. However, whilst the potential is being changed a measurable charge $|\Delta q|$ will be injected into the interface. For a non-blocking interface the initial potential will frequently be the equilibrium potential, i.e. the initial current will in this case be zero.

(2) The potential difference is changed linearly with time (potential sweep) and the current is again recorded as a function of time.

(3) The potential difference is changed in a sinusoidal manner with time with an amplitude < 10 mV (i.e. $\Delta E = \Delta E_0 \sin(\omega t)$). In this case the current response is also sinusoidal, but in general is out of phase with the exciting potential by an angle θ so that $\Delta i = \Delta i_0 \sin(\omega t + \theta)$. This response is expressed as an impedance ($\mathbf{Z}$) which is a vector quantity, the magnitude of which is the ratio of the amplitude of the potential to the current, i.e. $|\mathbf{Z}| = \Delta E_0/\Delta i_0$, whilst the two components of the vector $\mathbf{Z}$ are $Z' = |\mathbf{Z}| \cos \theta$ and $Z'' = |\mathbf{Z}| \sin \theta$. Impedance diagrams are produced by representing in the complex plane a series of Z values (points where Z' is plotted against Z'') corresponding to a range of frequencies. Where an interface behaves electrically like a capacitance (C_{dl}), which is generally the case for a blocking interface, then $Z = 1/j\omega C_{dl}$ where $j = \sqrt{(-1)}$. In this case the impedance plane display consists of a series of points on the Z'' axis each a distance $1/\omega C_{dl}$ from the origin (Fig. 10.3).

Often a non-blocking interface will behave like a resistance (R_{ct}) and capacitance (C_{dl}) in parallel. This leads to a semicircle in the impedance plane which has a high frequency limit at the origin and a low frequency limit at $Z' = R_{ct}$ (Fig. 10.4). At the maximum of the semicircle if the angular frequency is ω_{max}, then $R_{ct} C_{dl} \omega_{max} = 1$, from which C_{dl} can be evaluated.

Unfortunately it is always impossible to change the potential only across a single interface – potentials are always changed between two points in two bulk phases. This means that the observed electrical response is always that of the interface *plus* a part of the bulk phases. In most circumstances where one of the bulk phases is a metal, the only significant bulk contribution to the electrical response is due to the non-metallic phase and is in series with the interface response. For example if the interface is behaving as R_{ct} in parallel with C_{dl} the experimentally measured response will in addition have R_{Ω} in parallel with C_g, where R_{Ω} is the electrical resistance of the bulk phase contribution and C_g its (geometric) capacity (Figs. 10.3 and 10.4).

When a metal electrode is in contact with a non-metallic phase which has *at least* two mobile charge carriers present so that a non-blocking interface is formed, e.g. Li/PEO–$LiCF_3SO_3$, there is the possibility that the interface impedance has a *Warburg* impedance in series with R_{ct} due to the slow diffusion of the species crossing the interface. A Warburg impedance, generally given the symbol W, has the characteristic that $Z' = Z'' = A_W/\omega^{1/2}$, where A_W is called the Warburg coefficient. It arises from the fact that the concentration of a species at an electrode surface is out of phase by 45° with the flux of the same species across the surface.

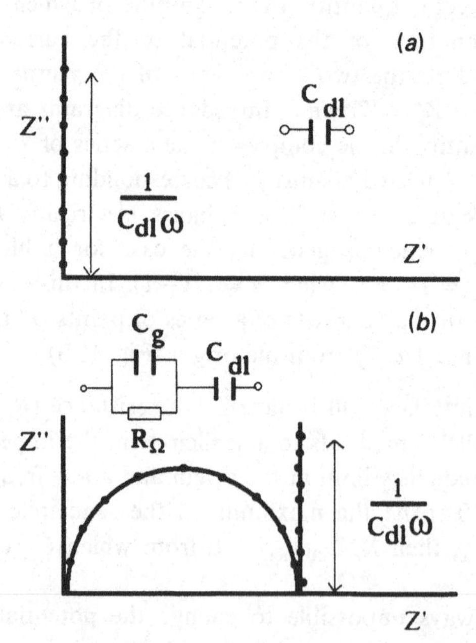

Fig. 10.3 Impedance diagrams for: (*a*) a blocking interface; (*b*) a blocking interface when the associated bulk impedance is taken into account.

10.1 The double layer at blocking interfaces

The structure of the charged layers which exist at the interface between conducting materials is best explored in the first instance by considering in detail (idealised) blocking interfaces. As stated above there is generally a charge q (C cm^{-2}) on one phase which is exactly balanced by a charge $-q$ on the contacting phase. In some systems q can be varied between as much as $+10$ μC cm^{-2} and -10 μC cm^{-2}. The limits on these values are determined by the point at which the charge starts to cross the interface as a current, causing an electrochemical reaction to occur. These charges, together with any dipole layers which exist at the interface, are responsible for the establishment of the interfacial potential $\Delta\phi$ between the two phases. The surface charge q can be varied (either increased or decreased) and its variation causes a systematic change in $\Delta\phi$. This leads to the possibility of defining an interfacial differential capacity via the

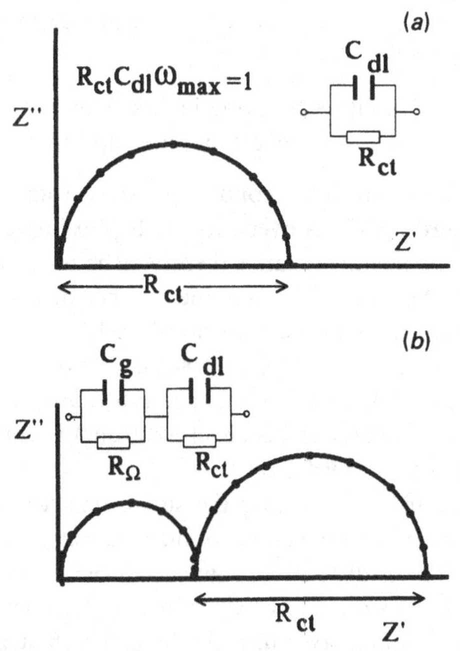

Fig. 10.4 Impedance diagrams for a non-blocking interface when: (*a*) bulk effects are neglected; (*b*) bulk effects are taken into account. The impedance diagrams are different if diffusional effects are significant.

relationship;

$$C_{dl} = dq/d\Delta\phi. \tag{10.1}$$

If $C_{dl} = 10\ \mu F\ cm^{-2}$ then a change of $1\ \mu C\ cm^{-2}$ in q leads to a change in $\Delta\phi$ of 0.1 V. It is necessary to use the concept of a differential capacity since the relationship between q and $\Delta\phi$ is generally not a linear one. The double layer capacitance (C_{dl}) is a directly measurable quantity whereas q and $\Delta\phi$ are not directly measurable, although changes in q and $\Delta\phi$ can be directly determined, and in some cases a situation where $q = 0$ can be inferred. It is important to note that the potential measured between two metal wires (E) is not related in any direct way to the interface potential difference $\Delta\phi$. Generally, C_{dl} is determined by verifying that the impedance of a (supposed) blocking interface is purely capacitive at limiting low ac frequencies or that it behaves as a resistance and capacitance in series (where the resistance is that of the bulk electrolyte) and assuming that the measured capacitance is C_{dl}. The lowest frequency used in practice may be 10^{-2} Hz, as it is difficult, experimentally, to make reliable measurements at lower frequencies. If the low frequency impedance is not purely capacitive (or behaves as a resistance and capacitance in series) this may be because either:

(*a*) the interface is not in reality a blocking interface; or
(*b*) the interface is too rough to behave as a pure capacitance.

The charge on the surface of an ionic conductor arises from a local excess of cations over anions or anions over cations. For example if the surface of Na-β-Al_2O_3 has a positive charge there will be an excess of Na^+ ions in the surface of the electrolyte over the number of Na^+ ions which would be required to maintain electroneutrality. Likewise if the surface has a negative charge there will be an overall deficit of Na^+ ions compared with the number required for electroneutrality. For a metal surface the surface charge (an excess or deficit of electrons) is generally assumed to be within 10 pm of the surface.

The next question which arises is whether the surface charge of, say, excess sodium ions is in contact with the atoms of the adjacent phase (e.g. Au) or whether the excess Na^+ is distributed some distance into the bulk of the electrolyte. There are two reasons why the excess charge may be distributed beyond the first atomic layer into the bulk. The first is that there may not be sufficient vacant sites within one ionic radius of the interface to accommodate all the excess charge, in which case some of the charge will be found in sites more than one ionic radius from the surface. This is unlikely to be a situation which arises frequently. The second reason, which may more frequently occur, is that thermal agitation of the system may cause the excess charge to distribute itself, on a time averaged basis, some considerable distance (many ionic radii) into the bulk of the ionic conductor. Gouy and Chapman (independently) constructed a theory which took account of the balance between the electrostatic attraction of the charges in the other phase (causing the charge to be as near to the surface as possible) and the thermal motion (causing a spread into the bulk). The most important parameter in this theory is the Debye length (r_D). Whether or not distribution of the charge into the bulk occurs is determined by the magnitude of r_D compared with the size of an ionic radius. Only when r_D is greater than the size of an ionic radius will the charge distribute itself into the bulk of the solid. This situation is favoured by a low concentration of mobile charges. Quantitative calculations are made as follows. We have:

$$r_D{}^2 = \varepsilon_0 \varepsilon_r RT/2F^2 I, \tag{10.2a}$$

where I, the ionic strength, is defined as

$$I = \tfrac{1}{2} \sum_i c_i z_i{}^2, \tag{10.2b}$$

c_i is the concentration of the ith charge, $z_i e$ its charge and the summation is made over all *mobile* charges, ε_0 is the vacuum permittivity, ε_r is the relative permittivity or dielectric constant, F is Faraday's constant, R is the gas constant and T is the temperature. Thus in Na-β-Al_2O_3 the summation is over Na^+ ions and must be calculated only in relation to the conduction plane. Orthogonal to the conduction plane r_D is, of course, infinite. For Na-β-Al_2O_3 and Ag_4RbI_5 the calculated value of the Debye length is much less than the size of the mobile ion (e.g. in Ag_4RbI_5, $r_D = 2.5 \times 10^{-11}$ m whereas the radius of the Ag^+ ion is 12.6×10^{-11} m). This suggests that in materials such as these, provided that there are sufficient vacant sites, the excess charge will be confined to the first atomic layer.

This leads to the *Helmholtz* model of the interface (Fig. 10.5) when the other contacting phase is a metal. The Helmholtz model of the interface predicts that the value of the double layer capacity (C_{dl}) will be given by:

$$C_{dl} = \varepsilon_0 \varepsilon_r A/a_0, \tag{10.3}$$

where a_0 is the radius of the mobile ion and A is the interface area and that C_{dl} will be invariant with both the interfacial potential difference $\Delta\phi$ and the temperature. Eqn (10.3) is simply the formula for a parallel plate condenser with the plate separation taken to be the radius of the mobile ion. In Eqn (10.3) $\varepsilon_0 = 8.85 \times 10^{-12}$ F m^{-1} (a universal constant) and if $A = 10^{-4}$ m^2, $a_0 = 1 \times 10^{-10}$ m, and $\varepsilon_r = 2$, then $C_{dl} = 17.7$ μF cm^{-2}. A value for the effective dielectric constant (ε_r) of 2 is reasonable since there can be no rotational dipole contribution on the atomic scale involved.

The Helmholtz model of the metal/electrolyte interface seems to be appropriate for such interfaces as Au/Na-β-Al_2O_3. For example the interfacial potential across this interface can be varied by 8 V without appreciable continuous current flowing and over this potential range the measured value of C_{dl} changes by only 20% (Fig. 10.6) (Armstrong, Burnham and Willis, 1976). In addition we note that:

(1) The value of C_{dl} is also only slightly temperature dependent over the range 100–300 K, as would be predicted from Eqn (10.3) (Armstrong and Archer, 1978).
(2) The value of C_{dl} for a single crystal of Na-β-Al_2O_3 is rather smaller than would be estimated from Eqn (10.3), i.e. near to

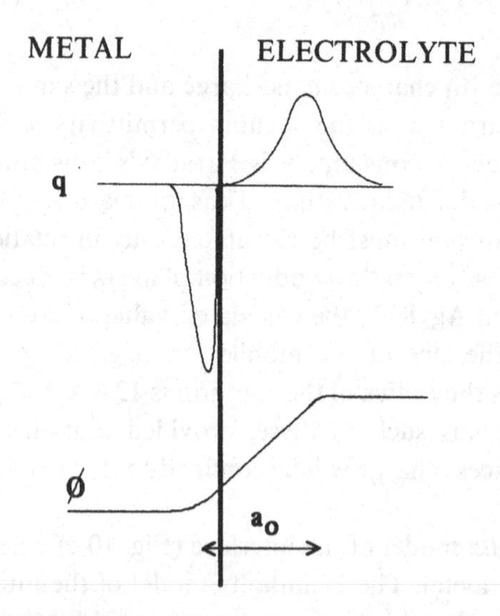

Fig. 10.5 Helmholtz model of the interface between a metal and an electrolyte. The metal is shown with a negative charge (excess of electrons) which is balanced by an excess of mobile cations, the centres of which are one atomic radius from the surface.

$0.4\,\mu F\,cm^{-2}$ rather than $10\text{–}20\,\mu F\,cm^{-2}$ expected from Eqn (10.3).

This may be an indication that not all of the surface change can be fitted into vacancies within the first ionic radius of the metal. It may also indicate that, as in any solid/solid contact, the true contact area is very much less than the geometric contact area when viewed on any atomic scale.

One complication which may be present, when the Helmholtz model is in other respects appropriate, is that of specific adsorption. If one of the mobile species is to some extent chemically bound rather than being simply electrostatically bound to the metal electrode, C_{dl} may show a dependence on the dc bias potential. Indeed this is the normal method of inferring specific adsorption. Another possibility in this case is that C_{dl} exhibits different high frequency and low frequency limits because at high frequencies the specific adsorption being an activated process is too slow to follow changes in interface potential. A further complication which is often present in real systems is the presence of an oxide layer on the surface of the metal electrode. Such an oxide layer can generate a potential

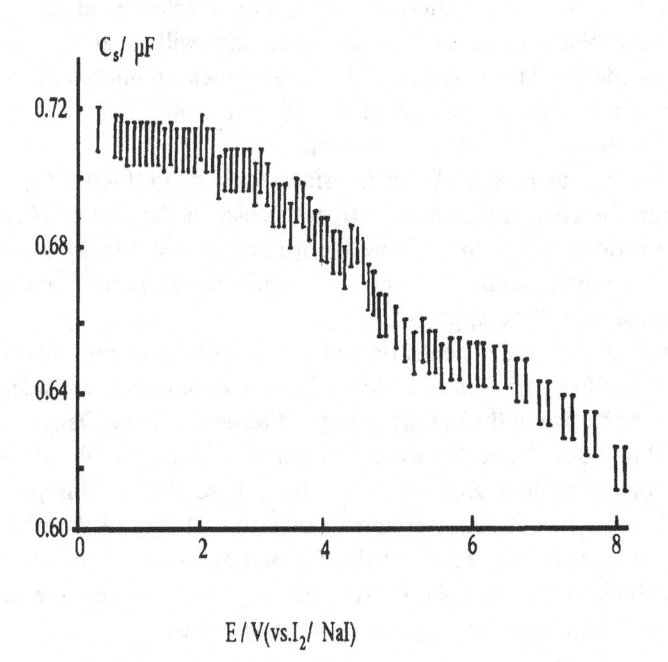

Fig. 10.6 C_{dl} for the Au/Na-β-Al_2O_3 interface as a function of the interfacial potential. Electrode area is 1.7 cm^2.

independent capacity, which, however, is often orders of magnitude smaller than a true Helmholtz double layer capacity.

Measurements of the value of C_{dl} using a two electrode cell such as Au/Na-β-Al_2O_3/Au do not allow the surface charge on the interfaces to be varied – it is necessary to accept the q which happens to be present when the cell is prepared (and to note that the properties of *two* identical interfaces are being measured). However, if measurements are made on a cell such as C/Ag_4RbI_5/Ag, the potential (E) between the C and Ag electrodes can be changed experimentally and since the potential across the interface Ag/Ag_4RbI_5 is fixed in this situation any change in E (ΔE) will cause a change in $\Delta\phi$ of $\Delta E = \Delta\phi_1 - \Delta\phi_2$ at the C/Ag_4RbI_5 interface. In this way the potential difference across a blocking interface can be systematically varied and with it q. It is important to note once again that the potential difference across a single interface can never be measured, though its magnitude can be varied to a known extent. If impedance measurements are made on an unsymmetrical two electrode cell it will be generally essential to ensure that the impedance of one interface is much less than the other so that the measured impedance can be ascribed to a particular interface. When one interface is blocking and the other non-blocking, the non-blocking interface will generally have the smaller impedance. The impedance of a single blocking interface can also be measured by using a three electrode cell containing a reference electrode as described in the next section.

The appropriate model of the interface for e.g. Pt/$LiCF_3SO_3$–PEO depends on the concentration of charged species in the PEO. When the salt:PEO ratio is less than 1:10 the Debye length will also be less than the size of a mobile charge. In this case again the Helmholtz model of the double layer will be appropriate.

However, when the concentration of salt dissolved in the polymer is comparatively low, e.g. 1 mM or less, the Debye length is considerably greater than the size of the mobile charges. For example the Debye length for a 10^{-4} mol dm^{-3} salt dissolved in a polymer with $\varepsilon_r = 10$ (and when ion pair formation is negligible) is 6×10^{-9} m so that in this polymer electrolyte the space charge will spread into the interior of the phase to about 60 ionic radii as a result of the thermal agitation of the atoms of the electrolyte. In this case the *Gouy–Chapman* model of the interface is required and Eqn (10.3) for C_{dl} has to be replaced by

$$C_{dl} = A(2z^2F^2\varepsilon_0\varepsilon_r c/RT)^{1/2} \cosh(zF\Delta\phi/2RT), \qquad (10.4)$$

where c is the salt concentration. Evidently C_{dl} no longer has a value independent of $\Delta\phi$ but has a minimum value of:

$$C_{dl} = A(2z^2F^2\varepsilon_0\varepsilon_r c/RT)^{1/2} \tag{10.5}$$

at $q = 0$ (the point of zero charge) and a value which sharply increases on either side of $q = 0$. Eqn (10.5) can be regarded as the parallel plate condenser formula again, but with the plate separation given by the Debye length, which is the effective charge separation distance in the Gouy–Chapman model at $q = 0$ (at q values different from 0 the effective plate separation is much less than this). This model of the interface between a metal and an electrolyte is illustrated in Fig. 10.7.

Of course, when the volume concentration of mobile charges is sufficiently high that the Debye length is comparable with the ionic radius of the mobile ion(s), a combination of the Helmholtz and Gouy–Chapman models is required. This is achieved by assuming that the measured C_{dl} value is a series combination of that due to the Gouy–Chapman model (C_{GC}) and that due to the Helmholtz model (C_H), i.e.

$$1/C_{dl} = 1/C_{GC} + 1/C_H. \tag{10.6}$$

This has the consequence that for low values of $|q|$, $C_{dl} = C_{CG}$ whilst for high $|q|$, $C_{dl} = C_H$.

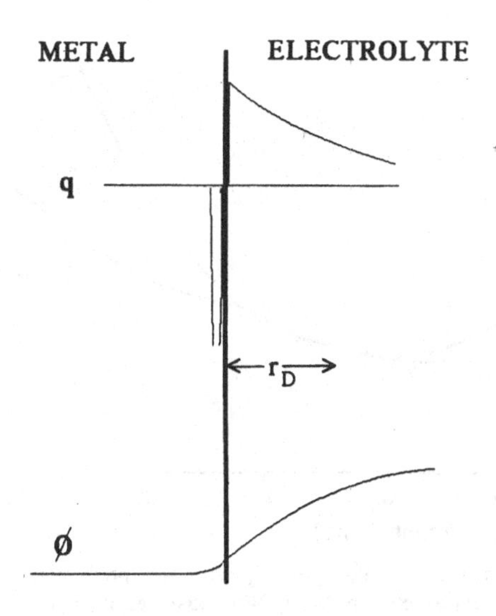

Fig. 10.7 Gouy–Chapman model of the interface between a metal and an electrolyte. The metal is shown with a negative charge on its surface.

In the discussion so far it has been assumed that one of the contacting phases is a metal and that the charge on the metal phase is confined to its geometric surface, so that it makes no contribution to C_{dl}. When neither of the contacting phases is a metal, e.g. when KCl–H_2O is contacted by PVC–$AsPh_4BPh_4$, we must consider the distribution of mobile charges in both phases. Of course, we once again have the charge $+q$ on phase 1 electrostatically balanced by $-q$ on phase 2. In each phase the Debye lengths will have to be calculated in order to see if the surface charge spreads into either or both phases. One possibility is that of back to back Gouy–Chapman models when the mobile charges in both phases are present in low volume concentrations. This is not the situation for the data shown in Fig. 10.8 (Armstrong, Proud and Todd, 1989) since the volume concentration of mobile charges in the aqueous phase is significantly higher than that in the polymer so that only the C_{GC} from the PVC needs to be considered.

Having discussed the way in which blocking interfaces behave we must now consider how blocking metallic contacts can be made on a given material, e.g. a ceramic electrolyte. Frequently a relatively inert metal such as Pt or Au is evaporated onto a ceramic material which has been polished

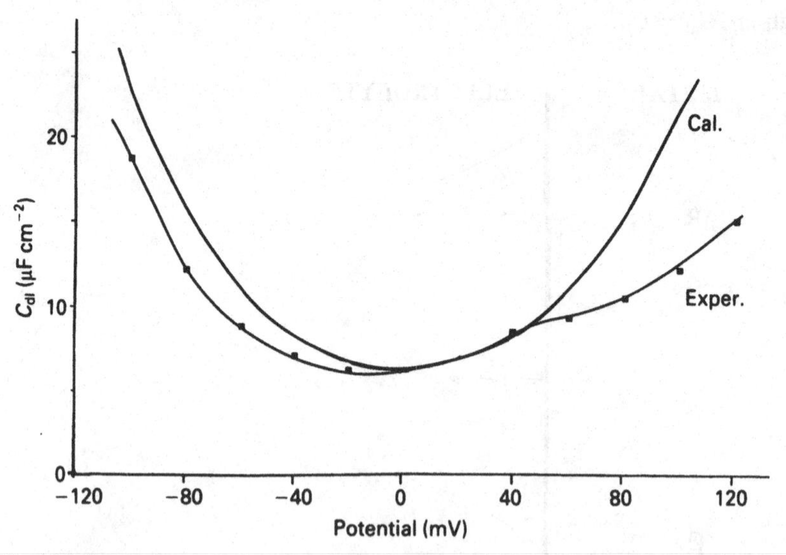

Fig. 10.8 Comparison of the experimental values of C_{dl} with the predictions of the Gouy–Chapman theory for the interface between $AsPh_4BPh_4$ dissolved in PVC and KCl dissolved in water. The concentration of KCl is much greater than that of the $AsPh_4BPh_4$ so that the double layer contribution from the aqueous phase can be neglected.

down to better than 1 μm with diamond paste and treated to remove the polishing materials. When precautions are not taken to obtain smooth, uncontaminated surfaces anomalous results are frequently found. In particular the interface may not behave as a pure capacitance at low frequencies, so that a unique experimental value of C_{dl} cannot be assigned to it. For a soft solid electrolyte which is easily deformed it is generally the contacting metal which is polished to a smooth finish. Contact is then established by the pressing of the two phases together.

10.2 Non-blocking metal electrodes – one mobile charge in the electrolyte

The interface structure for non-blocking interfaces is similar to that for related blocking interfaces. Thus the distribution of charge at the C/ Ag_4RbI_5 interface will be similar to that at the Ag/Ag_4RbI_5 interface. The major difference is that there is one particular interfacial potential difference $\Delta\phi_e$ at which the silver electrode is in equilibrium with Ag^+ ions in the bulk electrolyte phase. At this value of $\Delta\phi_e$, there is a particular charge q_e on the electrolyte balanced by an equal and opposite charge $-q_e$ on the metal. At any potential different from $\Delta\phi_e$ a value of q different from q_e will be on the electrolyte surface and a finite current of Ag^+ ions flows across the interface.

One of the major concerns which we have in relation to non-blocking electrodes is to understand the way in which the current crossing the interface varies with the overpotential η defined as:

$$\eta = \Delta\phi - \Delta\phi_e. \tag{10.7}$$

In order to determine the net current flowing, i, as a function of η (and in some cases time) it is generally necessary to work with three or four electrode cells where the electrode of interest (the working electrode) carries current into the cell and a second electrode (the auxiliary or subsidiary electrode) carries the current out of the cell. The third electrode is the reference electrode although in cases where one of the phases is not a metal a second reference electrode is required (four electrode cell).

The reference electrode(s) carries no *net* current and has a charged species in equilibrium between it and the electrolyte. The schematic form of such an arrangement is shown in Fig. 10.9 where the reference electrode is a silver wire inserted into Ag_4RbI_5 which is, in turn, between two silver sheets which constitute the working and subsidiary electrodes.

The reference electrode is placed as close as possible to the working electrode so as to minimise the part of the response due to the bulk phase. In order to measure the electrical characteristics of a blocking interface, e.g. C/Ag_4RbI_5, using a three electrode cell the Ag working electrode is simply replaced by a C electrode.

The overpotential is directly determined by measuring the potential difference between the working and reference electrodes in an arrangement such as that of Fig. 10.9. (In a four electrode cell the potential difference must be measured between the two reference electrodes.) Frequently a potentiostat is used to impose a known overpotential and the current flowing in the cell is measured as a function of η.

The relationship between current and overpotential at the non-blocking interface is generally dependent on both the interface structure and the number of mobile species in the contacting phases. The simplest situation is that represented by an interface of the type Ag/Ag_4RbI_5 where (i) the Helmoltz model of the interface is appropriate and (ii) there is only one mobile species in the electrolyte (Ag^+). In this case the relationship between i and η is a linear one at low values of η ($\eta < 10$ mV);

$$\eta = IR_{ct} \tag{10.8}$$

(provided that the ohmic resistance between the reference electrode and the interface is negligible). Here R_{ct} is the constant of proportionality. The relationship of R_{ct} to the exchange current for the process, i_0 (i.e. the

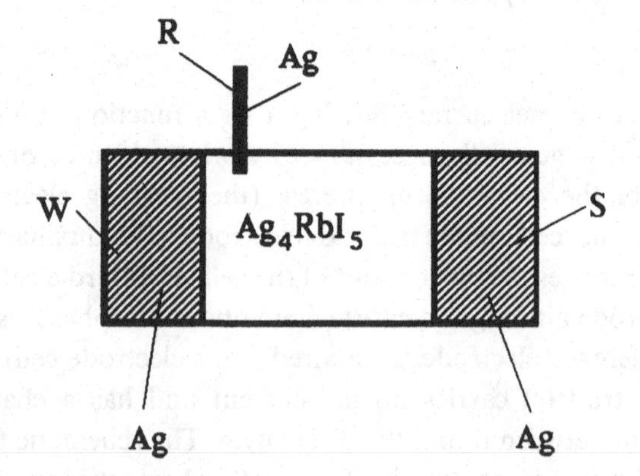

Fig. 10.9 Schematic representation of a three electrode cell. W – working electrode; R – reference electrode and S – subsidiary electrode. Note that the reference electrode is placed as close as possible to the working electrode.

current flowing across the interface in either direction of equilibrium), is given by

$$R_{ct} = RT/nFi_0 \tag{10.9}$$

Thus, measurement of R_{ct} is perhaps the simplest method of determining i_0.

For values of $\eta > 30$ mV the Tafel equation

$$i = i_0 \exp(\alpha nF\eta/RT) \tag{10.10a}$$

or

$$i = -i_0 \exp[-(1-\alpha)nF\eta/RT] \tag{10.10b}$$

is expected to hold, where η and i are taken to be positive for an anodic process, i.e. for $Ag = Ag^+ + e$ and n is the number of electrons involved in the reaction. The Tafel equation results from the fact that the overpotential changes the probability of a silver ion being transferred across the interface by changing the free energy of activation for the transfer by $\alpha nF\eta$. The quantity α can be shown to lie between 0 and 1. Applying transition state theory and assuming α is invariant with overpotential leads to the Tafel equation. There should be no transient currents when the potential is switched rapidly from $\eta = 0$ to a finite value of η. Values of i_0 can be found by plotting $\ln(|i|)$ vs η; the intercept on the $\ln(|i|)$ axis at $\eta = 0$ being the value of $\ln(i_0)$. The 'Tafel slope' is defined as $2.303\, d\eta/d\ln(|i|)$ and has the units of V decade^{-1}. From the Tafel equation this should have the value $2.303RT/\alpha nF$ for an anodic process or $2.303RT/(1-\alpha)nF$ for a cathodic process. Thus if $n = 1$ and $\alpha = 0.5$ the Tafel slopes should both have the value 0.120 V decade^{-1} at room temperature. If $n = 2$ and $\alpha = 0.5$ the Tafel slopes would have the value 0.060 V decade^{-1}.

The general relationship between η and i is obtained by combining Eqns (10.10a) and (10.10b) to give the Butler–Volmer equation:

$$i = i_0 \exp(\alpha nF\eta/RT) - i_0 \exp[-(1-\alpha)nF\eta/RT]. \tag{10.11}$$

This has the following limiting behaviour:

For low η: $i = i_0 nF\eta/RT$, i.e. a linear dependence of i on overpotential.

For high overpotentials the Tafel equations follow. The form of the Butler–Volmer equation is shown in Fig. 10.10.

The relatively simple equations which we expect to find in the Ag/Ag_4RbI_5 case result from the fact that $\Delta\phi$ can be changed without significantly changing the number of Ag^+ ions within one ionic radius on the electrolyte side of the double layer. Due to this the application of a finite value of η causes the rate of ion transfer across the interface to vary only as a result of a change in the value of $\Delta\phi$. Changes of Ag^+ concentration further than one ionic radius unit into the electrolyte are negligible because of the fact that electroneutrality must always hold in a material where the Debye length is small compared with the size of an ion. It is for this reason that the simple Tafel equation is expected to hold.

In the real situation there are, of course, numerous additional complications which arise from these factors such as:

(1) A finite ohmic resistance between the interface and the reference electrode (R_Ω). Thus, instead of $\eta = iR_{ct}$ for low overpotentials we will have

$$\eta = iR_{ct} + iR_\Omega, \tag{10.12}$$

so that in a situation where ion transfer at the interface is very rapid only the effect of R_Ω will be observable. Thus $\eta = iR_\Omega$ and no information is obtained in relation to interfacial processes. The sort of

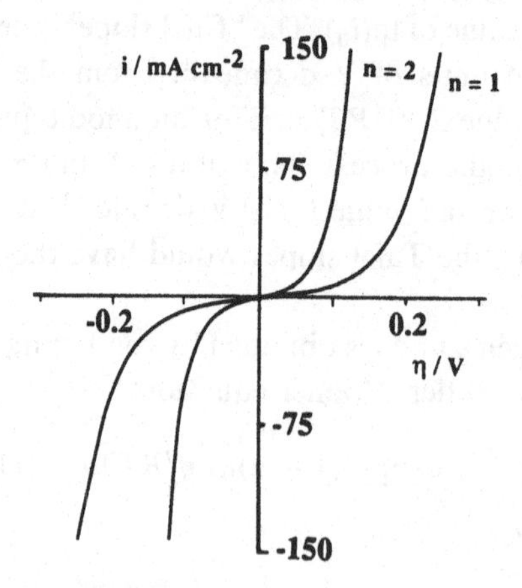

Fig. 10.10 Plot of the Butler–Volmer equation for $i_0 = 1$ ma cm^{-2}, $F = 9.65 \times 10^4$ C mol^{-1}, $\alpha = \beta = 0.5$, $T = 300$ K, $R = 8.31$ J K^{-1} mol^{-1}, and two different values of n.

distortion produced by R_Ω for higher overpotentials is shown in Fig. 10.11 (Armstrong, Dickinson and Willis, 1974).

(2) Electrocrystallisation effects in the metal. When Ag^+ ions are being deposited onto the preexisting Ag metal not all the available surface sites on the metal are energetically equivalent. There is, therefore, a tendency for Ag^+ ions to transfer from the high energy sites. Thus, in principle, only a liquid metal contact would behave as discussed above. In a solid metal, in addition to the charge transfer process we should consider a number of other processes such as adatom diffusion (Armstrong *et al.*, 1974). In the present context there is not the space to deal with these complexities. As a first step in the analysis of charge transfer at the metal/solid electrolyte interface the methods given above should be used and only if there is no possibility of fitting the experimental data to the simple scheme should more complicated mechanisms be invoked.

(3) The manner in which contact is maintained between the Ag metal and the electrolyte. Since the volumes of the two phases must change

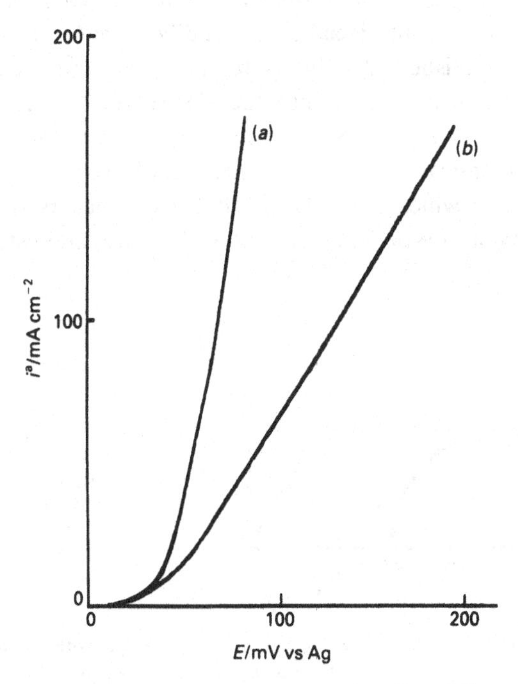

Fig. 10.11 Anodic current/overvoltage plot for silver foil at a temperature of 700 K: (*b*) original experimental data; (*a*) after correction for R_Ω.

when current passes, it is obvious that a situation could arise for example where contact is lost.

The interface impedance for a case such as Ag/Ag_4RbI_5 will consist of a capacitance C_{dl} (derived from the Helmholtz formula) in parallel with R_{ct} so that in the complex plane impedance a semi-circle will be found from which C_{dl} and R_{ct} can be evaluated. R_{Ω} will cause this semicircle to be offset from the origin by a high frequency semicircle due to the bulk impedance between the interface and the reference electrode (Fig. 10.12). *There can be no Warburg impedance* (a line at 45° to the real axis generally due to diffusion effects) *in this case.*

We must now consider the case where there is only one mobile charge in the electrolyte, but where the concentration of that mobile charge is so low that the Debye length is comparable with or greater than an ionic radius, but much less than the distance between the electrodes. This situation might arise where a metal is contacted by a polymer which has a low concentration of fixed anionic sites and a correspondingly low concentration of metal cations. In this case if we assume once more that the metal cations tend to be in equilibrium between the metal and the polymer there will again be an interfacial potential difference $\Delta\phi_e$ at which equilibrium will be established. At this potential, the charge on the electrolyte will extend some distance into the electrolyte. At distances which are greater than the Debye length from the interface, the cation concentration will be identical with the concentration of fixed anion sites, giving overall a situation which is similar to that which occurs when a metal is contacted by a p-type semiconductor. We are interested in

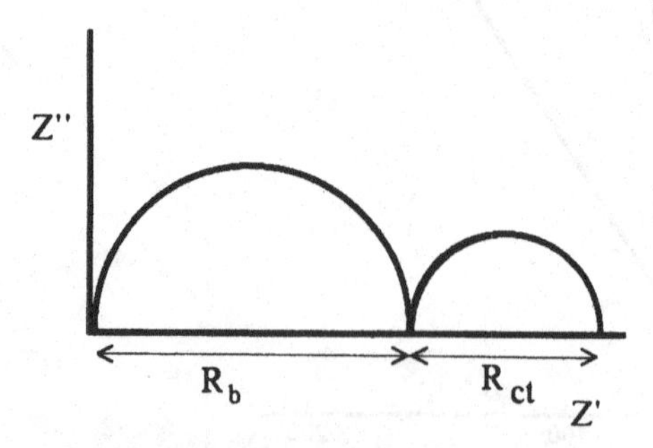

Fig. 10.12 Expected complex plane impedance diagram for an electrolyte with one mobile species which is contacted by two non-blocking metal electrodes, e.g. $Ag/Ag_4RbI_5/Ag$. R_b is the bulk resistance of the electrolyte and R_{ct} is the charge transfer resistance for the Ag/Ag_4RbI_5 interface.

evaluating the current which flows when a finite overpotential η is applied to the interface (on the assumption that the reference electrode is many Debye lengths distant from the interface). In this case for small values of η we would expect that:

$$\eta = iR_{ct} + iR_{\Omega} \tag{10.13}$$

with in most cases, because of the low conductivity of the bulk of the electrolyte, only the effect of R_{Ω} being observable.

R_{ct} will be no longer linked in a simple way to i_0, since it is impossible to change $\Delta\phi$ without simultaneously changing the concentration of cations at the interface to an appreciable extent. Thus we have the general relationship that

$$(R_{ct})^{-1} = \frac{di}{d\eta} = \left(\frac{\partial i}{\partial \eta}\right)_c + \left(\frac{\partial i}{\partial c}\right)_\eta \frac{dc}{d\eta}, \tag{10.14}$$

where the second term represents the fact that ΔE cannot be varied without changing the cation concentration (c) at the interface. The exchange current i_0 is related to $(i/\eta)_c = RT/nFi_0$ but not in a simple way to the experimental quantity R_{ct}.

The impedance of the system described above will generally consist of R_{Ω} in parallel with C_g, i.e. only the bulk electrolyte between the metal electrode and the reference electrode is likely to make a contribution to the impedance. Although C_{dl} in parallel with R_{ct} is theoretically present at sufficiently low frequencies it is only measurable in special circumstances – for example when the metal cations have an unusually high mobility so that R_{Ω} is comparatively small, enabling R_{ct} to be observed.

10.3 Non-blocking metal electrodes with more than one mobile charge in the electrolyte

The type of interface which will be considered in this section is typified by

$$Li/PEO–LiCF_3SO_3$$

As in the previous section there is a single charged species Li^+ which tends to equilibrate between the metal and the electrolyte, so that there is an equilibrium interfacial potential difference $\Delta\phi_e$ at which Li^+ is in equilibrium and this equilibrium is characterised by an exchange current i_0. The difference between this system and the systems considered in the previous section is that because there is more than one mobile charged species present in the electrolyte, as soon as current flows the composition of the electrolyte adjacent to the electrode starts to change. For example if a cathodic overpotential ($-\eta$) is applied so that Li^+ is deposited the concentration of $LiCF_3SO_3$ will start to decrease in the vicinity of the electrode. This means that instead of a current being established very quickly on the application of a finite value of η (starting with $\eta = 0$) and then remaining steady at that initial value, as in the case for electrolyte with a single mobile charge (provided that the resistance between the reference electrode and the working electrode is negligible), the value of the current generally falls with time, as concentration changes occur. The way in which the current falls with time, and also the final steady state current, depend on the detailed geometry.

In relation to describing the fundamental interfacial processes we are interested in determining the exchange current i_0. If the current which flows immediately after the application of an overpotential η is i_1 we have the relationship for $\eta < 10$ mV;

$$\eta = i_1 R_\Omega + i_1 R_{ct}. \tag{10.15}$$

For a highly conducting system, e.g., $LiCF_3SO_3$–PEO with high salt concentrations, R_Ω will be small which may enable R_{ct} to be evaluated and i_0 to be calculated from $R_{ct} = RT/nFi_0$. Similarly for high values of the overpotential if R_Ω can still be neglected the Tafel equation will hold in the form $i_1 = i_0 \exp(\alpha nF\eta/RT)$ for an anodic process (or in the equivalent form for a cathodic process) where i_1 is again the current flowing on the application of the overpotential, immediately after the charging of the double layer.

For low salt concentrations R_{ct} will be difficult to measure and will not be related in a simple way to i_0 (as discussed in the previous section).

Perhaps the best way to determine i_0 in this system is by measuring the impedance. The expected impedance plane display is shown in Fig. 10.13.

At high frequencies, a semicircle is expected as a result of a parallel combination of R_Ω and C_g. At low frequencies a Warburg impedance may be found as part of the interfacial impedance. In some cases it may dominate the interfacial impedance as in Fig. 10.13(a), in which case only the diffusion coefficient of the salt will be determinable. It should be noted that, in the absence of a supporting electrolyte, the electroactive species, in this case Li^+, cannot diffuse independently of the anions.

Electroneutrality demands that a concentration gradient exists for both lithium and triflate near the Li electrode and that the diffusion is coupled in a similar fashion to the coupled diffusion of ions and electrons in intercalation electrodes (Chapter 8). As in that case it is the coupled diffusion of both species which is observed, i.e. the salt diffusion coefficient. This may be evaluated by fitting the low frequency region of the curve to:

$$Z' = Z'' = A_W/\omega^{1/2}, \tag{10.16}$$

where A_W is the Warburg coefficient and ω is the angular frequency.

The relationship of A_W to the salt diffusion coefficient of the species crossing the interface (D_s) is:

$$A_W = RT/An^2F^2c(2D_s)^{1/2}. \tag{10.17}$$

In other cases as in Fig. 10.13(*b*) the interfacial impedance will show a semicircle due to R_{ct} and C_{dl} in parallel, with the Warburg impedance becoming apparent at significantly lower frequencies. In such cases R_{ct} can be evaluated without difficulty.

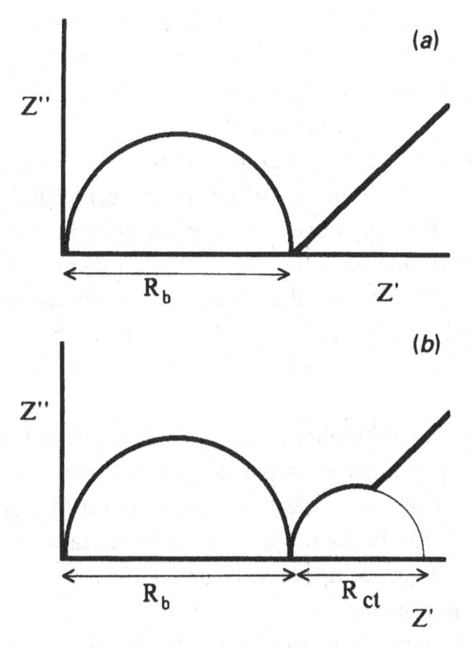

Fig. 10.13 Impedance plane diagrams for metal non-blocking electrodes with two mobile species in the electrolyte. (*a*) Interfacial impedance is only a Warburg impedance. (*b*) Interfacial impedance shows a charge transfer resistance semicircle.

As stated earlier the currents which flow in systems such as Li/ $LiCF_3SO_3$–PEO are generally time dependent (after the inevitable initial time dependent current which flows to charge the interface) after the application of an overpotential η. In some situations a steady state current will eventually appear, such as in the example of a thin slice of $LiCF_3SO_3$–PEO which is sandwiched between two lithium electrodes.

10.4 The effect of surface films on interfacial measurements

In most cases where an alkali metal is contacted by an electrolyte a thin layer of a third phase is formed between the two bulk phases so that the end result is a three phase system with two separate interfaces. Frequently the thin third phase is formed from O_2 or H_2O vapour which is present in the electrochemical cell before or after contact is made between the two initial phases, so that the thin layer present is often an alkali metal oxide or hydroxide or a mixture of these substances. However, there is considerable evidence that in the system Li/$LiCF_3SO_3$–PEO the lithium reacts chemically both with the PEO and with the $CF_3SO_3^-$ anion (Fauteux, 1985). The question therefore arises as to how the electrode kinetic behaviour, which is observed, is changed by the presence of such phases. If the phase is in the oxide or hydroxide form of an alkali metal it is likely to have a low ionic conductivity, generally via a defect mechanism, together with a very low electronic conductivity. Therefore if an overpotential is applied, for example, to the interface Li/$LiCF_3SO_3$–PEO some of the overpotential may have to be used to drive ions through a thin layer of lithium salt (oxide or hydroxide) if such a layer is present. In many cases the ohmic resistance of the thin layer of lithium salt will dominate the response leading to a linear i–η relationship so that the true characteristics of the Li/$LiCF_3SO_3$–PEO interface may be difficult to obtain.

The situation is analogous to the phenomenon of passivation found in aqueous electrochemistry where, for example, we are concerned with the formation of Fe_2O_3 on an iron electrode from the reaction of iron with water. It is, therefore, difficult in such systems to be certain that any apparent R_{ct} is due to the interface apparently under investigation. One indication as to what is happening is that passivating layers generally increase in thickness with time after contact, so that if R_{ct} appears to increase slowly with time it is most likely that it is not R_{ct} which is being measured but the ohmic resistance of the passivating layer.

10.5 The effect of surface roughness on interfacial measurements

The ideal interface would be one which was smooth on an atomic scale. The only interfaces which can be produced which are as smooth as it is theoretically possible are interfaces between a single crystal of a high melting metal, e.g. platinum, and a deformable polymer. All interfaces between two solids will tend to be very rough on an atomic scale. The question therefore arises – *How do real interfaces differ from idealised smooth interfaces*?

One important way in which they differ is that currents (either ac or dc) at real interfaces are not uniform across the interface. Therefore, a measured macroscopic dc current density i (A cm^{-2}) will not in general be a microscopic current density i on a small part of the interface (say 1 μm by 1 μm) on a rough or non-uniform electrode. In the formation of each interface it is necessary therefore to ensure that the surface is as smooth as is reasonably practicable. In most cases codes of best practice have been evolved and should generally be followed unless radical improvements are possible. In this way results should at least be comparable from one laboratory to another.

There is a particular and well-known problem when ac impedance measurements are made on cells containing rough electrodes. Let us consider a simple two electrode cell with identical blocking electrodes. The impedance plane displays for both rough and smooth electrodes are shown in Fig. 10.14. At frequencies which are sufficiently low, the smooth electrodes will give, at limiting frequencies, a line in the complex plane which is almost parallel to the Z'' axis. In real experiments angles against the real axis of 89° have been observed (90° is expected for the perfectly

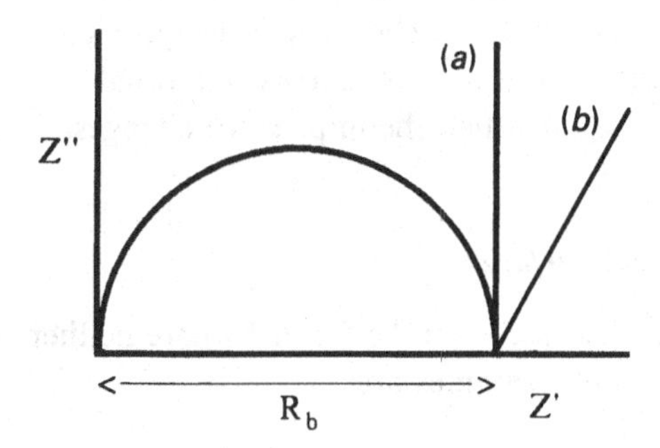

Fig. 10.14 Impedance diagram due to a blocking interface: (*a*) a perfectly smooth interface; (*b*) rough electrode.

smooth electrode). By contrast, for the rough electrode a slope as low as 45° may be observed. With such a slope it is impossible to assign unambiguously a value of C_{dl} to the interface. Of course, with the interface which is behaving as a perfectly smooth interface and exhibiting a slope close to 90° it is impossible to know if the value of C_{dl} which is calculated from $Z'' = 2/\omega C_{dl}$ is the same as the value which would be obtained if a surface as smooth on an atomic scale (rather than one smooth on a micron scale) had been used. Nevertheless a slope close to 90° is an indicator that gross roughness has been avoided.

When impedance measurements are made on non-blocking electrodes, different impedance spectra are again to be expected when rough electrodes are used instead of smooth electrodes. Let us once again consider a two electrode cell with identical non-blocking electrodes. The smooth electrodes are more likely to show the simple theoretical behaviour shown in Fig. 10.12 i.e. a high frequency semi-circle due to the bulk electrolyte followed by an interfacial semi-circle at low frequencies. With rough electrodes we would expect the high frequency behaviour to be identical but with a severe distortion in the low frequency semicircle.

One of the major causes of the distorted behaviour seen when rough electrodes are used is due to the fact that different parts of the interface have different bulk resistance pathways to them, in some cases causing transmission line type behaviour.

One constant problem which arises in the interpretation of impedance spectra is the question of the explanation of interfacial impedances which exhibit lines intercepting the real axis at slopes close to 45°. *Ab initio* it is not possible to say whether the observed behaviour is from a blocking electrode which is rather rough or from a non-blocking electrode which is smooth and exhibiting a *Warburg* impedance due to the slow diffusion of a reactant species to the interface. The only safe approach to this problem is deliberately either to increase or decrease the roughness of the interface and observe the way in which the impedance changes.

10.6 Other non-blocking interfaces

A number of non-blocking interfaces can be formed where neither of the phases is a simple metal. Some examples are:

> $LiCF_3SO_3$–PEO/$Li_xV_6O_{13}$ (Bruce and Krok, 1988) where the Li^+ ion tends to equilibrate between the phases.

$AsPh_4BPh_4$–PVC/$AsPh_4Cl$–H_2O where the $AsPh_4^+$ ion equilibrates between the phases.

There will always be a charge transfer resistance (R_{ct}) associated with the ion exchange across the interface. Where there are very small Debye lengths in each phase (compared with the size of an ion) the exchange current i_0 can be evaluated from the relationship

$$i_0 = RT/nFR_{ct}. \tag{10.18}$$

In general it will be necessary to measure R_{ct} via impedance measurements using a four electrode cell. A schematic diagram of the cell which would be used for such measurements is shown in Fig. 10.15. The expected behaviour will be as described in Eqn (10.3) except that Warburg impedances can arise from either or both phases. An example of an impedance spectrum of the H_2O/PVC interface is shown in Fig. 10.16. The application of a constant overpotential will, in general, lead to a slowly decaying current with time due to the concentration changes which occur in both phases, so that steady state current potential measurements will be of limited use.

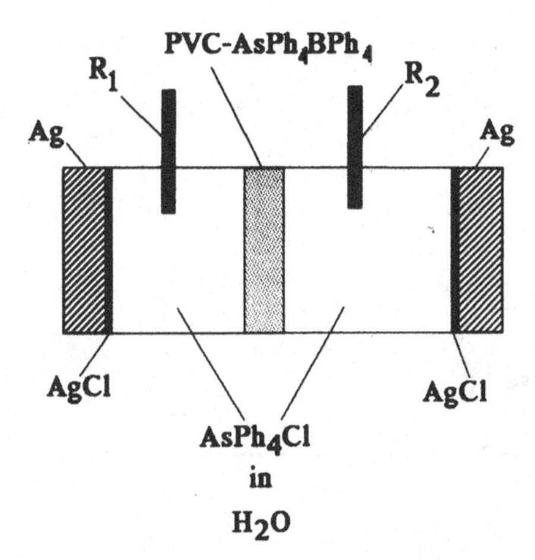

Fig. 10.15 Cell used for studying the transfer of $AsPh_4^+$ across the water/PVC interface. R_1 and R_2 are two Ag/AgCl wire electrodes. The cell current is carried in and out of the cell by the two massive silver electrodes at each end of the cell.

10.7 Two step charge transfer reactions

When an Li^+ ion is transferred between two phases an intermediate valency species is not possible. However, in the transfer of Cu^{2+} from a metal electrode into an electrolyte it is possible that the reaction mechanism is:

$$Cu = Cu^+ + e$$

$$Cu^+ = Cu^{2+} + e,$$

so that there is a Cu^+ intermediate species. In principle such an intermediate species is detectable from electrical measurements. For example, for the dissolution of Cu into an electrolyte under conditions where the Tafel law holds if Cu^+ were formed in a preequilibrium we would expect:

$$i = i_0 \exp[(\alpha + 1)Fh/RT] \tag{10.19}$$

rather than

$$i = i_0 \exp(2\alpha F\eta/RT), \tag{10.20}$$

i.e. the Tafel slope would be 0.040 V per decade rather than the 0.060 V per decade expected in the absence of a Cu^+ intermediate. Where there

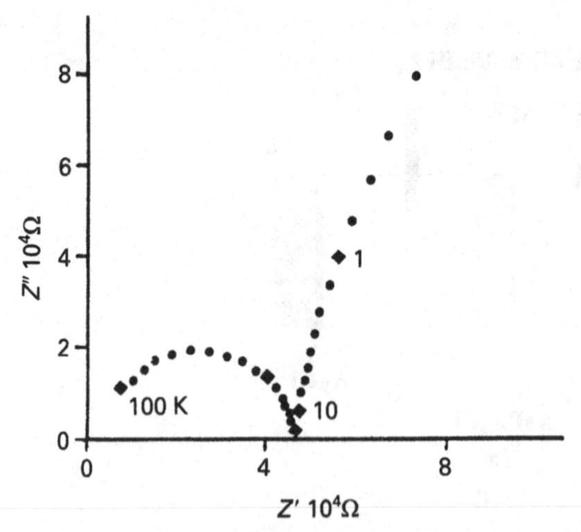

Fig. 10.16 Impedance diagram for the $AsPh_4BPh_4$/PVC/NOPE membrane in contact on one side with 0.01 M $AsPh_4Cl$ in 0.1 M NaCl and on the other side with 0.1 M $CaCl_2$ at $E = 0.34$ V. Contact area of each of the two interfaces is 0.78×10^{-4} m^2.

is a clearly defined intermediate species the impedance diagram may show two semicircles rather than one for the interface impedance.

References

Armstrong, R. D., Dickinson, T., Thirsk, H. R. and Whitfield, R. (1971) *J. Electroanal. Chem.*, **29**, 301–7.
Armstrong, R. D., Dickinson, T. and Willis, P. M. (1974) *J. Electroanal. Chem.*, **57**, 231–40.
Armstrong, R. D., Burnham, R. A. and Willis, P. M. (1976) *J. Electroanal. Chem.*, **67**, 111–20.
Armstrong, R. D. and Archer, W. I. (1978) *J. Electroanal. Chem.*, **87**, 221–4.
Armstrong, R. D. and Archer, W. I. (1980) in *Electrochemistry*, Vol. 7, Ed. H. R. Thirsk, The Chemical Society, London.
Armstrong, R. D., Proud, W. G. and Todd, M. (1989) *Electrochim. Acta*, **34**, 977–9.
Bruce, P. G. (1987) in *Polymer Electrolyte Reviews – 1*, Eds. J. R. MacCallum and C. A. Vincent, Elsevier, London.
Bruce, P. G. and Krok, F. (1988) *Electrochim. Acta*, **33**, 1669.
Fauteux, D. (1985) *Solid State Ionics*, **17**, 133–6.
Macdonald, J. R. (1987) *Impedance Spectroscopy*, Wiley, New York.

11 Applications

O. YAMAMOTO
Department of Chemistry, Mie University

11.1 Introduction

In the early part of this century, many types of solid electrolyte had already been reported. High conductivity was found in a number of metal halides. One of the first applications of solid electrolytes was to measure the thermodynamic properties of solid compounds at high temperatures. Katayama (1908) and Kiukkola and Wagner (1957) made extensive measurements of free enthalpy changes of chemical reactions at higher temperatures. Similar potentiometric measurements of solid electrolyte cells are still made in the context of electrochemical sensors which are one of the most important technical applications for solid electrolytes.

Another application of solid electrolytes is to be found in the field of power sources. Baur and Preis (1937) proposed a fuel cell system with an oxide ion conductive solid. The solid oxide fuel cells (SOFC) are attractive electric power generation systems, and over the last decade research devoted to their development has become intense. In 1967, Yao and Kummer found that β-alumina exhibited high sodium ion conductivity, and Weber and Kummer (1967) proposed a sodium-sulphur battery with β-alumina. This type of battery may be attractive as a power source for electric vehicles and for electric energy storage as part of a load-levelling system in consumer power distribution. Solid electrolyte cells, which operate at room temperature, have been developed over the past two decades. Also cells based on ionically conducting polymers have been developed and will be discussed in this chapter. Batteries based on electronically conducting polymer electrodes are more problematical but work continues (see Chapter 9). With the developments in electronics, miniature batteries with high reliability are required in many fields such as medical equipment.

Insertion electrodes also represent an important group of compounds for applications. Typical applications of these are in batteries, electrochromic devices and electrochemical memory devices. In only one chapter, it is impossible to deal comprehensively with the entire range of applications in the field of solid state ionics. This chapter will focus on the basic principles of several major applications for solid electrolytes and insertion compounds, namely batteries, fuel cells, chemical sensors, electrochemical memory devices and electrochromic displays.

11.2 Solid electrolyte batteries

Galvanic cells consist of three components: cathode, anode and electrolyte. Conventional batteries, such as Leclanché cells and the lead–acid accumulator, which were developed in the middle of the last century, have a liquid electrolyte. Cells with a solid electrolyte operate in a similar fashion to those with a liquid electrolyte. They may contain either liquid or solid electrodes. All-solid-state cells have many advantages such as miniaturisation, long storage life, operation over a wide temperature range, rugged structure, no volatilisation and no leakage, because all the cell components are solids. In the 1950s, various solid electrolyte cells were reported (see the review by Owens (1971)). In that period, most of them were unable to discharge more than a few microamperes per square centimetre at room temperature. This is due to the poor conductivity of the electrolytes used. At that time, silver halides and doped silver halides were mainly used for the electrolyte. The ionic conductivity of these electrolytes is less than 10^{-6} S cm^{-1} at room temperature. To reduce the resistance of the electrolyte, the silver halides were prepared in thin film form with a thickness of only several microns. The resistance of the electrolyte was still over several hundred ohms per square centimetre. Such silver ion conducting electrolytes were typically combined with silver metal anodes. However, the high electrode polarisation associated with silver dissolution restricted the high current drain. In the 1960s, high silver ion conducting solid electrolytes such as Ag_3SI (Reuter and Hardel, 1961) and $RbAg_4I_5$ (Owens and Argue, 1967; Bradley and Green, 1967) were found (the reasons for such high conductivity in these Ag^+ conductors are discussed in Chapters 2 and 3). Takahashi and Yamamoto (1966) reported a solid electrolyte cell,

$$Ag - Hg/Ag_3SI/I_2 - \text{acetylene black}, \qquad [11.1]$$

where the solid electrolyte Ag_3SI has a high silver ion conductivity,

0.015 S cm^{-1}, at room temperature. The cell resistance, with an electrolyte which was 0.15 cm thick with a diameter of 1.2 cm was about 9Ω at room temperature. When a silver plate was used as the anode, anode polarisation was dominant, and the overvoltage was several hundreds of millivolts at several hundred milliamperes per square centimetre. Whereas, when a silver amalgam anode was used, the polarisation was greatly decreased. At the cathode, the iodine and acetylene black mixture had the lowest polarisation. A typical discharge curve of cell [11.1] is shown in Fig. 11.1. The voltage dropped by only 100 mV after 3 h at a steady current discharge of 1 mA cm^{-2}. The internal resistance of the cell did not increase during the discharge. The reaction of cell [11.1] is essentially the same as that of the $Ag/AgI/I_2$ system,

$$Ag(S) + \tfrac{1}{2}I_2(G) = AgI(S), \tag{11.1}$$

where $I_2(G)$ is iodine in the vapour phase and AgI(S) and Ag(S) are solid silver iodide and silver. The observed open circuit voltage of cell [11.1] was 0.685 V at 20 °C, which is in good agreement with the calculated emf from the free energy change of the reaction, Eqn (11.1), 0.6865 V. If the reaction product AgI were produced at the electrolyte/cathode interface during discharge, a great increase in resistance and drop in the cell voltage would be observed. No change of the cell resistance suggests that the AgI produced at the cathode dissolves into the Ag_3SI. The phase diagram study confirmed the dissolution of AgI in Ag_3SI.

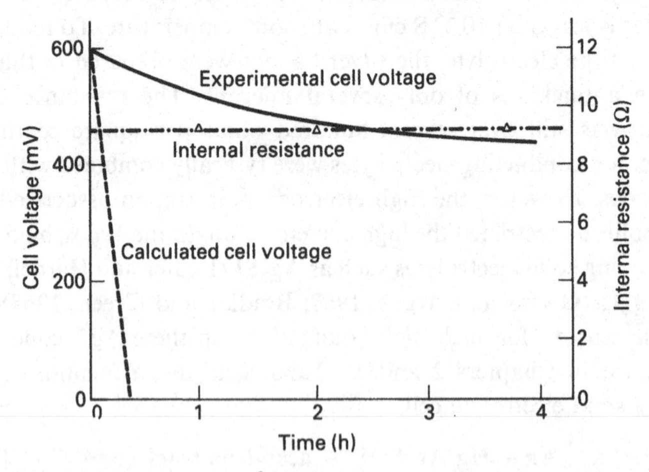

Fig. 11.1 Steady current (1 mA cm^{-2}) discharge curve and the change of internal resistance at 25 °C for the cell $Ag–Hg/Ag_3SI/I_2$–acetylene black (Takahashi and Yamamoto, 1966).

In 1968, Argue and Owens reported the advanced solid electrolyte cell,

$$Ag/RbAg_4I_5/RbI_3, C, \qquad [11.2]$$

which contains the high silver ion conducting solid $RbAg_4I_5$. RbI_3 was used as the cathode. Based upon the RbI–AgI phase diagram, the cell reaction below 27 °C should be

$$4Ag + 2RbI_3 = 3AgI + Rb_2AgI_3. \qquad (11.2)$$

Above 27 °C, the reaction would be

$$14Ag + 7RbI_3 = 3RbAg_4I_5 + 2Rb_2AgI_3. \qquad (11.3)$$

The X-ray diffraction analysis of the discharge products substantiated the two reactions, Eqns (11.2) and (11.3). In Fig. 11.2, constant-load discharge curves for cell [11.2] are shown. In all cases there is an initial *IR* drop, followed by a voltage plateau that extends out to 80–90% of discharge. No appreciable increase in internal resistance occurs during discharge of the cell. The low increase in the resistance may be attributed to the high effective 'porosity' of the solid electrodes, in which the electrolyte, the active cathode material and the discharge products are in close contact with the electronically conducting carbon. The solid state battery is capable of operation at both high and low current density and over a wide range of temperatures. However, the energy density of these batteries, 5 W h kg^{-1}, is quite low.

More recently, Takada, Kanbara, Yamamura and Kondo (1990) have

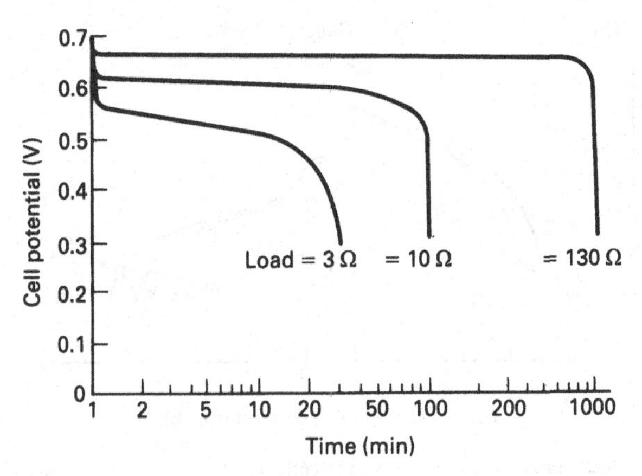

Fig. 11.2 Constant-load discharge at 25 °C for the cell $Ag/RbAg_4I_5/RbI_3$, C (Owens, 1971).

reported a rechargeable solid electrolyte cell with the silver ion conductor, $Ag_6I_4WO_4$, which is stable under ambient atmosphere and has a high conductivity of 0.05 S cm^{-1} at 25 °C. The proposed cell is

$$Ag_xV_2O_5/Ag_6I_4WO_4/Ag_xV_2O_5. \qquad [11.3]$$

$Ag_xV_2O_5$ is a vanadium bronze with a composition range of $0.3 < x < 1.0$, the bronze with $x = 0.29–0.41$ has the β-phase structure and that with $x = 0.67–0.89$ the δ-phase one. The δ-phase shows good reversibility for silver intercalation and deintercalation. The typical charge–discharge curves at a constant current of 0.3 mA cm^{-2} for cell [11.3] are shown in Fig. 11.3. No significant deterioration was observed for several hundred

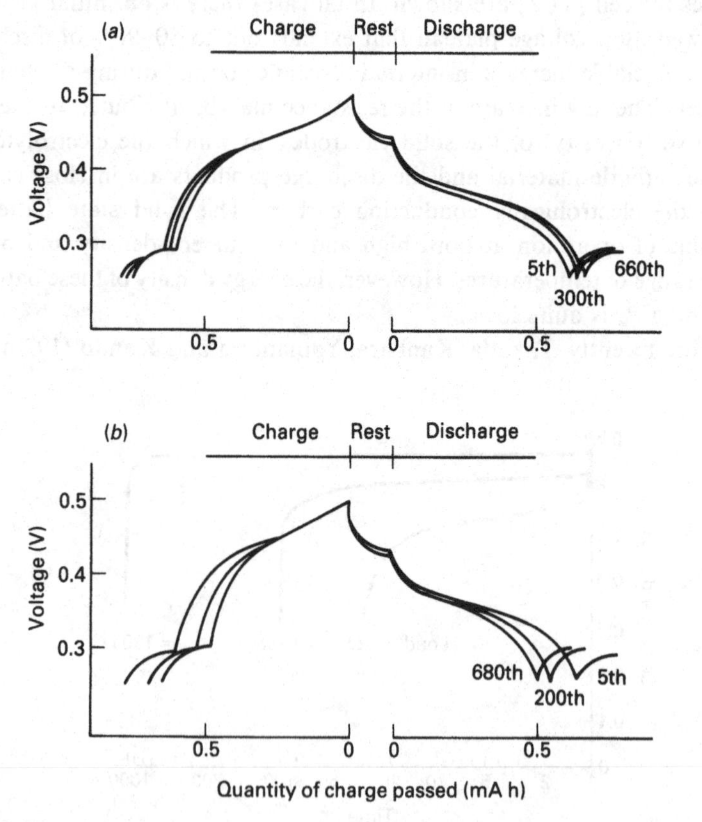

Fig. 11.3 Charge–discharge cycles at 25 °C for the cell $Ag_{0.7}V_2O_5|Ag_6I_4WO_4|Ag_{0.7}V_2O_5$: (*a*) in dry air; (*b*) in open air (Takada *et al.*, 1990).

charge–discharge cycles in open air. This cell has some attractive features in the context of practical applications.

Batteries with silver compounds are limited as far as practical application are concerned, because of the high cost of these materials. On the other hand, copper ion conducting solids are promising electrolytes for batteries. In 1979, Takahashi, Yamamoto, Yamada and Hayashi found that $Rb_4Cu_{16}I_7Cl_{13}$ has the extremely high copper ion conductivity of 0.34 S cm^{-1} and low electronic conductivity at room temperature. The temperature dependence of the conductivity of $Rb_4Cu_{16}I_7Cl_{13}$ is shown in Fig. 11.4. Over a wide temperature range, $Rb_4Cu_{16}I_7Cl_{13}$ has a high conductivity and the ionic transport number is unity. The crystal structure of the compound was analysed by Geller (1976). The structure is similar to that of $RbAg_4I_5$. The physical properties of $Rb_4Cu_{16}I_7Cl_{13}$ are summarised in Table 11.1 along with those of $RbAg_4I_5$, which indicates that the properties are comparable. The copper ion conducting solid was used as the electrolyte in an all-solid-state rechargeable cells. Kanno, Takeda, Oda, Ikeda and Yamamoto (1986) have proposed a cell of the type

$$Cu_4Mo_6S_8/Rb_4Cu_{16}I_7Cl_{13}/\text{intercalation compound}, \quad [11.4]$$

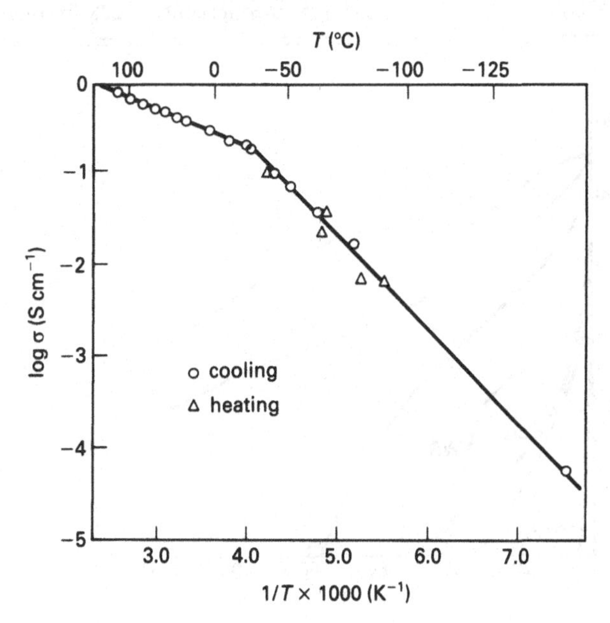

Fig. 11.4 Temperature dependence of the conductivity for $Rb_4Cu_{16}I_7Cl_{13}$ (Takahashi *et al.*, 1979).

where the intercalation compounds are metal dichalcogenides. The Chevrel phase $Cu_4Mo_6S_8$ is an excellent rechargeable anode material, because of the high diffusion rate of intercalated copper among the Mo_6S_8 blocks. The constant-current discharge curves are shown in Fig. 11.5. The highest

Table 11.1. *Physical properties of $Rb_4Cu_{16}I_7Cl_{13}$ and $RbAg_4I_5$ (Takahashi et al., 1979)*

	$Rb_4Cu_{16}I_7Cl_{13}$	$RbAg_4I_5$
Crystal structure (Å)	Cubic $a = 10.02$	Cubic $a = 11.24$
X-ray density (g cm^{-3})	4.47	5.38
Electrical conductivity at 25 °C (S cm^{-1})	0.34 ∞ Hz ac 0.26 10 kHz ac 0.31 dc	0.28 dc
Activation energy for conduction (kJ mole^{-1})	7.0	7.1
Electronic conductivity (S cm^{-1})	10^{-12} at 60 °C	10^{-11} at 25 °C
Copper or silver ion transport number	1.0	1.00
Decomposition potential (V)	0.69	0.67
Melting point (°C)	234 ± 5 (incongruent)	228 (incongruent)

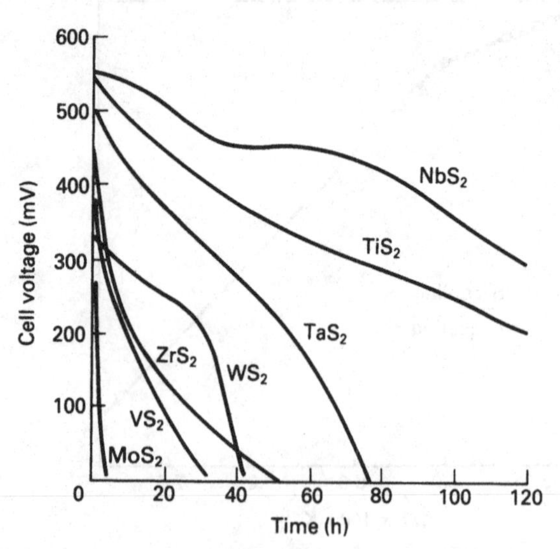

Fig. 11.5 Constant-current discharge curves (100 μA) at 25 °C for the cell $Cu_4Mo_6S_8$/$Rb_4Cu_{16}I_7Cl_{13}$/MS_2 (Kanno *et al.*, 1986).

discharge performance was obtained for the cell with the NbS_2 cathode. The charge–discharge cycles of the cell with NbS_2 at 0.75 mA cm^{-2} are shown in Fig. 11.6. The charge–discharge depth is 0.003 in e/NbS_2. The charge–discharge profiles slightly change with cycling. The dichalcongenide compounds show a volume change during the insertion and removal of ions. These morphological changes could have a profound effect upon the integrity of the electrolyte/electrode interface. The reversibility tends to be best when structural differences between the host and insertion compounds are minimised. Kanno, Takeda, Oya and Yamamoto (1987) proposed the following rechargeable solid electrolyte cell, in which both anode and cathode are Chevrel phases,

$$Cu_2Mo_6S_{7.8}/Rb_4Cu_{16}I_7Cl_{13}/Cu_2Mo_6S_{7.8}. \qquad [11.5]$$

The structure of the copper Chevrel phase is shown in Fig. 7.7. It is based on Mo_6S_8 building blocks which stack in three dimensions. A

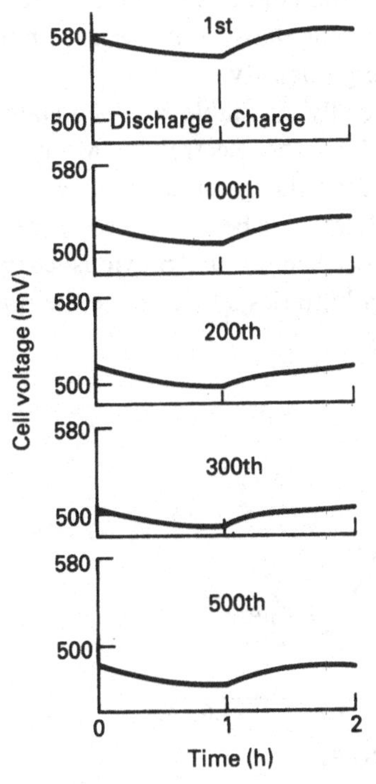

Fig. 11.6 Charge–discharge cycles at 25 °C for the cell $Cu_4Mo_6S_8/Rb_4Cu_{16}I_7Cl_{13}/NbS_2$ (Kanno *et al.*, 1986).

three-dimensional network of pathways is thus formed along which copper ions can move (more details are presented in Chapter 7). The copper content x in $Cu_xMo_6S_{8-y}$ changes as the cell is charged and discharged. The cell is charged up to the composition

$$Cu_{3.8}Mo_6S_{7.8}/Rb_4Cu_{16}I_7Cu_{13}/Cu_{0.2}Mo_6S_{7.8}. \qquad [11.6]$$

Fig. 11.7 shows the constant-current discharge curves of cell [11.6] at room temperature. The cell yields an open circuit voltage of 0.55 V and a current of several hundreds of microamperes with high cathode efficiency at room temperature. Fig. 11.8 shows the charge–discharge cycles of cell [11.6] at a current density of 0.375 mA cm^{-2}. The composition ranges of the cathode and anode were $0.2 < x < 0.3$ and $3.7 < x < 3.8$, respectively. No significant deterioration was observed during 1400 test cycles. This is considerably superior to other types of solid-state cell with Cu/TiS_2 and $Cu_4Mo_6S_8/NbS_2$ couples, which can sustain a larger number of cycles only at the lower current densities of 0.01 mA cm^{-2} and 0.075 mA cm^{-2}, respectively. The rechargeable silver and copper solid electrolyte cells are quite interesting. However, these cells have not been commercially developed, because of their low energy density.

At this time the only commercially available all-solid-state cell is the lithium battery containing LiI as the electrolyte. Many types of solid lithium ion conductors including inorganic crystalline and glassy materials as well as polymer electrolytes have been proposed as separators in lithium batteries. These are described in the previous chapters. A suitable solid electrolyte for lithium batteries should have the properties

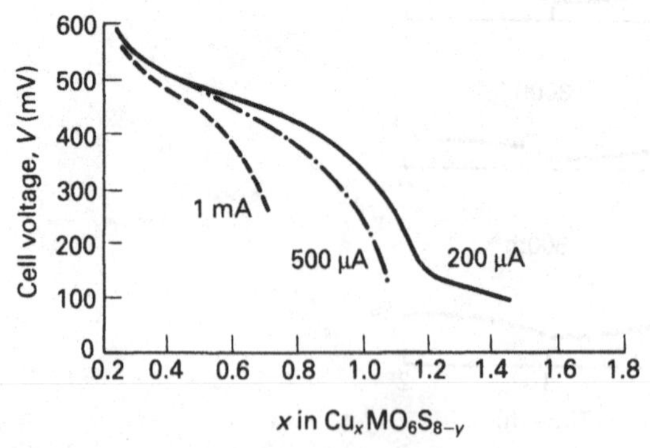

Fig. 11.7 Constant-current discharge curves at 25 °C for the cell $Cu_{3.8}Mo_6S_{7.8}/Rb_4Cu_{16}I_7Cl_{13}/Cu_{0.2}Mo_6S_{7.8}$ (Kanno *et al.*, 1987).

of (i) high lithium ion conductivity, (ii) compatibility with lithium metal or carbon anodes, (iii) a high decomposition potential and (iv) ease of forming a thin film. The highest lithium ion conductivity was found to be 0.001 S cm^{-1} at 25 °C in Li_3N. The decomposition potential of Li_3N was observed to be 0.44 V vs lithium metal at room temperature. Because of this low value, the utilisation of this electrolyte in batteries is limited. In Table 11.2, several lithium solid electrolyte batteries are summarised. Lithium iodide fulfils requirements (ii)–(iv) for a battery electrolyte but not (i). Batteries which demonstrate high reliability are required for implantable cardiac pacemakers. It was recognised that the cell

lithium/LiI/iodine-poly(2vinyl pyridine)(PVP) [11.7]

showed promise in fulfilling the low current, long life and high reliability

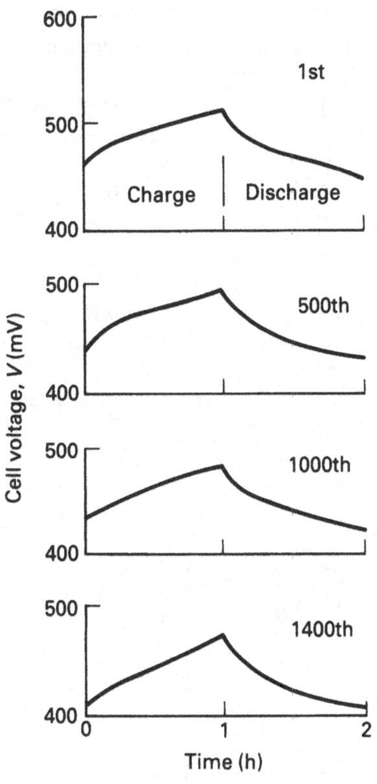

Fig. 11.8 Charge–discharge cycles at 375 μA cm^{-2} for the cell $Cu_{3.8}Mo_6S_{7.8}$/$Rb_4Cu_{16}I_7Cu_{13}$/$Cu_{0.2}Mo_6S_{7.8}$ (Kanno *et al.*, 1987).

requirements (Holmes, 1986). Today, nearly 90% of all pacemakers are powered by lithium/I_2–PVP cells. The thin film LiI electrolyte is formed by *in-situ* reaction of the lithium metal anode and the cathode. A charge transfer complex of I_2 with PVP is used as the cathode. In Fig. 11.9, the proposed structure of the complex (Holmes, 1986) is shown. At values of the I_2/PVP ratio above 1.25 all iodine in the complex is at unity activity. The basic cell reaction in cell [11.7] is quite simple, namely,

$$\mathrm{Li} + \tfrac{1}{2}\mathrm{I}_2 = \mathrm{LiI}. \tag{11.4}$$

The Gibbs free energy change of this reaction is $-270\ \mathrm{kJ\ mol^{-1}}$ and the open circuit voltage is 2.8 V at 25 °C. A major problem in cell [11.7] is to reduce the resistance of the LiI electrolyte. The conductivity of LiI is

Table 11.2. *Solid electrolyte lithium batteries*

System	Observed open circuit voltage (V)	Practical volumetric energy density ($\mathrm{W\,h\,cm^{-3}}$)
Li/LiI/AgI, Ag	2.10	
Li/LiI/PVP, I_2	2.80	0.4–1
Li/LiI(Al_2O_3)/PbI_2, I_2	1.91	0.2
Li/LiI(Al_2O_3)/TiS_2, S	2.14	
Li/PEO–$LiCF_3SO_3$/V_6O_{13}	3.3–3.6	
Li/2.5LiI–$Li_4P_2S_7$/ TiS_2, 2.5LiI–$Li_4P_2S_7$	2.5	0.15–0.21
Li/Li_2O–V_2O_5–SiO_2/MnO_x	2.5	

PVP: Poly(2vinyl pyridine)
PEO: Poly(ethylene oxide).

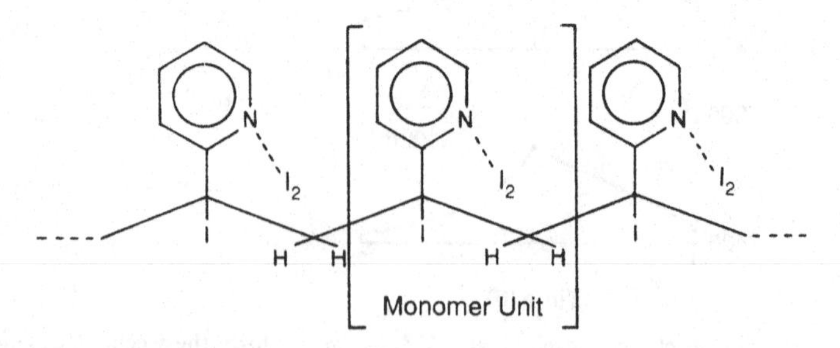

Fig. 11.9 Proposed structure of I_2–PVP reaction product (Holmes, 1986).

around 10^{-7} S cm^{-1} at room temperature. It was found that the cell impedance was reduced considerably when the lithium anode was pre-coated with PVP. The rate with which the resistance of the coated anode cells increased was about two orders of magnitude lower than that of the uncoated anode cells (Owens and Skarstad, 1979). The exact mechanism of this improved behaviour is not well understood. Owens and Skarstad suggested that the phenomenon appears to be less related to the LiI itself than to the expanded area of the interfacial region on a microscopic level. Fig. 11.10 shows typical discharge curves at 37 °C for the Wilson Greatback Ltd Model 755 cell, where the cell dimensions are 33 × 9 × 40 mm^3 and the capacity is 3 A h. At a drain rate of 20 μA (140 kΩ load), over 80% of the 3 A h rated capacity can be utilised without a significant loss of cell voltage. A cutaway view of the WGL cell is shown in Fig. 11.11. The cell consists of a central lithium anode around which molten cathode material is poured through a small fill port which is later hermetically sealed.

The composite of LiI and Al_2O_3 shows a high lithium ion conductivity, 10^{-5} S cm^{-1} at 25 °C. A primary battery with this electrolyte was proposed

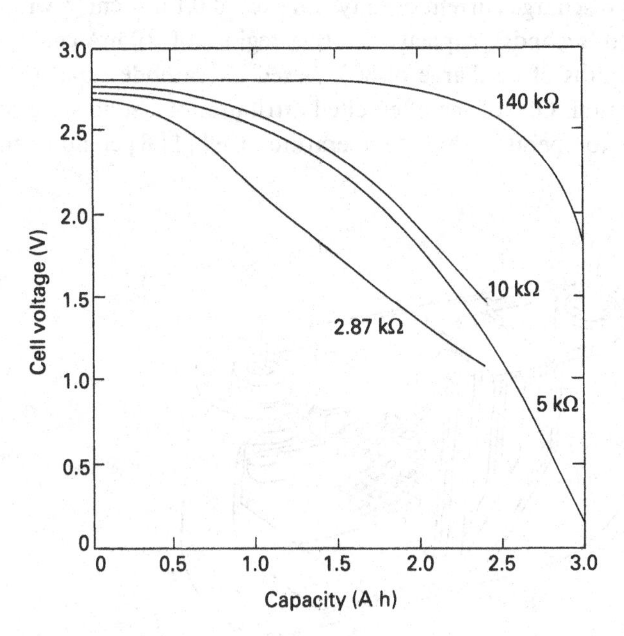

Fig. 11.10 Discharge characteristics of the WGL Model 755 Li/I_2–PVP cell at different constant loads and 37 °C (Shahi, Wagner and Owens, 1983).

by Liang (1973). As it is difficult to make a thin film of the electrolyte, the battery cannot pass a high current. Applications have been limited to areas where current requirements do not exceed about 5–20 μA cm^{-2}. The discovery of high lithium ion conducting glass provided materials having both high decomposition potential and high conductivity at room temperature. Akridge and Vourlis (1986) reported lithium solid state batteries incorporating 2.5 $LiI{:}Li_4P_2S_7$ as the glass electrolyte, and TiS_2 as the cathode. The conductivity of the glass is around 10^{-3} S cm^{-1} at room temperature. In Fig. 11.12, typical constant-current discharge curves at 25 °C for the cell

$$Li/2.5LiI{:}Li_4P_2S_7/(TiS_2 + 2.5LiI{:}Li_4P_2S_7), \qquad [11.8]$$

are shown. The discharge reaction of cell [11.8] is

$$x\mathrm{Li} + \mathrm{TiS_2} = \mathrm{Li}_x\mathrm{TiS_2}. \qquad (11.5)$$

The TiS_2–glass composite cathode was used to establish an electrolyte/cathode interface which will remain intact during discharge. The anode capacity of the cell was 46 mA h while that of the cathode was 37 mA h. Continuous discharge current density can exceed 0.1 mA cm^{-2} with pulse current (two seconds) capacity in the region of 10 mA cm^{-2} while achieving depths of discharge of 80% based on cathode capacity, down to a 1.4 V cutoff. One of the often cited attributes of a solid state battery is the ability to operate at high temperatures. Cell [11.8] could operate at

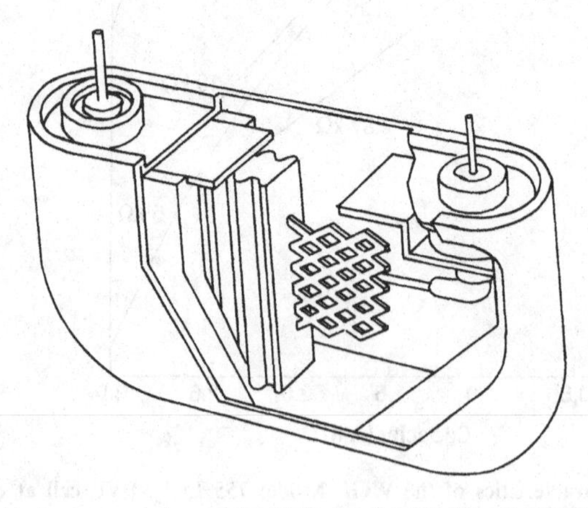

Fig. 11.11 A cutaway view of the WGL Model 761/23 cell (Shahi *et al.*, 1983).

150 °C without serious degradation, where nonaqueous and aqueous batteries have difficulties.

The rechargeable all-solid-state lithium batteries are more attractive for practical applications such as memory back-up than are the silver or copper systems, Ohtsuka, Okada and Yamaki (1990) have demonstrated secondary lithium cells with a Li_2O–V_2O_5–SiO_2 film. The film was fabricated by an rf-sputtering method (Ohtsuka and Yamaki, 1989). The conductivity of the film was 1×10^{-6} S cm^{-1} at 25 °C. The cells were fabricated by depositing MnO_x cathode, solid electrolyte, and lithium anode on stainless steel substrates. The film thicknesses of the cathode, electrolyte and anode were 0.5, 1.0, and 4.0 μm, respectively. The cell was cycled between 1.0 and 3.0 V with a current density of 10 μA cm^{-2}. The change in the discharge and charge capacity of the cell on cycling at room temperature is shown in Fig. 11.13 for the 10th and 70th cycles. The cell reaction is

$$y\text{Li} + \text{MnO}_x = \text{Li}_y\text{MnO}_x. \qquad (11.6)$$

The cell has good rechargeability, though the discharge capacity was relatively small at 14 μA h cm^{-2}.

More recently, solid state batteries with lithium conducting polymer electrolytes have been extensively studied. The development has focused on secondary batteries for an electric vehicle, because lithium polymer batteries have a theoretical energy density that approaches 800 W h kg^{-1}

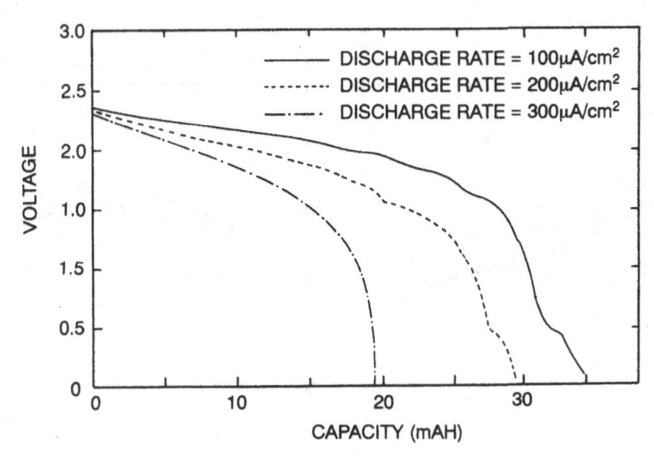

Fig. 11.12 Constant-current discharge curves for the cell $Li/2.5LiI{:}Li_4P_2S_7/(TiS_2 + 2.5LiI{:}Li_4P_2S_7)$ at 25 °C (Akridge and Vourlis, 1986).

and an achievable performance of 425 W h kg^{-1} has been anticipated (Tofield, Dell and Jensen, 1984). A major disadvantage of polymer electrolytes based on poly(ethylene oxide) is that the conductivity hardly reaches 10^{-6} S cm^{-1} until the temperature is increased above the melting point (60 °C); the operating temperature of lithium polymer batteries based on this electrolyte is generally above 100 °C. However, by adding low molecular weight solvents to the high molecular weight polymers high ionic conductivity, above 10^{-3} S cm^{-1}, may be achieved at room temperature. Ambient temperature batteries based on such plasticised polymers are under development. A typical lithium polymer battery designed for commercial application is shown in Fig. 11.14, where poly(ethylene oxide) (PEO) with $LiCF_3SO_3$ is used as the electrolyte and V_6O_{13} as the cathode. The cell reaction is

$$x\mathrm{Li} + \mathrm{V_6O_{13}} = \mathrm{Li}_x\mathrm{V_6O_{13}}. \tag{11.7}$$

Li^+ ions travel across the electrolyte, the electrons leave the anode and proceed along a wire to a current collector to which the cathode is attached. IREQ in Canada (Kapfer, Gauthies and Belanger, 1990) has designed lithium polymer batteries with a Li anode and VO_x cathode for electric vehicle applications. The proposed system comprises 144 cells in two banks each containing 72 cells in series in order to satisfy the minimum requirement for 120 V. Each cell has a nominal capacity of 280 W h. This battery design results in a weight and a volume of 407 kg and 0.506 m^3. The operating temperature is in the range 60–100 °C. In

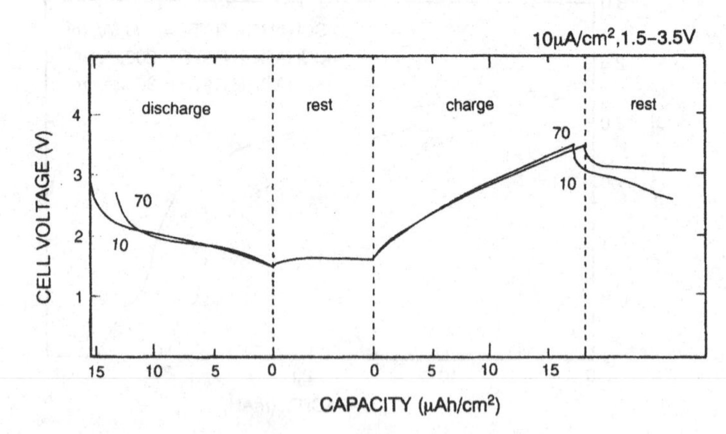

Fig. 11.13 Typical discharge–charge curves of the cell $Li/Li_2O–V_2O_5–SiO_2/MnO_x$ at room temperature (Ohtsuka *et al.*, 1990).

Fig. 11.15, the loss of capacity with cycle life is shown. The available energy capacity can be calculated as 35 kW h at the beginning of the battery's life and 21 kW h at the end. The energy density of 86 W h kg^{-1} at the beginning is attractive compared with the conventional secondary batteries of 40 W h kg^{-1} or less. The energy capability with cycling must be improved for practical applications.

During the past two decades, the sodium–sulphur battery which

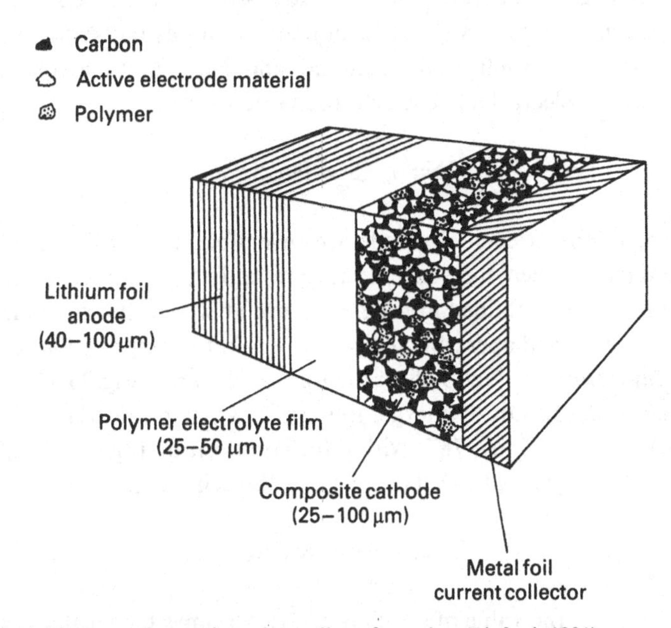

Fig. 11.14 Lithium polymer electrolyte cell configuration (Linford, 1991).

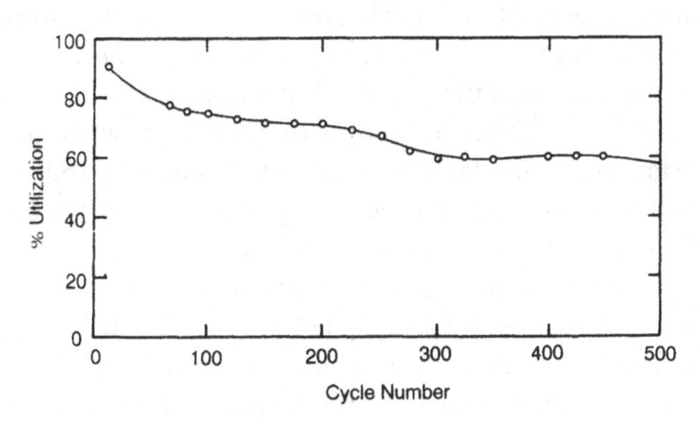

Fig. 11.15 Loss of capacity with cycle life for polymer batteries (Kapfer *et al.*, 1990).

incorporates a sodium ion conducting solid electrolyte, Na-β-alumina, has been developed. This type of battery is quite attractive as a means of storing the electricity generated by power stations at times of low consumer demand then feeding this power back into the distribution system at times of peak demand, i.e. load levelling. It is also attractive for electric vehicles, because of its high ratio of energy to weight and volume compared with conventional lead–acid or nicad batteries (Fischer, 1989). So far this chapter has dealt with all-solid-state cells, in contrast the sodium–sulphur battery consists of molten sodium as the anode, molten sulphur with carbon felt as the cathode, and Na-β-alumina (or Na-β''-alumina) as the electrolyte. The cell discharge reaction is

$$2Na + xS = Na_2S_x \tag{11.8}$$

and the reactions at each electrode are reversible. In order to maintain good contact between the solid electrolyte and the electrode materials, the cell must be maintained at a temperature at which the electrodes are molten. The practical operating temperature is in the range of 300–400 °C, in accordance with the Na–S phase diagram as shown in Fig. 11.16 (Gupta and Tischer, 1972). On discharge, the emf is at first constant and then declines from 2.076 V to 1.78 V within the composition range of $Na_2S_{5.2}$–$Na_2S_{2.7}$. The theoretical specific energy of the cell reaction

$$2Na + 3S = Na_2S_3 \tag{11.9}$$

is 760 W h kg^{-1}, the value of which is about 4.5 times higher than that of the conventional lead–acid battery. A tubular cell design for a practical battery is shown in Fig. 11.17. In this type of cell, sodium is located in the centre, and the current collector for the cathode is the outer steel case. A cell with sulphur at the centre has also been developed. The advantage of this cell is that the positive current collector can more easily be provided with a corrosion protection layer than can the casing of a sodium core cell. On the other hand, it has a lower energy density than the cell with sodium at the centre. The cell performance is shown in Table 11.3. The most important properties are energy and power per weight and volume and cycle life. Practical energy density is 132 W h kg^{-1}, which is approximately three times as high as that of lead–acid cells (40 W h kg^{-1}). Continuous degradation is caused by effects such as impurities migrating into the β''-alumina electrolyte, and corrosion products migrating into the

Table 11.3. *Characteristics of the sodium–sulphur cell (Hitachi)*

Total cell size	ϕ75 mm × 398 mm
β''-Al_2O_3 tube	ϕ49.3 mm × 350 mm
Total weight	4.0 kg
Current density	72 mA cm^{-2}
Capacity	280 A h
Power	66 W
Energy	528 W h
Cell voltage	1.85 V
Cell resistance	4.4 mΩ
Energy efficiency	86%
Energy density	132 W h kg^{-1}
	300 W h l^{-1}

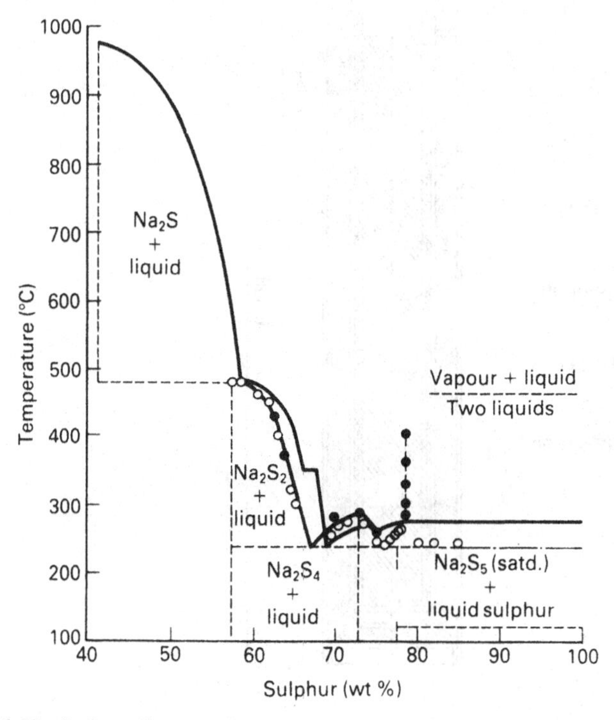

Fig. 11.16 Na–S phase diagram (Gupta and Tischer, 1972).

front layer of the sulphur electrode (Fischer, 1989). Abrupt failures are mainly correlated to rupture of the ceramic components. The abrupt rupture of β''-alumina occurs as a consequence of thermal stress, inhomogeneities within the ceramic or uneven current distribution as a consequence of inhomogeneous electrodes. For load levelling at power stations a lifetime of several thousand cycles is required, and for an electric vehicle 500–1000 cycles are necessary. The maximum cycle life of the present generation of Na–S batteries is several hundred and this is already close to the requirement for electric vehicles, but reliability has to be improved (Fischer, 1989).

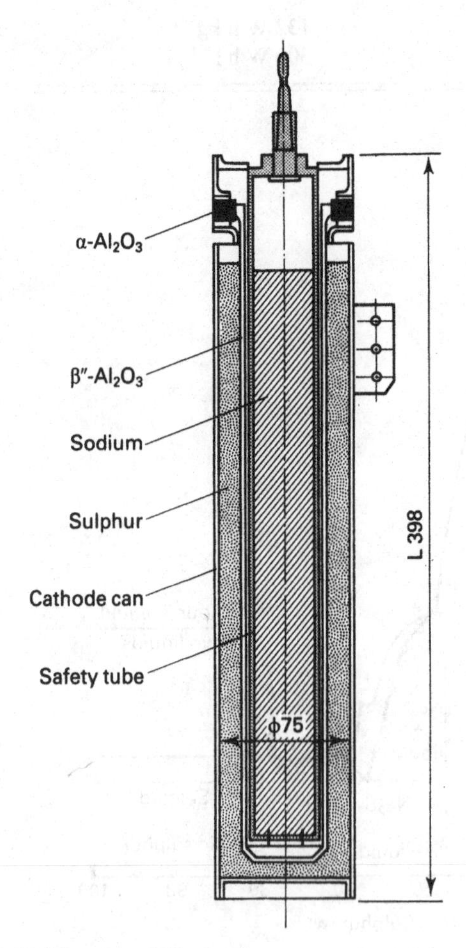

Fig. 11.17 Schematic of Na–S cell.

11.3 Intercalation electrodes for batteries

The term intercalation describes the reversible insertion of guest species into a host. The term can equally be applied to one-, two- and three-dimensional solids. The fundamental electrochemistry of these electrodes is discussed in detail in Chapters 7 and 8. They have also been mentioned in the previous section. Intercalation compounds are very attractive as the electrode in rechargeable batteries. Titanium disulphide has been extensively studied as the cathode for lithium batteries, because it has many positive attributes such as high reversibility of the chemical reaction and a high free enthalpy change of reaction. It adopts a lamellar structure (Chapter 7), where guest atoms like Li can move freely between the layers. The chemical diffusion coefficient of the lithium ion in TiS_2 is 10^{-9} cm^2 s^{-1} (Whittingham and Silbernagel, 1977). The emf of the cell,

$$\text{Li/propylene carbonate–LiPF}_6\text{/TiS}_2 \qquad [11.9]$$

is shown as a function of x in Li_xTiS_2 at 25 °C in Fig. 11.18. There is a single phase over the entire Li composition range, $0 < x < 1$, so that no

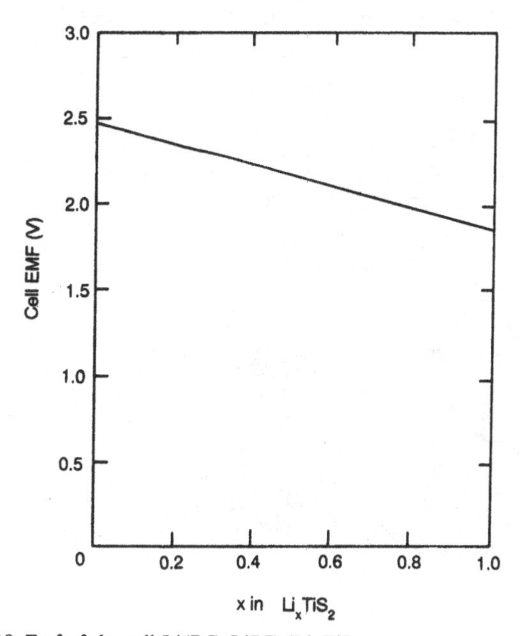

Fig. 11.18 Emf of the cell $Li/PC–LiPF_6/Li_xTiS_2$.

energy is expended in nucleating a new phase, although careful examination of the variation of emf with composition does reveal fine structure associated with local ordering (Chapter 7). The cell reaction is

$$x\mathrm{Li} + \mathrm{TiS_2} \underset{\text{charge}}{\overset{\text{discharge}}{\rightleftharpoons}} \mathrm{Li}_x\mathrm{TiS_2}. \tag{11.10}$$

The high free energy change of the reaction leads to a theoretical energy density of 481 W h kg^{-1} and 455 W h kg^{-1} at a load of 10 mA cm^{-2}. The intercalation and deintercalation of lithium has been accomplished more than 600 times. Most of the work on this cell type was done by research groups at EIC Laboratories and the Exxon Corporation in the USA. Large prismatic Li/TiS_2 cells of approximately 5.5 A h capacity were built and tested at EIC (Brumer, 1984). These cells had practical energy densities of less than 100 W h kg^{-1} at medium to high rate, with high cycle life. Brumer has suggested that the technology still requires a positive electrode with a higher voltage if it is to achieve its market potential.

A number of transition metal oxides can also be intercalated by lithium. One of the best known examples is V_6O_{13}. The V_6O_{13} structure, shown in Fig. 11.19, consists of alternate double and single layers of V_2O_5 ribbons. The layers are connected by vertex sharing of octahedral sites, and this leads to a relatively open framework structure (Wilhelmi, Waltersson and Kihlburg, 1971). Insertion of lithium into the oxide matrix

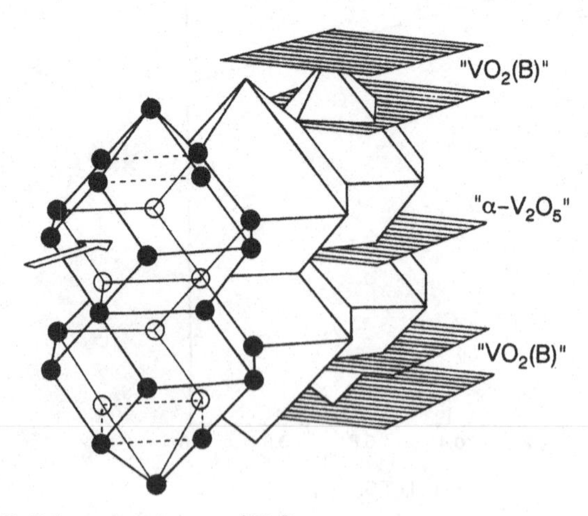

Fig. 11.19 Schematic structure of V_6O_{13}.

causes small changes in the lattice parameters, but the basic framework is preserved. As there are a number of different sites in the perovskite-like cavities, the emf curve for $Li/Li_xV_6O_{13}$ shows several distinct plateaux, reflecting the sequential filling of inequivalent sites. The excellent rechargeability of the material is indicated by the charge–discharge cycles shown in Fig. 11.20 for an Li/2Me–THF, $LiAsF_6/V_6O_{13}$ cell. The material also has good rate capability and a high theoretical energy density of 890 W h kg^{-1}. The high reversibility and rate capability of V_6O_{13} make it an attractive positive electrode.

For consumer applications, alternative high energy density rechargeable cathode materials have been developed. Mizushima, Jones, Wiseman and Goodenough (1980) reported that the open circuit voltage of Li/Li_xCoO_2 was 3.8 V at $x = 1.0$ and gradually increased to about 4.7 V with decreasing amount of lithium as shown in Fig. 11.21. When charged to 4.0 V the cathode has a chemical composition of $Li_{0.5}CoO_2$ and remains isostructural with $LiCoO_2$. $LiCoO_2$ has a layered structure analogous to $LiTiS_2$ except that the anions in the former case adopt cubic close packing instead of hexagonal close packing. The idealised structures of $LiCoO_2$ and $LiTiS_2$ are shown in Fig. 11.22. Lithium ions can be reversibly extracted from between the layers with a corresponding oxidation of Co^{3+} to Co^{4+}. $LiCoO_2$ has a high theoretical energy density of 766 W h kg^{-1} and is an attractive cathode material for lithium secondary batteries. This material is used as the cathode in the Li secondary cells now commercially available from Sony Co., Japan (Mashiko, Yokokawa

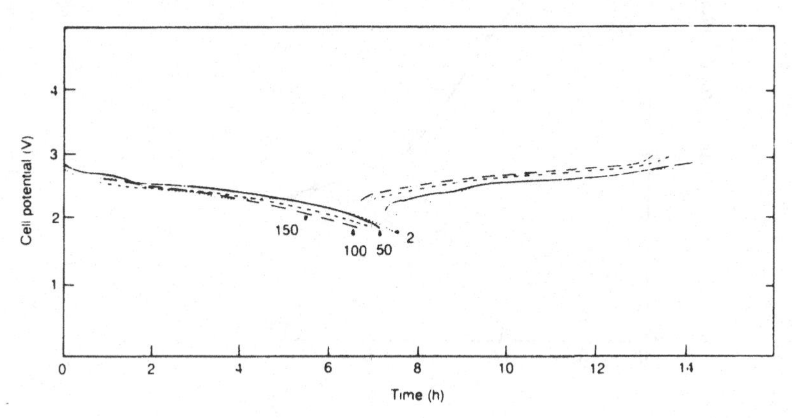

Fig. 11.20 Typical charge–discharge plot for the Li/V_6O_{13} cell (Abraham, Goldman and Dempsey, 1981).

and Nagaura, 1991). The cell consists of a $LiCoO_2$ cathode and carbon anode,

$$Li_xC/PC + DME + LiPF_6/Li_{1-x}CoO_2. \quad [11.10]$$

On first charging the cell the lithium in $LiCoO_2$ is moved to the carbon anode,

$$LiCoO_2 + C = Li_{1-x}CoO_2 + Li_xC. \quad (11.11)$$

It has been known for some time that lithium can be intercalated between the carbon layers in graphite by chemical reaction at a high temperature. Mori *et al.* (1989) have reported that lithium can be electrochemically intercalated into carbon formed by thermal decomposition to form LiC_6. Sony has used the carbon from the thermal decomposition of polymers such as furfuryl alcohol resin. In Fig. 11.23, the discharge curve for a cylindrical cell with the dimensions ϕ 20 mm × 50 mm is shown, where the current is 0.2 A. The energy density for a cutoff voltage of 3.7 V is 219 W h l^{-1}, which is about two times higher than that of Ni–Cd cells. The capacity loss with cycle number is only 30% after 1200 cycles. This is not a lithium battery in the spirit of those described in Section 11.2.

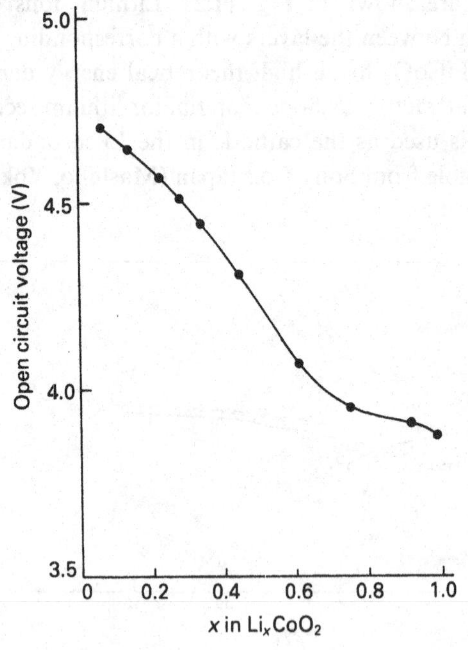

Fig. 11.21 Open circuit voltage versus composition x for Li/Li_xCoO_2 (Mizushima *et al.*, 1980).

Instead it is a known as a lithium ion battery since it is free from lithium metal and hence free from the safety and stability problems of lithium cells. The commercialisation of this cell by Sony represents one of the most important breakthroughs in battery technology for many years and is a major success for solid state electrochemistry.

11.4 Solid oxide fuel cells

High temperature solid oxide fuel cells (SOFCs) have become of great interest as a potentially economical, clean and efficient means of producing electricity in a variety of commercial and industrial applications (Singhal, 1991). A SOFC is based upon the ability of oxide ions to be conducted through a solid at elevated temperatures. Oxide ion conductivity was observed in ZrO_2–9 mol% Y_2O_3 by Nernst as early as 1899. In 1937, Bauer

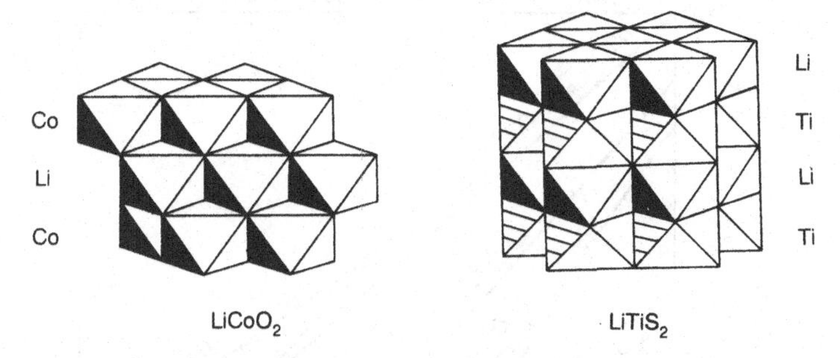

Fig. 11.22 Idealised structure of $LiCoO_2$ and $LiTiS_2$ (Mizushima *et al.*, 1980).

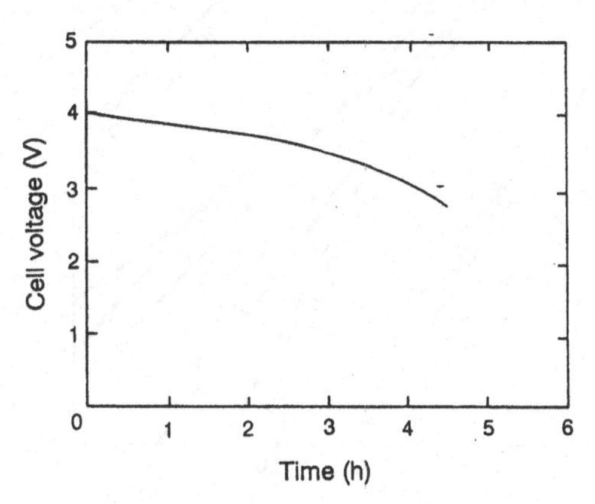

Fig. 11.23 Constant-current (0.2 A) discharge curve for the $Li_xC/Li_{1-x}CoO_2$ cell. Cell size: ϕ 20 mm × 50 mm.

and Preis constructed the first SOFC using this electrolyte, the oxide ion conductivity of which is around 0.1 S cm^{-1} at 1000 °C. Since the 1960s, many oxide systems have been examined as oxide ion conductors. The electrolyte for a SOFC has to meet the following requirements: high ionic conductivity, low electronic conductivity, chemical and physical stability under reducing and oxidising atmospheres at high temperature, and ease of preparation in the form of a dense film. In Fig. 11.24, the temperature dependence of the conductivity for several oxide ion conductors is shown. The systems based on CeO_2 and Bi_2O_3 have a high ionic conductivity, but Ce^{4+} ions and Bi^{3+} ions are easily reduced to Ce^{3+} and Bi^{2+} valence states at low oxygen pressures. Measurement of the ionic transport number, t_i, in the CeO_2–11 mol% La_2O_3 system indicated that $t_i = 0.92$

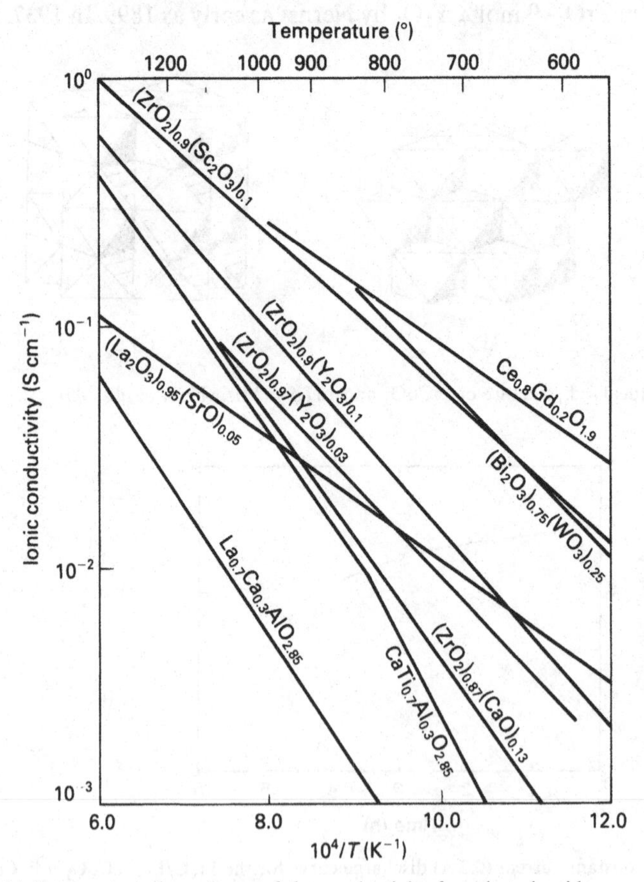

Fig. 11.24 Temperature dependence of the conductivity for selected oxides.

Table 11.4. *Mechanical properties and electrical conductivity of 3 mol% Y_2O_3–ZrO_2(TZP) and 8 mol% Y_2O_3–ZrO_2(FSZ)*

	TZP	FSZ
Bending strength (MPa)	1200	300
Fracture toughness ($MN\ m^{-1.5}$)	8	3
Electrical conductivity at 1000 °C ($S\ cm^{-1}$)	6.5×10^{-2}	1.6×10^{-1}

at $P_{O_2} = 0.1$ atm, and was only 0.54 at $P_{O_2} = 10^{-8}$ atm and 1000 °C. Despite much research since the early 1960s stabilised zirconia, the first material discovered to exhibit oxide ion conductivity, remains one of the best materials for a SOFC. The fluorite phase of ZrO_2 is stabilised when it is doped with di- or tri-valent metal oxides like CaO, Y_2O_3 or Sc_2O_3. More details on the doped zirconia electrolytes and the efforts which have been made to develop alternative O^{2-} ion conductors, which can operate at lower temperatures, are given in Chapters 2 and 3. To obtain a high performance SOFC, the contribution of the electrolyte resistance should be less than 0.2–0.3 $\Omega\ cm^2$ at the operating temperature. Therefore, the thickness of the cubic stabilised zirconia must be around 0.2–0.3 mm, because the conductivity at 1000 °C is about 0.1 $S\ cm^{-1}$. The tetragonal phase of zirconia has been stabilised by taking advantage of fine particle technology (Gupta *et al.*, 1977). This phase is expected to find an application as the electrolyte in SOFCs, because of its high mechanical strength as shown in Table 11.4.

The operating principle of a SOFC is schematically shown in Fig. 11.25. When an external load is applied to the cell, oxygen is reduced at the porous air electrode to produce oxide ions. These ions migrate through the solid electrolyte to the fuel electrode, and oxide ions react with the fuel, H_2 or CO, to produce H_2O or CO_2.

SOFCs of several different configurations are presently under investigation, such as planar, monolithic and tubular geometries (Singhal, 1989; Yamamoto, Dokiya and Tagawa, 1989). The typical configurations of each design are shown in Fig. 11.26. Most progress to date has been achieved with the Westinghouse tubular cell. In this design, the active cell components are deposited in the form of thin layers (to keep the overall cell resistance low) on a porous calcia-stabilised zirconia support tube. The materials for the cell components, and their fabrication methods are summarised in Table 11.5. Using the electrochemical vapour deposition technique (Isenberg, 1977), gas-tight thin layers of electrolyte and interconnect were fabricated. A voltage–current plot at different temperatures for the Westinghouse SOFC single cell is shown in Fig. 11.27. To construct an electric generator, individual cells are connected in both parallel and series. A 3 kW SOFC generation system was delivered for field testing in

Table 11.5. *Cell components, materials and fabrication process for the Westinghouse SOFC (Singhal, 1991)*

Component	Material	Thickness	Fabrication process
Support tube	$ZrO_2(CaO)$	1.2 mm	Extrusion-sintering
Air electrode	$La(Sr)MnO_3$	1.4 mm	Slurry coat-sintering
Electrolyte	$ZrO_2(Y_2O_3)$	40 μm	Electrochemical vapour deposition
Interconnection	$LaCr(Mg)O_3$	40 μm	Electrochemical vapour deposition
Fuel electrode	$Ni–ZrO_2(Y_2O_3)$	100 μm	Slurry coat-electrochemical vapour deposition

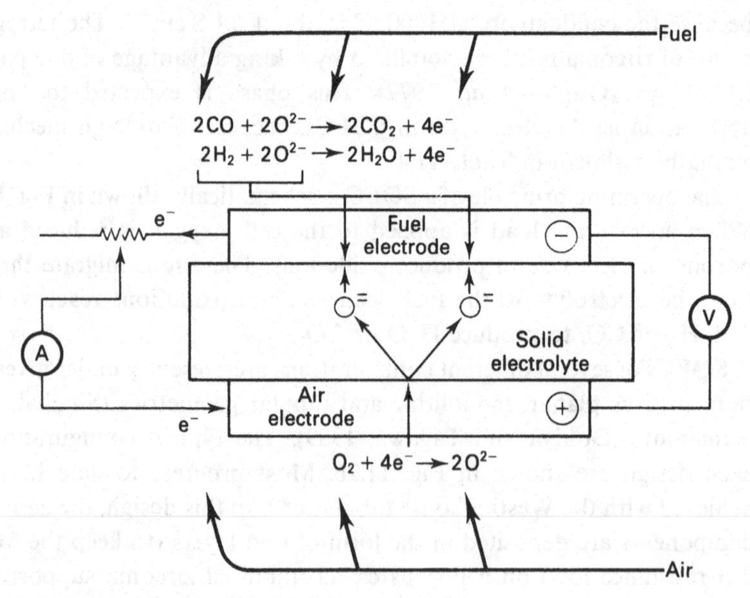

Fig. 11.25 Operating principle of a SOFC (Singhal, 1989).

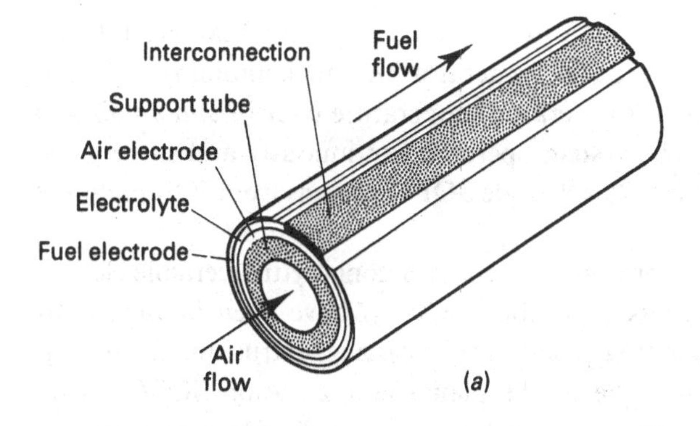

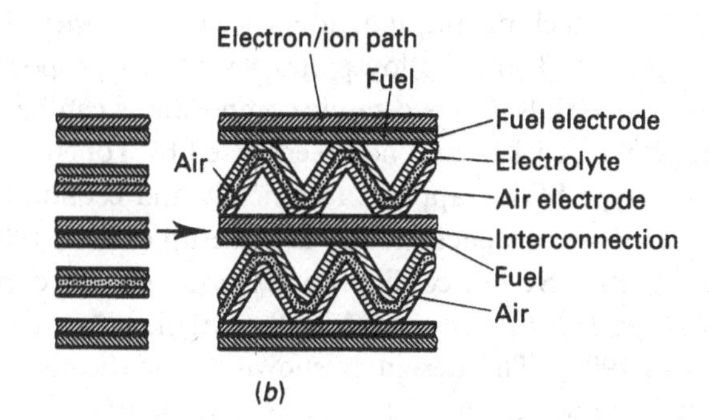

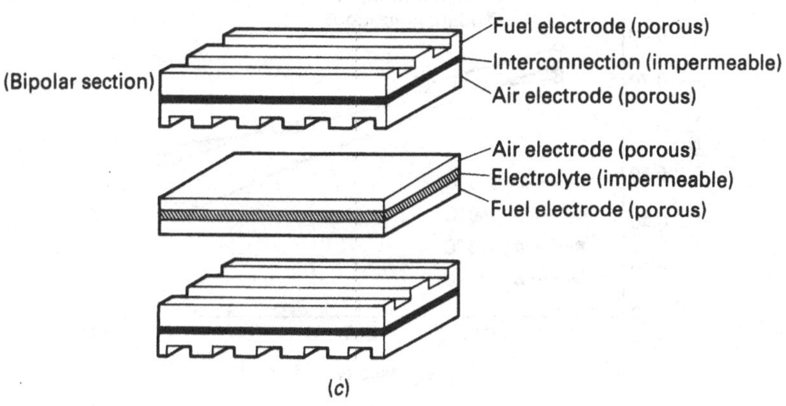

Fig. 11.26 SOFC configuration: (*a*) Westinghouse single tube configuration; (*b*) Argonne multichannel monolithic configuration; (*c*) planar configuration.

1987 (Harada and Mori, 1988; Yamamoto, Kaneko and Takahashi, 1988). The system consisted of a SOFC generator module containing 144 cells, electrical air preheaters for generator temperature control and an air and fuel handling system. The system operated continuously and successfully for over six months. Now, 25 kW scale SOFC systems from Westinghouse are being tested.

Strong, dense and impermeable sheets of zirconia–yttria ceramic electrolytes with thicknesses less than about 300 μm have been fabricated by advanced ceramic technology such as tape-casting, extrusion, hot-rolling, etc. Using these zirconia sheets, the planar configuration SOFC shown in Fig. 11.26(*c*) has been developed (Steele, 1987). Electrodes can be deposited by appropriate techniques and then multicell units can be assembled by stacking the individual sheets together. Doped $LaCrO_3$ sintered plates and metal alloy plates have been proposed as the interconnection materials. Glass ceramic compositions can be used to seal the sheets together. A 1 kW unit has been tested by Tonen (Sakurada, 1991). The planar type SOFC appears technically and economically attractive, but problems may be encountered in scale-up (Steele, 1987).

An alternative SOFC configuration, based on the corrugated monolithic concept, was first proposed by the Argonne National Laboratories (Fee *et al.*, 1986). The design is shown schematically in Fig. 11.26(*b*). The cells incorporate two triple layer subassemblies each prepared by

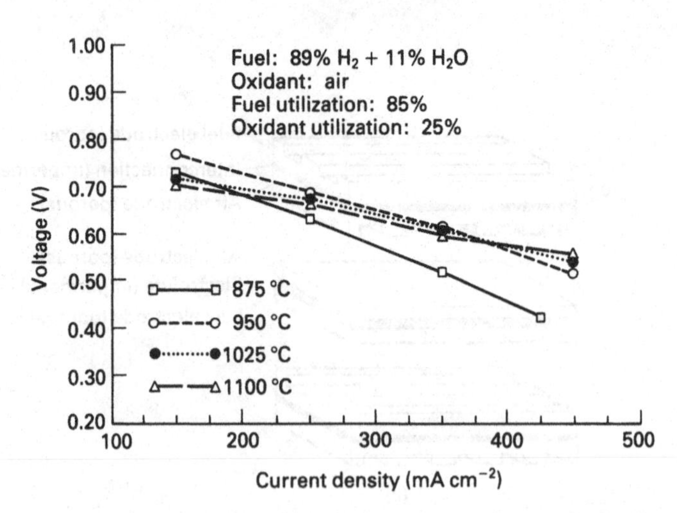

Fig. 11.27 Voltage–current curves at different temperatures for the Westinghouse SOFC single cell (Singhal, 1991).

sequentially tape-casting the anode material, electrolyte or interconnecting material and finally the cathode material. After drying, the subassemblies are stacked on top of each other and cofired. The principal problem with the monolithic configuration is associated with the cofiring stage. The electrodes must remain porous whilst at the same time the electrolyte, YSZ, and interconnector, doped $LaCrO_3$, must be impermeable. A specific problem relates to $LaCr(Mg)O_3$. To obtain dense and impermeable doped $LaCrO_3$, many approaches have been investigated. Argonne have developed a fabrication procedure which involves the addition of sintering aids to $LaCr(Mg)O_3$ in order to allow this component to be sintered to a dense impermeable layer in an oxidising environment at moderate temperatures. Whether these sintering aids will diffuse and cause degradation of the cell will only become apparent during long term performance tests (Steele, 1987). Sakai *et al.* (1990) have reported that a slight deficit of chromium drastically enhanced densification in calcium-doped lanthanum chromates.

Consequently, high temperature SOFCs can achieve extremely high current densities, as high as $1\ \mathrm{A\ cm^{-2}}$, because of the high operating temperature of 1000 °C. However, performance data for the large scale tubular configuration SOFC developed by Westinghouse are not as good as expected. To obtain a high performance SOFC, many technical and materials problems remain to be solved.

11.5 Solid electrolyte sensors

Different types of sensor based on solid electrolytes have been developed following a report by Kiukkola and Wagner (1957). These sensors are based on one of two principles: (*a*) the chemical potential difference across the solid electrolyte (potentiometric sensor), or (*b*) the charge passed through the electrolyte (amperometric sensor). In the following galvanic cell,

$$\mu_X'/\mathrm{MX}/\mu_x'', \qquad [11.11]$$

emf, E, is given by the relation

$$E = \frac{1}{Z_X F}\int_{\mu_{X'}}^{\mu_{X''}} t_{\mathrm{ion}}\, d\mu_X = -\frac{1}{Z_M F}\int_{\mu_{M'}}^{\mu_{M''}} t_{\mathrm{ion}}\, d\mu_M, \qquad (11.12)$$

where M and X denote the metal and nonmetal of the MX electrolyte, respectively, μ' and μ'' are the chemical potentials of the anode and the cathode, respectively, Z is the valence, F is Faraday's constant and t_{ion} is the ionic transport number in MX. For a solid electrolyte with $t_{ion} = 1.0$, Eqn (11.12) is simplified to

$$E = (\mu_X'' - \mu_X')/Z_X F = -(\mu_M'' - \mu_M')/Z_M F. \qquad (11.13)$$

Generally, in solid electrolytes, ionic conductivity is predominant ($t_{ion} = 1$) only over a limited chemical potential. The electrolytic conductivity domain is an important factor limiting the application of solid electrolytes in electrochemical sensors.

The most thoroughly developed sensor based on a solid electrolyte is the oxygen sensor using a stabilised zirconia electrolyte. This type of sensor is one of the most successful commercial sensors to date. They are widely used in industry, especially in the analysis of exhaust gases from combustion engines. The following configuration is used in the O_2 sensors:

$$\text{Pt}, P_{O_2}'/ZrO_2(Y_2O_3)/P_{O_2}'', \text{Pt}. \qquad [11.12]$$

The emf of the cell is given by

$$E = -\frac{RT}{4F} \ln \frac{P_{O_2}''}{P_{O_2}'}, \qquad (11.14)$$

where P_{O_2}' is the partial pressure of oxygen in the test gas and P_{O_2}'' is the partial pressure in the reference gas (generally, air). By measuring the emf of the cell, the chemical potential and hence O_2 concentration in the test gas may be determined from Eqn (11.14) since the O_2 content in the reference compartment is fixed. Operation of automobile engines at the stoichiometric air/fuel ratio, A/F, improves the engine efficiency as well as reducing the content of toxic gases like CO and NO_x in the exhaust. To reduce the NO_x gas content, A/F is controlled within several per cent of the stoichiometric value and then NO_x is reduced by the three-way catalytic system. A typical emf of this sensor is shown as a function of A/F in Fig. 11.28 (Fleming, 1977). The electrodes are platinum black backed onto the stabilised zirconia tube. The platinum at the anode and cathode plays an important role in catalysing the reactions. A sharp step in the A/F vs emf curve is not obtained in the absence of an effective catalyst.

Several oxygen sensors based on oxygen pumping with stabilised zirconia have been reported (Hetrick, Fate and Vassell, 1981). This type of oxygen sensor is able to measure the oxygen partial pressure in the exhaust gas from the engine in lean burn. The operating principle of the sensor proposed by NGK (Soejima and Mase, 1985) is shown in Fig. 11.29. The sensor consists of two electrochemical cells: a pumping cell and a sensing cell. By controlling the pumping current flowing through the pumping cell, excess oxygen or fuel in the exhaust reaches the internal electrode of the pumping cell through the gap having a specified diffusion

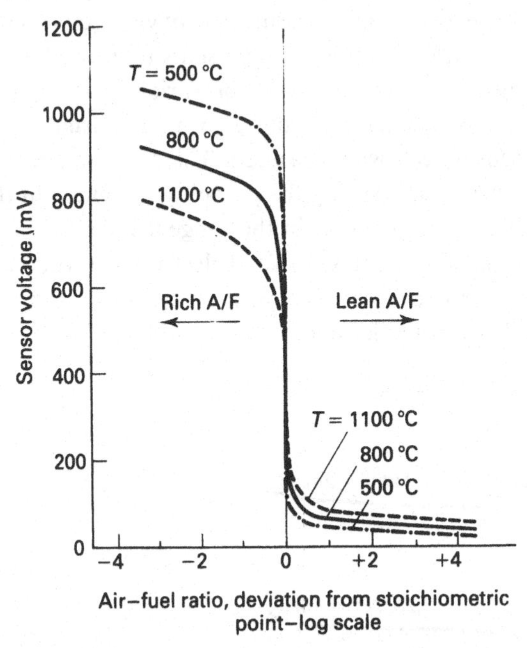

Fig. 11.28 Ideal sensor voltage curves for the following cell: exhaust gas, Pt stabilised ZrO_2/Pt, air (Fleming, 1977).

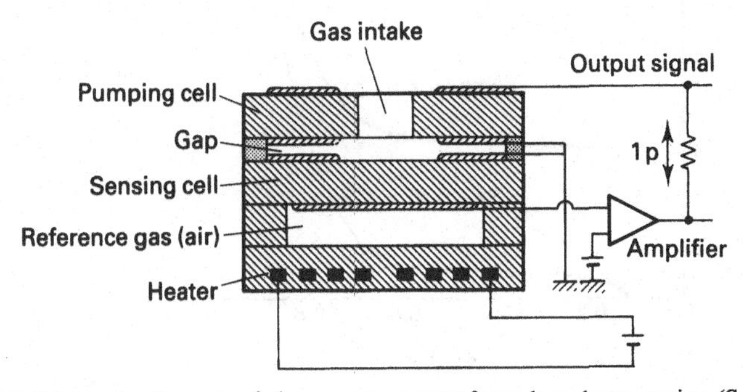

Fig. 11.29 Schematic diagram of the oxygen sensor for a lean burn engine (Soejima and Mase, 1985).

resistance and is consumed completely on the internal electrode surface of the pumping cell. In this case, the amount of oxygen converted through the pumping cell is proportional to the pumping current. Therefore, by measuring the pumping current, the amount of excess oxygen or fuel in the exhaust can be obtained. The other cell serves as an oxygen sensing cell; it detects the oxygen partial pressure of the gas between the gap and reference gas (air). This new type of amperometric oxygen cell can be used with lean and rich A/F ratios. Figure 11.30 presents typical relationships between the pumping current and the emf for various A/F ratios. The sensor was placed in the exhaust gas from a propane burner at 310 °C and the emf of the sensing cell was measured. The rich–lean region can be detected continuously by measuring the pumping current with the emf of the sensing cell being kept constant in the range 0.2–0.6 V.

To reduce the working temperature for solid electrolyte oxygen sensors, fluoride electrolytes have been proposed. Such sensors are composed of the following cell (Sibert, Fouletier and Vilminot, 1983),

$$Sn,SnF_2/PbSnF_4/Pt,P_{O_2}, \quad [11.13]$$

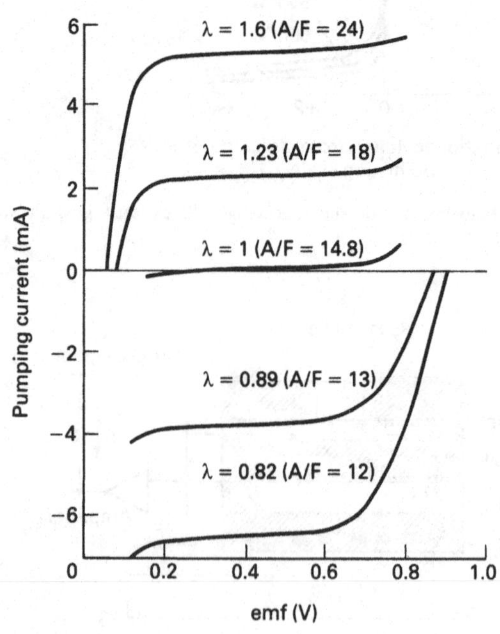

Fig. 11.30 Relation between pumping current and emf of the lean burn sensor for various λ (= (air/fuel ratio)/(theoretical air/fuel ratio)) (Soejima and Mase, 1985).

where Sn/SnF_2 is the reference electrode and $PbSnF_4$ is a good fluoride ion conductor. Pseudoequilibrium conditions for oxygen diffusion through the interface between $PbSnF_4$ and the Pt electrode give a Nernst emf which is added to the constant one appearing at the $Sn/SnF_2/PbSnF_4$ due to the equilibrium of F^- ions. A potential difference, E, can be measured depending on the partial pressure of oxygen:

$$E = E^\circ + \frac{kT}{2e} \ln P_{O_2} - \frac{1}{e}(\delta\mu_{O^{2-}} - \delta\mu_{F^-}), \qquad (11.15)$$

where $\delta\mu_{O^{2-}}$ and $\delta\mu_{F^-}$ represent possible variations of the chemical potential of the oxide and fluoride ions in the region close to the $PbSnF_4/Pt$ interface.

Sensors have also been developed which are capable of detecting a variety of other gases including H_2, SO_2, CO_2 etc. (see, for example, the *Proceedings of the 8th International Conference on Solid State Ionics*).

11.6 Electrochromic devices (ECDs)

Electrochromism can be defined as a colour change induced in a material by an applied electric field or current. Some ions in solid compounds can be reduced or oxidised (redox) electrochromically with a consequent change in colour. WO_3 and MoO_3 solid films have been extensively used for this purpose. The electrochromic reaction is expressed by

$$xA^+ + MO_y + xe^- = A_xMO_y \quad (A = H, Li,), \qquad (11.16)$$

where the cation A^+ and electron e^- are inserted into the host oxide, MO_y, which results in the nonstoichiometric compounds A_xMO_y. For WO_3, the colour of A_xWO_3 is blue. While most of the developmental work to date has focused upon the use of proton or lithium injection into WO_3 using aqueous or organic electrolytes, $RbAg_4I_5$, β-alumina, $HUO_2PO_4 \cdot 4H_2O$ and polymer electrolytes have been examined as solid electrolytes in an ECD. The ECD with $RbAg_4I_5$ developed by Green and Richman (1974) has the geometry;

$$Ag/RbAg_4I_5/WO_3,ITO \qquad [11.14]$$

The cell is initially white when viewed from the WO_3 cathode. When a dc potential is applied across the cell with WO_3 on ITO (electrically conducting but transparent Sn-doped In_2O_3), a blue colouration develops over the region covered by $RbAg_4I_5$. Even after removal of the voltage, the film remains coloured.

Interest has developed in electrochromic light transmission modulators, which are called 'smart windows', for control of temperature and lighting in buildings and automobiles. A cross section of an electrochromic light transmission modulator is shown in Fig. 11.31 (Rauh and Cogan, 1988). The two electrochromic elements of the structure are designated EC1 and EC2, and are sandwiched between two thin film, optically transparent, electrodes of ITO and separated by an electrolyte. The EC1 layer should colour when a negative potential is applied and the EC2 layer should either colour under positive potentials or remain in a transparent state. This is indicated by the chemical reactions:

$$\left.\begin{array}{l} \underset{\text{(bleached)}}{(\text{EC1})} + xM^{+} + xe^{-} \rightleftharpoons \underset{\text{(coloured)}}{M_x(\text{EC1})} \\ \underset{\text{(bleached)}}{M_x(\text{EC2})} \rightleftharpoons xM^{+} + xe^{-} + \underset{\text{(coloured or bleached)}}{(\text{EC2}).} \end{array}\right\} \quad (11.17)$$

Polycrystalline films of WO_3 and MoO_3 are good candidates for the EC1 layer (Goldner *et al.*, 1988). WO_3 has been used extensively for electrochromic displays. Of the identified candidate materials for the EC2 layer, V_2O_5, Nb_2O_5 and In_2O_3 (Goldner *et al.*, 1988), and IrO_2 and NiO (Nagai, 1990) are promising.

The electrolyte layer should have relatively high ionic (H^+ or Li^+) and low electronic conductivity at room temperature over the operating spectral range of the window (approximately, $0.35\ \mu m < \beta < 1.5\ \mu m$). Of the number of identified candidate materials for the solid electrolyte layer, glassy films of $LiNbO_3$ (Goldner *et al.*, 1988), hydrated Ta_2O_3 (Nagai, 1990), and poly(ethylene oxide)–H_3PO_4 (Pedone, Armand and Deroo,

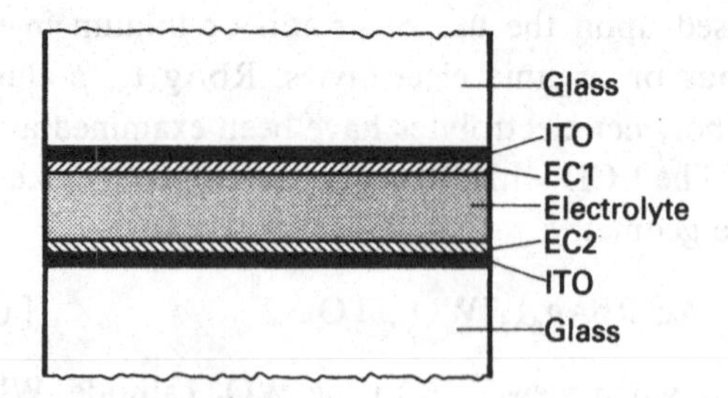

Fig. 11.31 Cross section of an electrochromic light transmission modulator (smart window) (Rauh and Cogen, 1988).

1988) are promising. In Fig. 11.32, a typical transmission vs wavelength curve reported by Goldner *et al.* (1988) is shown. Materials used for the layers were: EC1 = WO_3, electrolyte = $LiNbO_3$ and EC2 = In_2O_3. The devices exhibit good spectrally selective transmission in the near IR and visible portions of the spectrum. These devices have been switched between +3 V more than 3000 times with little detectable change in their optical and electrochemical properties.

Electronically conducting polymers also have an important role to play as electrodes in electrochromic devices. This is described in Chapter 9.

11.7 Electrochemical potential memory device

Electrochemical potential memory devices were proposed by Takahashi and Yamamoto (1973); they have been extensively developed by Ikeda and Tada (1980) of the Sanyo Electric Company, Japan, and are commercially available. The device construction is

$$\begin{array}{c} Ag(1)/Ag_6I_4WO_4/Pt(Ag_2Se)_{0.925}(Ag_3PO_4)_{0.075}/Ag_6I_4WO_4/Ag(2), \\ \uparrow \\ Pt \end{array} \quad [11.15]$$

in which $(Ag_2Se)_{0.925}(Ag_3PO_4)_{0.075}$ is a mixed conductor exhibiting high ionic (0.13 S cm^{-1} at 25 °C) and electronic (10^3 S cm^{-1} at 25 °C)

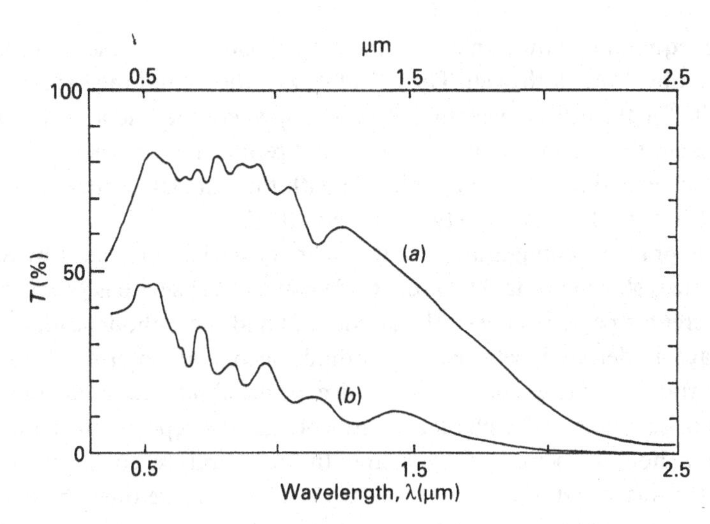

Fig. 11.32 Transmission, T(%) vs wavelength, λ, for the smart window, ITO/WO_3/$LiNbO_3$/V_2O_5/In_2O_3: (*a*) bleached state; (*b*) coloured state (Goldner *et al.*, 1988).

conductivity (Takahashi and Yamamoto, 1972). $Ag_6I_4WO_4$ has a silver ion conductivity of 0.047 S cm^{-1} at 25 °C and extremely low electronic conductivity (Takahashi, Ikeda and Yamamoto, 1972). Upon passing a direct current from the platinum electrode to the Ag(1) electrode in cell [11.15], a definite amount of silver is transferred from the mixed conductor phase to the silver electrode, decreasing the cation-to-anion ratio in the mixed conductor. If the direction of the current is reversed, the ratio increases. The chemical potential of silver, μ_{Ag}, in the mixed conductor is related to the potential difference, E, between the reference Ag(2) electrode and the platinum electrode by

$$-EF = \mu_{Ag} - \mu_{Ag}^{\circ}, \tag{11.18}$$

where μ_{Ag}° is the chemical potential of pure silver and F is Faraday's constant. According to the theoretical calculation by Takahashi and Yamamoto (1973), the relation between the number of coulombs passed through the cell, q, and E is expressed by

$$q/x = 16.0[f(8.48) - f(-39.3E + 8.48)], \tag{11.19}$$

where x is the number of moles of anion in the mixed conductor and f is the Fermi–Dirac function. The Fermi–Dirac function $f(\eta)$ may be reduced to a linear function of η for small deviations of η, and Eqn (11.19) may then be expressed by the following simple formula,

$$q/x = 1810 \times E. \tag{11.20}$$

This equation shows that E increases linearly with the number of coulombs through the cell. The calculated isotherm and the experimental results for the cell are shown in Fig. 11.33, where the line and the circles represent the calculated and experimental results, respectively. It is found that the experimental values coincide with the calculated ones showing a good linear relationship between 0 and 10 mV.

The practical configuration of the memory devices produced by Sanyo Electric is shown in Fig. 11.34 (the trade name of the device is Memoriode). A current is passed between the anode lead and the cathode lead, and the voltage is detected between the cathode lead and potential lead. The important characteristic of the device is its ability to hold and thus memorise voltage. The change in the voltage was reported to be 0.3 mV after a holding period of 48 h, and the temperature coefficient of the voltage was found to be −10 to +40 μV $°C^{-1}$. The life time characteristics showed hardly any capacitance change after more than 10^6 cycles of use, indicating a very high reliability.

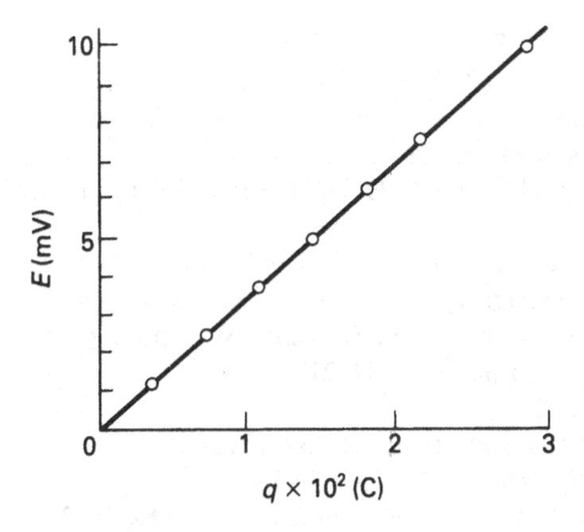

Fig. 11.33 E vs q curve of the cell $Ag/RbAg_4I_5/(Ag_2Se)_{0.925}(Ag_3PO_4)_{0.075}$(0.484 g) at 25 °C. The circles represent the experimental values and the line is the theoretical isotherm (Takahashi and Yamamoto, 1973).

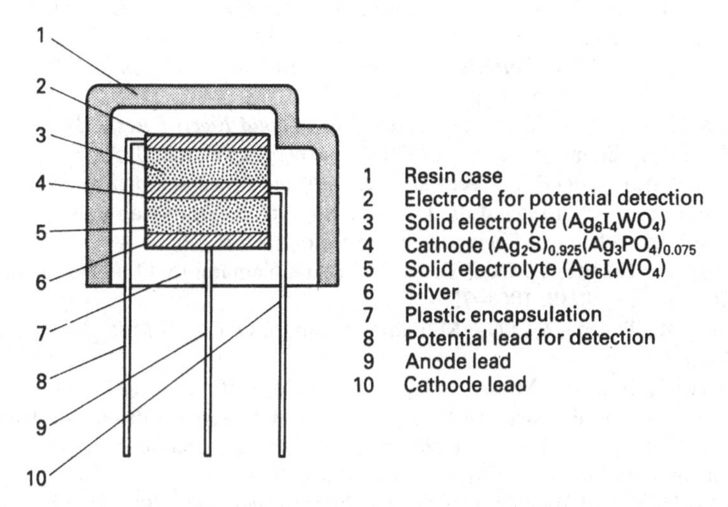

Fig. 11.34 Schematic diagram of an electrochemical potential memory device (Ikeda and Tada, 1980).

References

Abraham, K. M., Goldman, J. L. and Dempsey, M. D. (1981) *J. Electroch. Soc.*, **128**, 2493–2500.

Akridge, J. R. and Vourlis, H. (1986) *Solid State Ionics*, **18/19**, 1082–7.

Argue, G. R. and Owens, B. B. (1968) *Abstracts, 133rd National Meeting of the Electrochemical Society, Boston, Mass*, No. 281, Electrochemical Society.

Bauer, E. and Preis, H. (1937) *Z. Elektrochem.*, **44**, 727–32.
Bradley, J. N. and Green, P. D. (1967) *Trans. Farad. Soc.*, **63**, 424–30.
Brummer, S. B. (1984) in *Lithium Battery Technology*, Ed. H. V. Venkatasetty, John Wiley and Sons, New York, p. 159.
Dudley, G. J., Cheung, K. Y. and Steele, B. C. H. (1980) *J. Solid State Chem.*, **32**, 269.
Fee, D. C. *et al.* (1986) *Fuel Cell Seminar Abs., Tuscon, USA*, p. 40.
Fischer, W. (1989) in *High Conductivity Solid Ion Conductors: Recent Trends and Applications*, Ed. T. Takahashi, World Scientific, Singapore, p. 595.
Fleming. W. J. (1977) *J. Electrochem. Soc.*, **124**, 21–8.
Geller, S. (1976) *Science*, **157**, 21–8.
Goldner, R. B. *et al.* (1988) *Solid State Ionics*, **28/30**, 1715–21.
Green, M. and Richman (1974) *Thin Solid Films*, 24S45.
Gupta, N. K. and Tischer, R. P. (1972) *J. Electrochem. Soc.*, **119**, 1033–7.
Gupta, T. K., Bechtold, J. H., Kuznickl, R. C., Gadoff, L. H. and Rossing, B. R. (1977) *J. Mat. Sci.*, **12**, 2421.
Harada, M. and Mori, Y. (1988) *National Fuel Cell Seminar Abstract*, Courtesy Associates, Inc., Washington DC, p. 18.
Hetrick, R. E., Fate, W. A. and Vassell, W. C. (1981) *Sensors and Actuators Engineering*, Technical Paper Series No. 810433.
Holmes, O. F. (1986) *Batteries for Implantable Biomedical Devices*. Ed. B. B. Owens, Plenum Press, New York, p. 130.
Ikeda, H. and Tada, K. (1980) *Applications of Solid Electrolytes*, Eds. T. Takahashi and A. Kozawa, JES Press Inc., Cleveland, USA.
Isenberg, A. O. (1977) in *Electrode Materials and Processes for Energy Conversion and Storage*, Eds. J. D. E. McIntyre, S. Srinivasan and F. G. Will, The Electrochemical Society, Inc., Princeton, NJ, p. 572.
Kanno, R., Takeda, Y., Oda, Y., Ikeda, H. and Yamamoto, O. (1986) *Solid State Ionics*, **18/19**, 1068–72.
Kanno, R., Takeda, Y., Oya, M. and Yamamoto, O. (1987) *Mat. Res. Bull.*, **22**, 1283–90.
Kapfer, B., Gauthies, M. and Belanger, A. (1990) in *Proceedings of the Symposium on Primary and Secondary Lithium Batteries*, Eds. K. M. Abraham and M. Salomon, The Electrochemical Society, Inc., Pennington, p. 227.
Katayama, M. (1908) *Z. Phys. Chem.*, **61**, 566–87.
Kiukkola, K. and Wagner, C. (1957) *J. Electrochem. Soc.*, **104**, 379–87.
Liang, C. C. (1973) *J. Electrochem. Soc.*, **120**, 1289–92.
Linford, R. G. (1991) in *Solid State Material*, Eds R. Radhakrishna and G. Daud, Narosa Pub., New Delhi, p. 30.
Mashiko, E., Yokokawa, M. and Nagaura, T. (1991) in *Extended Abstract of the 32nd Battery Symposium in Japan*, pp. 31–2.
Mizushima, K., Jones, P. C., Wiseman, P. J. and Goodenough, J. B. (1980) *Mat. Res. Bull.*, **15**, 783–9.
Mori, M. *et al.* (1989) *J. Power Sources*, **26**, 545.
Nagai, J. (1990) *Solid State Ionics*, **40/41**, 383–7.
Ohtsuka, H., Okada, S. and Yamaki, J. (1990) *Solid State Ionics*, **40/41**, 964–6.

Ohtsuka, H. and Yamaki, J. (1989) *Solid State Ionics*, **35**, 201–6.
Owens, B. B. (1971) in *Advances in Electrochemistry and Electrochemical Engineering*, Vol. 8, Eds. P. Delahay and C. W. Tobias, Wiley-Interscience, New York, p. 1.
Owens, B. B. and Argue, G. R. (1967) *Science*, **157**, 308–9.
Owens, B. B. and Skarstad, P. M. (1979) in *Fast Ion Transport in Solid*, Eds. P. Vashishta *et al.*, North-Holland, New York, p. 61.
Pedone, D., Armand, M. and Deroo, D. (1988) *Solid State Ionics*, **29/30**, 1729–32.
Proceedings of the 8th International Conference on Solid State Ionics (1991) Lake Louise, Canada, Eds P. S. Nicholson, M. S. Whittingham, G. C. Farrington, W. W. Smeltzer and J. Thomas, North-Holland, Amsterdam.
Rauh, R. D. and Cogan, S. F. (1988) *Solid State Ionics*, **28/30**, 1707–14.
Reuter, B. and Hardel, K. (1961) *Naturwissenschaften*, **48**, 161.
Sakai, N., Kawada, T., Yokokawa, H., Dokiya, M. and Iwata, T. (1990) *Solid State Ionics*, **40/41**, 394–7.
Sakurada, S. (1991) in *Proceedings of the Second International Symposium on Solid Oxide Fuel Cells*, Eds. F. Grosy, P. Segers, S. C. Singhal and O. Yamamoto, pp. 45–54.
Shahi, K., Wagner, J. B. and Owens, B. B. (1983) in *Lithium Batteries*, Ed. J. P. Gabano, Academic Press, London, p. 407.
Sibert, E., Fouletier, J. and Vilminot, S. (1983) *Solid State Ionics*, **9/10**, 1291–4.
Singhal, S. C. (Ed.) (1989) *Solid Oxide Fuel Cell*, The Electrochemical Society, Inc., Pennington, NJ.
Singhal, S. C. (1991) in *Proceeding of the Second International Symposium on Solid Oxide Fuel Cells*, Eds F. Grosy, P. Segers, S. C. Singhal and O. Yamamoto, p. 25.
Soejima, S. and Mase, S. (1985) *Sensors and Actuators Engineering*, Technical paper series No. 850378.
Steele, B. C. H. (1987) in *Ceramic Electrochemical Reactors*, Ceramic, London.
Takada, K., Kanbara, T., Yamamura, Y. and Kondo, S. (1990) *Solid State Ionics*, **40/41**, 988–92.
Takahashi, T. and Yamamoto, O. (1966) *Electrochem Acta*, **11**, 779–89.
Takahashi, T., Ikeda, S. and Yamamoto, O. (1972) *J. Electrochem. Soc.*, **120**, 647–51.
Takahashi, T. and Yamamoto, O. (1972) *J. Electrochem. Soc.*, **119**, 1735–40.
Takahashi, T. and Yamamoto, O. (1973) *J. Appl. Electrochem.*, **3**, 129–35.
Takahashi, T., Yamamoto, O., Yamada, S. and Hayashi, S. (1979) *J. Electrochem. Soc.*, **126**, 1655–8.
Tofield, B. C., Dell, R. M. and Jensen, J. (1984) *Energy Conservation Industry*, **2**, 120–4.
Yamamoto, O., Dokiya, M. and Tagawa, H. (Eds.) (1989) *Solid Oxide Fuel Cells*, Science House Co. Ltd., Tokyo, Japan.
Yamamoto, O., Kaneko, S. and Takahashi, H. (1988) *National Fuel Cell Seminar Abstracts*, Courtesy Associates, Inc., Washington DC, p. 25.

Yao, Y. F. and Kummer, J. K. (1967) *J. Inorg. Nucl. Chem.*, **29**, 2456.
Weber, N. and Kummer, J. T. (1967) *Advanced Energy Conversion.* ASME conference, p. 916.
Whittingham, M. S. (1982) in *Intercalation Chemistry*, Ed. M. S. Whittingham, Academic Press, New York, p. 1.
Whittingham, M. S. and Silbernagel, B. G. (1977) in *Solid Electrolyte*, Eds. W. van Gool, and P. Hagenmuller, Academic Press, New York.
Wilhelmi, K. A., Waltersson, K. and Kihlburg, L. (1971) *Acta Chem. Scand.*, **25**, 2675.

Index

References which refer to figures in the text are shown in *underlined italics*.
References which refer to tables are **underlined bold**.
References which are textual footnotes have an 'n' after the page number e.g. 45n.

Index

Index

Printed in the United States
By Bookmasters